Environmental Earth Sciences

Series editor

James W. LaMoreaux, Tuscaloosa, USA

More information about this series at http://www.springer.com/series/8394

Wei Dong · Yanqing Lian · Yong Zhang
Editors

Sustainable Development of Water Resources and Hydraulic Engineering in China

Proceedings for the 2016 International Conference on Water Resource and Hydraulic Engineering

Springer

Editors
Wei Dong
Hebei Engineering University
Handan
China

Yong Zhang
Hebei Engineering University
Handan
China

Yanqing Lian
University of Illinois at Urbana-Champai
Champaign, IL
USA

ISSN 2199-9155 ISSN 2199-9163 (electronic)
Environmental Earth Sciences
ISBN 978-3-030-09663-2 ISBN 978-3-319-61630-8 (eBook)
https://doi.org/10.1007/978-3-319-61630-8

Printed on acid-free paper

This Springer imprint is published by the registered company Springer International Publishing AG part of Springer Nature
The registered company address is: Gewerbestrasse 11, 6330 Cham, Switzerland

Preface

The "2016 International Conference on Water Resource and Hydraulic Engineering" was held at the Hebei University of Engineering, Handan, China, on October 9–12, 2016. More than 160 professionals attended. This conference was sponsored by the Hebei University of Engineering.

The central theme of the conference focused on sustainable development of water resources and the environment. Papers for conference presentation were selected from a wide variety of research areas, including watershed hydrology, river hydraulics, groundwater hydrology, water resource management and sustainability development, water supply planning under climate change, water quality analysis and water pollution, Sponge City Development and urban watershed management, environment and sustainability, global connection of air and water, irrigation and drainage issues for agricultural engineering. Out of 72 submitted papers, 34 were selected for this proceedings.

Global climate change and variability have had great impacts on the hydrologic cycle and subsequently on our living environment. Human activities have placed significant roles in an altered hydrologic environment. China has already experienced some environmental impacts from its rapid economic development in recent years. Issues such as air quality, surface water and groundwater environment, flooding especially in big cities, sustainable water resources, etc., have become major concerns by the Chinese government and general public. The central government of China is promoting the Sponge City or Low Impact Development concept to address these issues to sustain the continuing economic growth in China and at the same time to create a healthier and eco-friendly environment. Papers selected from this conference for the proceedings covered research related to these on advanced technology for air quality and water quality monitoring, and research on sustainable water resource development under global climate change and variability.

Papers in this proceedings shall be of interest to a worldwide audience in addressing emerging problems a developing country might face and by research and practice to successfully deal with these issues for a greener and eco-friendly living environment.

Handan, China — Wei Dong
Champaign, USA — Yanqing Lian
Handan, China — Yong Zhang
Tuscaloosa, USA — Jim LaMoreaux

The original version of the book was revised: A new chapter has been included. The erratum to the book is available at https://doi.org/10.1007/978-3-319-61630-8_36

Contents

Establishment of Groundwater Level Warning for Covered Karst Areas in Northern China

Xiaowei Wang, Lizhi Wang, Jingli Shao and Zhiwei Zhao

Abstract In view of the resources shortage and geological environment problems caused by over-exploitation of groundwater in northern China, the delineation of the groundwater level warning (GLW) is urgently needed. Based on the comprehensive consideration of the management objectives of water resources supply guarantee and karst collapse hazards warning, this paper proposed a method for the delineation of karst GLW in covered karst areas. Specific steps include: (1) establish the criterion of GLW based on statistical analysis of groundwater level using the groundwater depth criterion model of Pennsylvania as a reference; (2) establish the criterion of GLW for karst collapse in line with the experience and mechanism research; (3) establish the comprehensive judgment criterion model to determine the GLW of a single observation well; and (4) determine the GLW divisions based on observation wells cluster and issue and renew the warning afterward. The method was verified in the Chengqu-Jiuxian Karst System (CJKS) of Taian City in Shandong Province. It is demonstrated that the GLW conducted by the method in the study area was reasonable and reliable, and was appropriate for decision making for water authorities and governments.

Keywords Groundwater level warning · Covered karst areas · Statistical analysis Karst collapse

X. Wang (✉) · J. Shao
School of Water Resource and Environment, China University of Geosciences (Beijing), Beijing, China
e-mail: wangxw@cugb.edu.cn

J. Shao
e-mail: jshao@cugb.edu.cn

L. Wang
Beijing Geotherm Research Institute, Beijing, China
e-mail: wlz4@163.com

Z. Zhao
Shandong 5th Geo-mineral Prospecting Institute, Taian, Shangodng, China
e-mail: 363100255@qq.com

W. Dong et al. (eds.), *Sustainable Development of Water Resources and Hydraulic Engineering in China*, Environmental Earth Sciences,
https://doi.org/10.1007/978-3-319-61630-8_1

1 Introduction

As the most important water resources of covered karst areas in northern China, groundwater is severely overexploited due to rapidly increasing demand in recent years. The groundwater level is continuing to decrease. A series of karst collapse hazards were initiated by the intense groundwater level fluctuation when the over-exploitation occurred in strong flow areas of karst groundwater. It is necessary to research groundwater warning to prevent a water supply crisis and geo-environmental problems.

Groundwater Warning (GW) is identified as the evaluating model of groundwater status based on the integrated analyses of factors of groundwater benefits, qualities and geo-environmental hazards initiated utilizing groundwater monitoring data [1]. The groundwater level is composed of Groundwater Warning Level (GLW) and Groundwater Quality Warning (GQW). Common methods in current GLW research consist of numerical simulation [2, 3], multiple regression [4], combining relevant indicators [5] and statistical approaches. The statistical approach can warn and predict groundwater level states rapidly and dynamically compared with other methods due to its available analyses of groundwater monitoring data which is relatively easy to gather. A typical research was conducted to monitor dynamic warning based on the mathematical statistics method in Pennsylvania, USA by the United States Geological Survey (USGS) in 2002, which was adapted for the GLW in the Beijing plain areas [1].

There is not enough research on GLW of covered karst areas in northern China, especially on comprehensive consideration of groundwater supply and geo-environmental problems in an integral karst system. This study promoted a method for GLW in covered karst areas using the groundwater depth criterion model of Pennsylvania as a reference and combining the GLW of karst collapse. The method was verified in Chengqu-Jiuxian Karst System (CJKS) of Taian City in Shandong Province. It is demonstrated that this method was rational and feasible and was suitable for decision making of water authorities and governments.

2 Method

2.1 Establishment of the Criterion Model of GLW

An observation well cluster would be selected for analysis. It is stipulated that the selected wells should: (1) have monitoring data with a long-term time series which is more than 15 years; (2) have high quality monitoring data with the deficit no longer than half a year; and (3) be equally distributed in the area.

The percentile of long-term monitoring data for each observational day was calculated with the centile of 5, 25 and 75. The criterion model of GLW is established by the percentile curves of the whole hydrological year consist of minimum curve, 5 centile curve, 25 centile curve and 75 centile curve.

2.2 Establishment of the Criterion for GLW of Karst Collapse

It is proven that the karst collapse is initiated commonly due to the rapid fluctuation of groundwater level around the roof of the bedrock in covered karst areas in northern China [6]. The criterion for GLW of karst collapse can be established with the expression h ± i (m), where h is the roof of karst bedrock, i the threshold of water level wave of karst collapse initiation according to mechanism research or experiences.

2.3 Evaluation of GLW for Single Observation Well

The comprehensive judgment criterion model for single observation well would be elaborated as follows: (1) area above centile 75 curve is safe area; (2) area between centile 25 and 75 curve is green area; (3) area between centile 5 and 25 curve is blue area; (4) area between minimum curve and centile 75 curve is orange area; (5) area beneath minimum curve is red area; and (6) area between the threshold i is red area overlapping other types of areas. The GLW for single observation well can be obtained by verifying the designated area in the criterion model of updated monitoring data. It is necessary that the criterion model be updated every year in each hydrological year.

2.4 Issue of Warning

Warning divisions for the entire area that are based on the geological hazard susceptibility partitions will be determined by the combination of GLW results of observation well clusters and illustrated. A warning should be issued for different divisions, respectively, as well as suggestions such as reducing groundwater abstraction amount or more circumspect on karst collapse hazard.

3 Verification

3.1 Study Area

The Chengqu-Jiuxian Karst System (CJKS) is located in and to the southeast of the urban area of Taian City in Shandong Province. The CJKS is a monoclinic karst system [7] with an area of 112 km^2. The system is restrained and divided by faulted structures and stratigraphic, the main in which are Cambrain-Ordovician stratum. The depth of the aquifer is about 70–80 m. The aquifer has karst grown strongly and good water retention properties as the flow of a single-well can be up to 1000–5000 m^3/d. This system contains two water-rich areas—Chengqu and Jiuxian, both of which have been developed to groundwater-source locations. The main recharge of the karstic groundwater consists of inter-aquifer flow from Quaternary deposits through "skylights", lateral flow and surface water from the Muwhen River in the south. The flow path is from east to west in the east side of the Daidaoan Fault and from north to south in the west. Groundwater discharges to surface or Quaternary around Jiuxian village and Xujiapu village at the southernmost area of CJKS in natural station. Groundwater abstraction formed wells are currently the most important discharge. The Quaternary deposits overlaying on CJKS are thicker from northeast to southwest with a depth of 0–40 m.

The Chengqu groundwater-source location was constructed in the 1950s. The abstraction amount was up to the peak of 5–6 $\times$ 10^4 m/d during 1982. Karst collapses were initiated by overexploitation. Abstraction from the water supply company was prohibited in 1993 and from enterprises had been reduced to 2.1 $\times$ 10^4 m/d since 2012. The Jiuxian groundwater-source location was set up for operation in 1982 by the water supply company. The peak pumping amount was about 5 $\times$ 10^4 m/d around 2002. Karst collapse occurred in a large area soon afterwards. Although the pumping was presently limited, there still are more than 100 wells for municipal water supply and rural drinking and irrigation supply in the location with the abstraction amount up to 5 $\times$ 10^4 m/d during extreme drought.

3.2 Results

Seven observation wells were selected along with the groundwater flow direction in CJKS (Table 1). Monitoring frequency is 3/month or 6/month. The percentile curve, including minimum 5, 25 and 75, were illustrated as the criterion model using long-term monitoring data. The roof of bedrock h for each observation well was from drilling data. Karst collapse intensively occurred when the water level was fluctuating around the roof of bedrock when the inter-aquifer recharge was occurring [6]. Therefore, h $\pm$ 2 m is determined to be the karst collapse warning criterion. According to the established comprehensive judgment criterion model, GLW for every observation well was obtained by taking the monitoring water level

Table 1 Information of selected observation wells and GLW for June 5, 2016

ID	Depth	Time series	Frequency/month	h ± i(m)	Water level June 5, 2016	GLW for single well on June 5, 2016
201	294.77	1997–2015	3	130.57 ± 2	118.54	Safe area
206	150.26	1990–2015	6	126.79 ± 2	118.82	Safe area
208	100.03	1990–2015	6	118.14 ± 2	112.66	Green area
212	175.00	1990–2015	6	109.19 ± 2	124.49	Safe area
225	294.77	1990–2015	3	110.41 ± 2	107.68	Green area
230	120.90	1990–2015	3	117.14 ± 2	115.85	Red area
236	150.05	1990–2015	3	132.44 ± 2	109.97	Red area

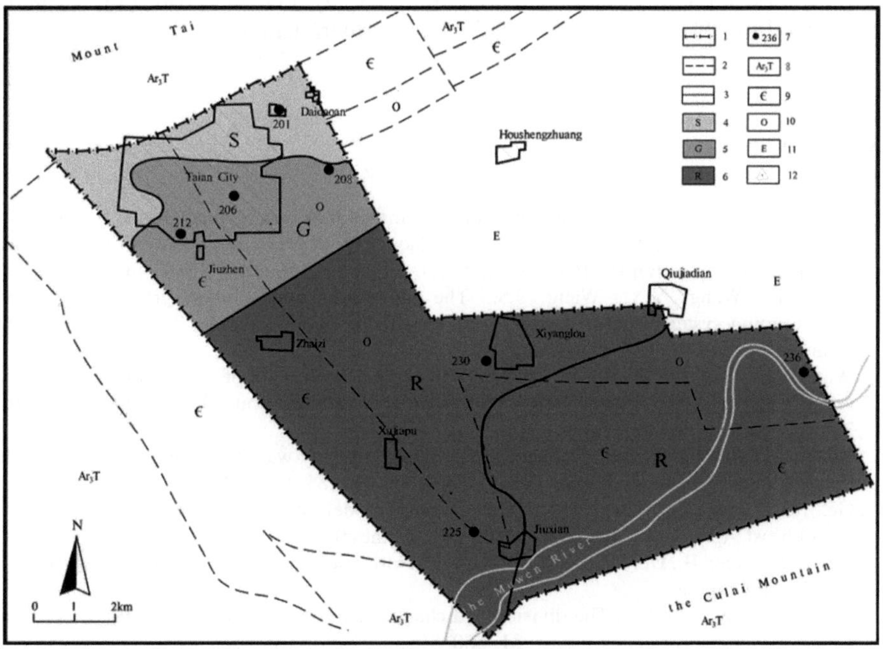

Fig. 1 Groundwater warning division of CJKS in June 5, 2016. 1-Study area boundary, 2-Stratigraphic boundary, 3-Warning divisions boundary, 4-Safe area, 5-Green area, 6-Red area, 7-Observation wells with ID, 8-Metamorphic rocks group, 9-Cambrian stratum, 10-Ordovician stratum, 11-Eogene stratum, 12-Groundwater-source location

on June 5, 2016 for example. Then the warning division of the whole study area was illustrated based on the geological hazard susceptibility partitions of Taian City (Fig. 1).

4 Conclusion

In this study, a method for groundwater level warning (GLW) in covered karst areas was put forward. The method was based on statistics of percentile and groundwater warning of karst collapse. This approach was verified in a typical study area of Taian City in Shandong Province. The study area identified as Chengqu-Jiuxian Karst System (CJKS) is a typical covered karst system in northern China where groundwater overexploitation coexists with serious karst collapse. The basis of this verification is a reasonably comprehensive groundwater observation net. Results showed that the warning divisions can be illustrated with four partitions in the study area including one safe area, one green area and two red areas on June 5, 2016. It is indicated that the method promoted is reasonably clear and easy to calculate. The verification result corresponds to the fact of the study area and powerfully sustains the method to be used for other locations in northern China.

References

1. Smith, T.F., Waterman, M.S.: Identification of common molecular subsequences. J. Mol. Biol. **147**, 195–197 (1981); Li, W.P., Zheng, Y.J., Hao, A.B.: A preliminary study of groundwater level pre-warning in Beijing Plain. J. Earth Sci. Front. **17**, 166–173 (2010). (in Chinese)
2. Bai, L.P., Wang, Y.Y., Wang, J.S.: The numerical model based groundwater level early-warning system: a case study of Linfen basin. J. Geol. China **36**, 246–253 (2009). (in Chinese)
3. Wang, K.J., Cao, J.F., Xu, L., et al.: Stablishment and application of early-warning system in groundwater resource management in the urban area of Changchun. J. Adv. Water Sci. **16**, 238–243 (2005). (in Chinese)
4. Adams, B., Bloomfield, J.P., Gallagher, A.J., et al.: An early warning system for groundwater flooding in the Chalk. Q. J. Eng. Geol. Hydrogeol. **43**, 185–193 (2010)
5. Ge, H.L., Jiao, Y., Ren, Y.T.: Early warning model of groundwater level in Harbin. J. Northeast Agric Univ. **42**, 77–83 (2011). (in Chinese)
6. He, K.Q., Wang, B., Du, R.L.: Karst Collapse in Northern China. Geological Publisher, Beijing (2005). (in Chinese)
7. Liang, Y.P., Wang, W.T.: The division and characteristics of karst water systems in northern China. J. Acta Geoscientica Sinica. **31**, 860–868 (2010). (in Chinese)

Flow Chocking Characteristics of Leak-Floor Flip Buckets

Shu-fang Li and Ji-wei Yang

Abstract Leak-floor flip bucket is a new type of flip bucket recently proposed. It has the advantages of decreasing flow choking on the bucket in small flow regimes and improving energy dissipation by a typical long-narrow nappe. However, if the structure parameters are designed unreasonably, flow choking may also occur on the bucket if the impact location of the lower jet trajectory is too near to the base of the structure, and will threaten the safety of the dam. The purpose of this paper is to study the critical conditions when flow choking begins to disappear or appear on the leak-floor flip bucket, during the increasing and decreasing discharge regimes, respectively. Five leak-floor flip bucket models were conducted, and one circular-shaped flip bucket was prepared for comparison. The critical conditions were investigated under a systematic variation of the approach flow depth, gap width and gap length. It concludes that the critical Froude numbers are primarily influenced by the relative bucket height and the area ratio of the gap; empirical equations for the prediction of critical conditions are obtained and conformed to the test data reasonably.

Keywords Leak-floor flip bucket · Flow choking · Critical condition Dissipater

S. Li (✉) · J. Yang
College of Water Conservancy and Hydropower Engineering,
Hebei University of Engineering, Handan 056000, China
e-mail: shuili_1981@126.com

J. Yang
e-mail: yangjw@hebeu.edu.cn

S. Li
College of Water Conservancy and Hydropower Engineering,
Hohai University, Nanjing 210098, China

W. Dong et al. (eds.), *Sustainable Development of Water Resources and Hydraulic Engineering in China*, Environmental Earth Sciences,
https://doi.org/10.1007/978-3-319-61630-8_2

1 Introduction

Ski jumps are a major element of high dam spillways or tunnel outlet for its satisfactory energy dissipation, especially when the velocity is larger than about 15–20 m/s [1, 2]. Many types of flip buckets were designed as ski jump generators. After the traditional continuous circular-typed (CCT) flip bucket [3–6], a series of different types of energy dissipaters such as slit-type flip bucket [7], triangular-shaped flip bucket [8–10] and deflector dissipaters [11] were proposed. However, a significant disadvantage of the mentioned bucket is the increased level of local flow choking, which is the breakdown of supercritical flow and a local hydraulic jump due to small approach Froude number and the presence of the bucket. When flow choking occurs, the water flow on the bucket is unstable, the jet trajectory impinges almost vertical and causes significant scour at the toe of the dissipater. Further, the choking makes the hydrodynamic and fluctuation pressures on the sidewalls much greater, thus flow choking must be carefully checked.

Leak-floor (LF) flip bucket has firstly been proposed by Deng [12] (Sichuan University, China) in 2009 to improve flow choking, energy dissipation and impact location. It is made up of a curved bed and two side walls, with a gap in the center axis of the bed, and that the length of the two beds which connect the two side walls can be designed to be the same or different, and the bed may be curved or distorted. As its plan view is like a swallowtail, it is also called the swallowtail-type flip bucket. Figure 1a gives a specific LF flip bucket with an equal side wall length proposed by Deng [12]. The LF flip bucket mainly has three advantages as a dissipater: (1) it can decrease flow choking on the bucket and reduce the incipient ski-jump discharge; (2) it makes the water jet diffuse in the longitudinal direction due to the existence of the gap, and reduces the pressure that effects on the side walls; and (3) by changing the length of the two side walls or the bed forms, it can adjust the jet direction into the downstream water flexibly, and adapt to complex terrain conditions to protect the banks of the downstream river.

The LF flip bucket has firstly been used in the right spillway tunnel of Jinping I hydropower project in China [13], and is also being used in the testing stage of Nam Ngiep II spillway [14]. In 2015, Deng [15] studied the flow pattern, the formation and the mechanism of the LF flip bucket based on experiments and numerical simulation. Until now, several questions have so far not been systematically addressed, such as flow choking characteristics, cavitations, energy dissipation, and so on. Although the flow choking characteristic is somewhat not as important as the other problems, the research on it will fill in the gaps in the systematic study of LF flip bucket. In this paper, the flow choking characteristics of LF flip bucket are experimentally investigated. As a preliminary research, this paper only considered a simple condition with equal side bucket length and an axis symmetric gap as shown in Fig. 1b.

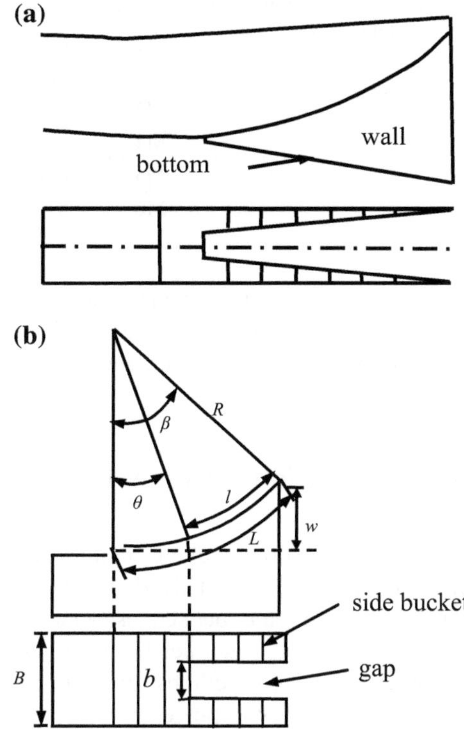

Fig. 1 Schematic view of leak-floor flip bucket **a** proposed by Deng [12]; **b** the present research

2 Experimental Setup

The experiments were conducted in a rectangular channel as described by Wu et al. [16]. It involved a horizontal approach channel to simplify the research. It is 1.25 m long, 0.15 m wide and 0.38 m high. Water was pumped from a laboratory sump to a water tank and then entered the horizontal approach channel. The maximum pump capacity was 400 Ls^{-1}, and the working head was about 1.50 m.

The discharge Q was measured by discharge measurement weir at the end of the tail water channel. The flow depth in the scope of 0.04 m $\leq h_o \leq$ 0.18 m was controlled by a radial sluice gate, which separates the pressure and the free-surface flow section at the inlet of the channel, approach Froude number $Fr = v_o/(gh_o)^{1/2}$ was generated by the jet box and the average approach flow velocity $v_o = Q/bh_o = q/h_o$ (q is the unit discharge), was adjusted by the working head.

The test model, made of Perspex, including five LF flip buckets inserted at the end of the channel, and one circular-shaped bucket was tested for comparison. Figure 1b gives a schematic view of LF flip bucket, where the width B = 0.15 m, the side bucket radius R = 0.50 m and deflection angle β = 45° were fixed. The gap deflection angle θ and the gap width b was changed for θ = 0°, 15° and 30°; b = 0.03, 0.05

Table 1 Test program with basic parameter variation

Cases	b (m)	θ (°)	S	h_o (m)
M1	0	0	0	0.04, 0.07, 0.10, 0.18
M2	0.05	30	0.11	
M3	0.05	15	0.22	
M4	0.05	0	0.33	
M5	0.03	0	0.20	
M6	0.07	0	0.47	

and 0.07 m, with the gap area ratio coefficient $S = lb/BL$ changed accordingly. Table 1 lists the experimental cases and geometric parameters in this paper, in which case M1 is a CCT flip bucket, and cases M2 to M5 are LF flip buckets.

3 Observations of Flow Choking

Figure 2 shows flow choking regimes for all the cases. From visual observations, several aspects can be derived: for the CCT flip bucket of case M1, a hydraulic jump occurs on the bucket, associated with significant air entrainment at the air-water interface, water depth on the bucket increasing dramatically and even part of the turbulent roller climbs over the side walls, the outlet flow flapping sharply, with a jet trajectory impinges almost vertical at the toe [2]. This situation must be avoided in practical engineering because it will endanger the hydraulic structure's foundation; As for the LF flip bucket, case M2 has almost the same flow choking regime as M1 except it is a little weaker, and has a small fin below the nappe because of the small gap on the bucket bed and a small part of water flow from the gap; as for case M3, when the gap deflection angle is decreased to $\theta = 15°$, a fully developed hydraulic jump still exists on the bucket, but the surface turbulence is much weaker than M1 and M2, and the fin extends longitudinal along the gap; when decrease θ to 0° as is the case for M4 (Fig. 2d), there is only a little water block near the bucket outlet. If seen from the plan view (Fig. 3), it can be found that, shock waves intersect on the axis when the water flow direction changed by the gap; observations from Fig. 2e, f found that, when the gap width is decreased to $b = 0.03$ m, a slight surface hydraulic jump appears again; but if the gap width is enlarged to $b = 0.07$ m, the hydraulic jump is then replaced by shock waves once again. It can be obtained from the experiments that the shock wave height is decreased with the increasing of the gap width b.

In conclusion, the flow choking regimes of LF flip bucket can be divided into three types by visual observation: strong hydraulic jump (SHJ) (Fig. 2a, b), weak hydraulic jump (WHJ) (Fig. 2c, e) and shock wave (SW) flow choking (Fig. 2d, f).

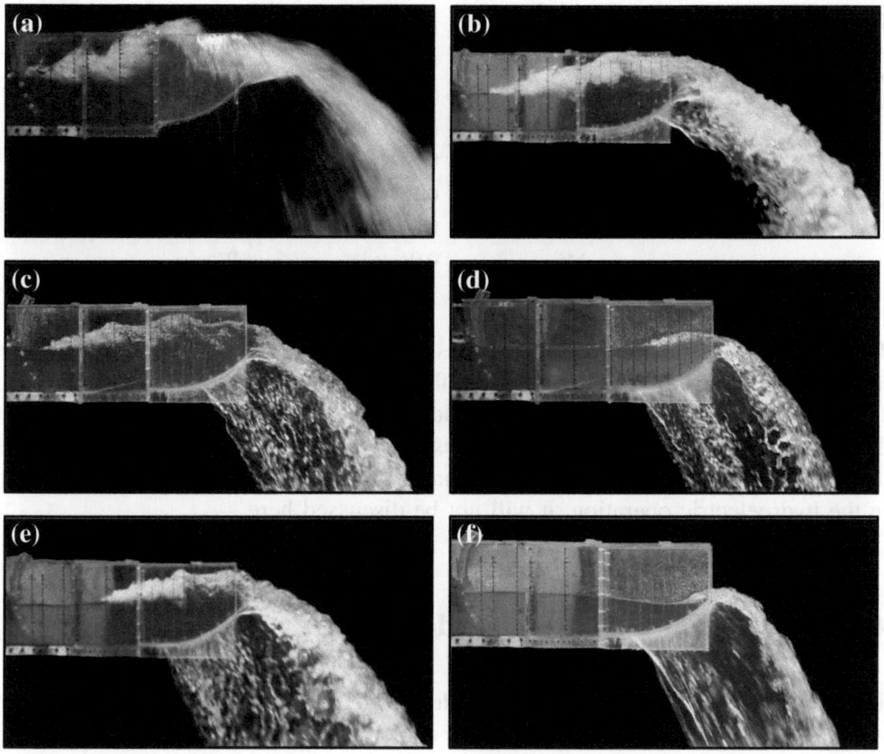

Fig. 2 Flow choking **a** M1: b = 0 m, $\theta = 0°$, Fr = 1.97; **b** M2: b = 0.05 m, $\theta = 30°$, Fr = 1.91; **c** M3: b = 0.05 m, $\theta = 15°$, Fr = 1.54; **d** M4: b = 0.05 m, $\theta = 0°$, Fr = 1.22; **e** M5: b = 0.03 m, $\theta = 0°$, Fr = 1.64; **f** M6: b = 0.07 m, $\theta = 0°$, Fr = 1.11

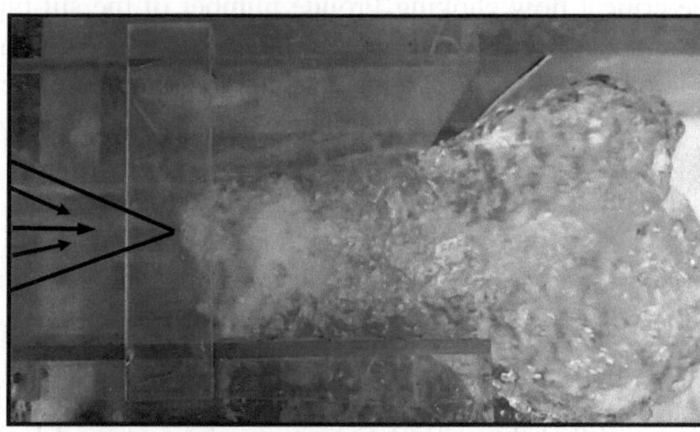

Fig. 3 Plan view of shock waves for M4: h_o = 0.04 m

Table 2 Experimental observations of flow choking regimes

Cases	h_o (m)	Flow choking types
M1	0.04–0.18	SHJ
M2	0.04–0.18	SHJ
M3	0.04–0.18	WHJ
M4	0.04–0.10	SW
	0.18	WHJ
M5	0.04–0.18	WHJ
M6	0.04–0.18	SW

The first two situations need special consideration in hydraulic operation, but the last can be ignored. Table 2 lists all the flow choking cases in the experiments.

Similarly, as with the CCT flip bucket, flow choking also occurs in the decreasing discharge regime, and it is just the opposite process as the increasing discharge regime. As the decreasing discharge process is always less important [17] in the hydroelectric operation, it will not be discussed here.

4 Critical Flow Choking Froude Number

The flow choking characteristics can be defined by the critical Froude number F_{cri}, where $i = 1$ and 2 represent that flow choking completely disappeared in the increasing discharge regime and appeared in the decreasing discharge regime, respectively. From Heller's [4] result it can be obtained that the relative bucket height w/h_o is the main influence parameter of the flow choking characteristics for the CCT flip bucket, and it can be noted from Wu's [16] experimental results that the outlet width, the contraction angle and the approach flow depth are important in the critical flow choking Froude number of the slit-type flip bucket. As for the LF flip bucket, the relative bucket height w/h_o and the area ratio $S = lb/LB$ are considered as the main parameters influencing the flow choking characteristics.

The critical Froude numbers F_{cri} were recorded during experiments. The experimental results were plotted as F_{cri} versus w/h_o (Fig. 4a, b). It can be found that both F_{cr1} and F_{cr2} are increasing with w/h_o, and F_{cr1} is obviously larger than F_{cr2}. This is reasonable as w enhances the flow depth choked on the bucket and a larger momentum is needed to push the choked flow jump out of the bucket. Otherwise, the depth below the hydraulic jump increases with the flow depth h_o and it is relatively easy to push the hydraulic jump out of the bucket. This is similar to the result of the CCT flip bucket proposed by Heller [4]. It can be obviously observed that the critical Froude numbers of the LF flip buckets are much smaller

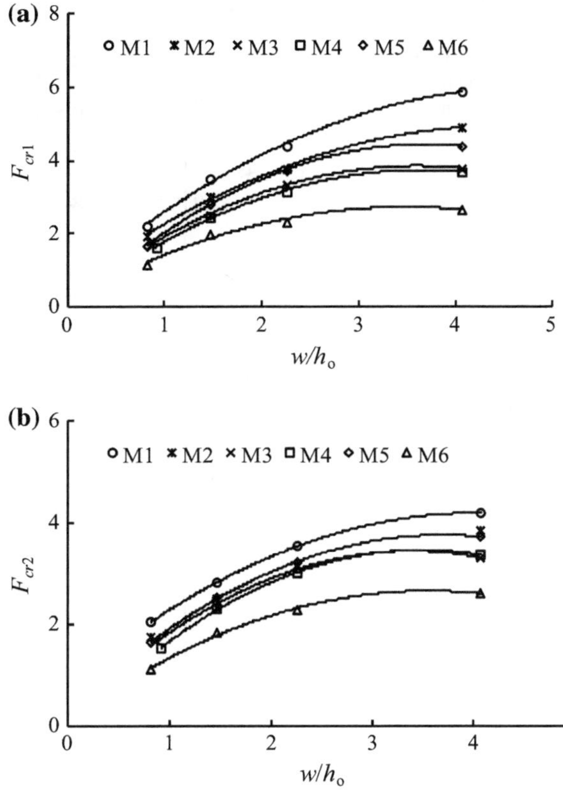

Fig. 4 Critical Froude numbers versus w/h_o: **a** increasing discharge regime; **b** decreasing discharge regime

than the corresponding CCT flip buckets (Fig. 4). Figure 5a, b relate to F_{cr1} and F_{cr2} versus the gap area ratio parameter S for LF flip buckets of cases M2 to M6. Both F_{cr1} and F_{cr2} are decreased with S when w/h_o are fixed.

Considering all of the parameters above, a combined parameter $K = (1-S)\ (w/h_o)$ was proposed and Fig. 6a, b relate to F_{cr1} and F_{cr2} versus K, in which, the dashed line represents results of $h_o = 0.04$ m, and the solid line represents $h_o = 0.07$, 0.10 and 0.18 m. For $h_o = 0.04$ m, the data can be expressed as:

$$F_{cr1} = 1.63K - 0.93 \tag{1}$$

$$F_{cr2} = 0.79K + 1.01 \tag{2}$$

with the correlation coefficients $R^2 = 0.95$ for F_{cr1} and $R^2 = 0.92$ for F_{cr2}.

For cases when $h_o = 0.07$, 0.10 and 0.18 m, the critical Froude number can be expressed as:

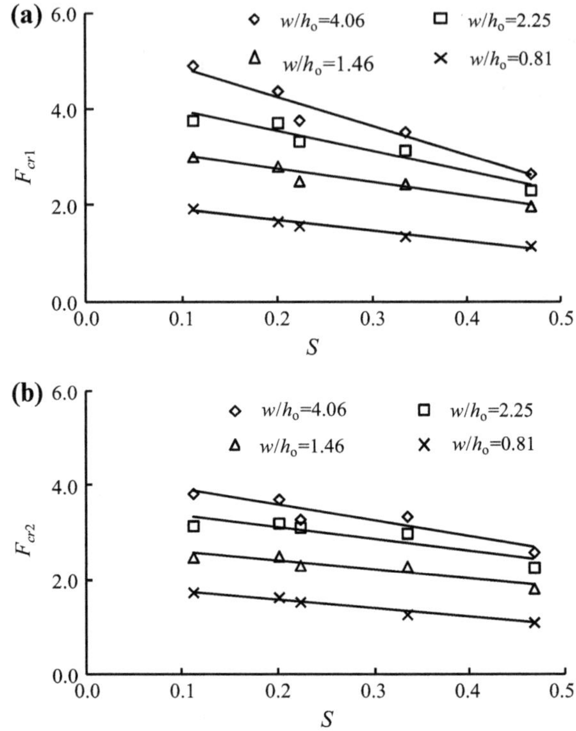

Fig. 5 Critical Froude numbers versus S: **a** increasing discharge regime; **b** decreasing discharge regime

$$F_{cr1} = 2.37K^{0.75} \tag{3}$$

$$F_{cr2} = 2.16K^{0.66} \tag{4}$$

with both correlation coefficients $R^2 = 0.97$.

The limitations of the above equations are $0 \le S \le 0.5$ for the bucket gap area ratio and $0.8 \le w/h_o \le 4.1$ for the relative bucket height.

Figure 7 is the comparisons of the calculated critical Froude number F_{cric} by Eqs. (1)–(4) with the experimental results; the dotted lines represent the ranges of 10% error. The results show that the error is mostly less than 10% and Eqs. (1)–(4) have high precision in the critical flow choking Froude number.

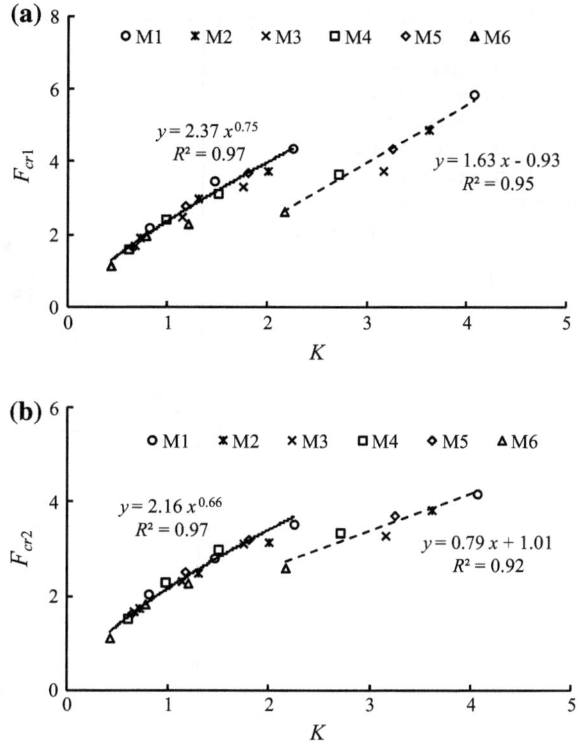

Fig. 6 Critical Froude number versus $K = (1-S)(w/h_o)$: **a** Increasing discharge regime; **b** decreasing discharge regime

5 Discussions

Scale effects may exist when the approach flow depth is too small, such as when $h_o = 0.04$ m, resulted in a different relationship of F_{cri} and K, thus Eqs. (1) and (2) actually have no practical meanings and just for reference only. Equations (3) and (4) can be expressed in a uniform format as:

$$F_{cri} = a_i \left(\frac{w}{h_0} (1 - S) \right)^{b_i} \quad (5)$$

where $a_1 = 2.37$, $b_1 = 0.75$ for F_{cr1} and $a_2 = 2.16$, $b_2 = 0.66$ for F_{cr2}.

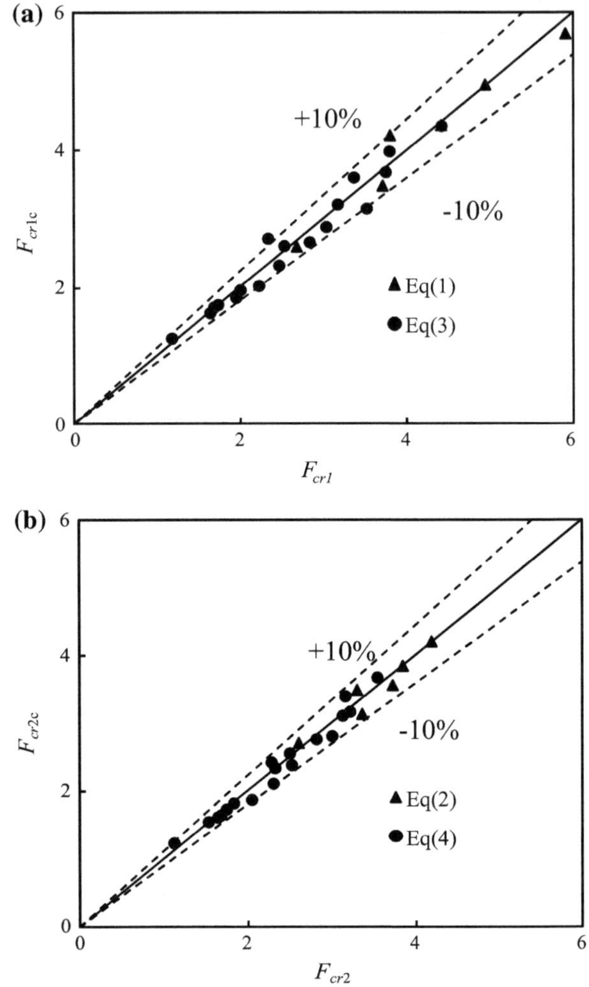

Fig. 7 Comparisons between the measured (*x*-axis) and calculated (*y*-axis) critical Froude number: **a** increasing discharge regime and F_{cr1c} is calculated by Eq. (1) and Eq. (3); **b** decreasing discharge regime and F_{cr2c} is calculated by Eqs. (2) and (4)

The experimental data of the CCT flip bucket was also included in Fig. 6 while considering $S = 0$. This represents that the critical flow choking Froude number, for both LF flip bucket and the CCT flip bucket, has the same tendency with Eq. (5).

From the experiments, it can be shown that the flow choking decreased with S but increased with w/h_o. Besides, it also can be concluded that when $S \geq 0.33$ and $w/h_o \leq 0.81$, or $S \geq 0.47$ and $w/h_o \leq 4.06$, only slightly a weak shock wave appeared, and these situations are far-fetched to be called flow choking, can be considered as reasonable situations in practical engineering. In addition, the design standard suggested that $4 \leq R/h_o \leq 10$ for CCT flip bucket [18]. Then, considering from the aspect of avoiding flow choking, the LF flip bucket can be designed as $S \geq 0.33$ and $w/h_o \leq 0.81$, or $S \geq 0.47$ and $w/h_o \leq 4.06$. Additionally, the trajectory distance must be avoided being too close to cause scour at the toe of bucket.

6 Conclusions

Flow choking regimes and critical conditions of LF flip buckets are explored experimentally. The critical flow choking Froude numbers F_{cr1} and F_{cr2} are focused on and empirical equations to calculate them are obtained. Comparisons between the empirical equations and the test data showed that the paper's equations are reasonable in critical flow choking prediction, both for LF and CCT flip buckets, and can be used for hydraulic design as a preliminary estimation. Furthermore, a preliminary design standard of $S \geq 0.33$ and $w/h_o \leq 0.81$, or $S \geq 0.47$ and $w/h_o \leq 4.06$ for LF flip bucket was proposed from the consideration of avoiding flow choking.

Acknowledgements This research was financially supported by the National Science Foundation. Thanks to the guidance of Professor Jianhua Wu and Fei Ma in Hohai University. Thanks to the proposal of Professor Jiwei Yang in Hebei University of Engineering.

References

1. Grigoryan, K.A.: Calculation of the regime of submergence of a ski-jump bucket behind a deep outlet. Hydrotech. Constr. **18**(7), 313–316 (1984). https://doi.org/10.1007/BF01427682
2. Juon, R., Hager, W.H.: Flip bucket without and with deflectors. J. Hydraul. Eng. ASCE **126** (11), 837–845 (2000). https://doi.org/10.1061/(ASCE)0733-9429(2000)126:11(837)
3. Butakov, A.N.: Determination of the optimal height of the bucket lip of a ski-jump spillway. Hydrotech. Constr. **15**(6), 335–341 (1981). https://doi.org/10.1007/BF01432568
4. Heller, V., Hager, W.H., Mior, H.E.: Ski jump hydraulics. J. Hydraul. Eng. ASCE **5**, 347–355 (2005). https://doi.org/10.1061/(ASCE)HY.1943-7900.0001178
5. Khatsuria, R.M.: Discussion of "ski jump hydraulics" by Valentin Heller, Willi H. Hager and Hans-Erwin Minor. J. Hydraul. Eng. **131**(5), 347–355 (2005). https://doi.org/10.1061/(ASCE)0733-9429(2006)132:10(1115)
6. Savic, L., Kuzmanovic, V., Milovanovic, B.: Ski jump design. Water Manag. **163**(WM110), 523–527 (2010). https://doi.org/10.1680/wama.900052
7. Wu, J.H., Yao, L., Ma, F.: Hydraulics of a multiple slit-type energy dissipater. J. Hydrodyn. **26**(1), 86–93 (2014). https://doi.org/10.1016/S1001-6058(14)60010-X
8. Steiner, R., Heller, R.V., Hager, W.H., et al.: Deflector ski jump hydraulics. J. Hydraul. Eng. **134**(5), 562–571 (2008). https://doi.org/10.1061/(ASCE)0733-9429(2008)134:5(562)
9. Pfister, M., Hager, W.H.: Deflector-generated jets. J. Hydraul. Res. **47**(4), 466–475 (2009). https://doi.org/10.3826/jhr.2009.3525

10. Pfister, M., Hager, W.H.: Deflector-jets affected by pre-aerated approach flow. J. Hydraul. Res. 1–11 (2012). https://doi.org/10.1080/00221686.2012.657875
11. Lucas, J., Hager, W.H., Boes, R.M.: Deflector effect on chute flow. J. Hydraul. Eng. ASCE **139**(4), 444–449 (2013). https://doi.org/10.1061/(ASCE)HY.1943-7900.0000652
12. Deng, J., Liu, S.J., Zhou, Z., et al.: A swallowtail—type of bucket. 2009.12.25, Application number: 200910263563.9. (in Chinese)
13. Wang, J.M., Duan, S.H., Zheng, J.: Key technical problems in the construction of Jinping I high arch dam. Technical Progress in the Dam Construction and Management, The Chinese Association of Dam, Sichuan China, 2012:11. (in Chinese)
14. Chen, X.M., Wang, Z.M., Wang, X.B., et al.: Key techniques in the Nam Ngiep II hydropower construction. Technical Progress of high dam construction and operation management, The Chinese Association of Dam, Guiyang, China, 2014.1016. (in Chinese)
15. Deng, J., Yang, Z.L., Tian, Z., et al.: A new type of leak-floor flip bucket. Sci. China Technol. Sci. 1–8 (2015). https://doi.org/10.1007/s11431-015-5925-x
16. Wu, J.H., Wan, B., Ma, F.: Flow choking characteristics of slit-type energy dissipaters. J. Hydrodyn. **27**(1), 159–162 (2015). https://doi.org/10.1016/S1001-6058(15)60468-1
17. Liu, Z.R., Jiang, Y.Y., Liu, G.C., et al.: Arc radius and critical jet flow discharges in the ski jump energy dissipaters. Proc. Energy Dissipation Eros. Control 441–448 (1980). (in Chinese)
18. Design specification for concrete gravity dams. Water Conservancy Industry Standard of the People's Republic of China, SL319-2005. (in Chinese)

Effects of Brackish Water Salinity on the Soil Salt and Water Movements and the Cotton Seedling Growth Under Film Hole Irrigation

Chao Ma, Haoyun Hu, Lihua Jia, Ce Zhang and Fei Li

Abstract Use of brackish water is becoming more and more important in agricultural irrigation, and it has not been effectively utilized. The effects of brackish water on the soil salt and water movements and the emergence rate of cotton with different salinities of brackish water (0 g/L (control group—CK), 2, 3, 5 and 7 g/L) through film hole irrigation were examined. The results showed that: (1) the cotton emergence rate significantly decreased under the salinity of brackish water more than 5 g/L, whereas the brackish water salinity of 0–5 g/L barely affected the cotton seedling growth; (2) the leaf net photosynthetic rate (Pn) and transpiration rate (E) gradually decreased with the increasing salinity brackish water; (3) the soil salinity increases with the increase of salinity brackish water; and, as time increased, soil salinity gradually moved downward. The results suggested that high salinity of brackish water may have impacts cotton seedling growth. The results may have important significance on cotton plantations with brackish water film hole irrigation.

Keywords Brackish water · Salinity · Soil water · Soil salt · Cotton seedling

C. Ma · H. Hu (✉) · C. Zhang · F. Li
Hydropower College, Hebei University of Engineering, Handan 056000, China
e-mail: hhaoyun@126.com

C. Ma
e-mail: 18231086956@163.com

C. Zhang
e-mail: 41325411917@qq.com

F. Li
e-mail: 18303099770lifei@sina.com

L. Jia
Agriculture Engineering Reconnaissance and Design College of Shanxi, Xi'an 710068, China
e-mail: 3375256152@qq.com

W. Dong et al. (eds.), *Sustainable Development of Water Resources and Hydraulic Engineering in China*, Environmental Earth Sciences,
https://doi.org/10.1007/978-3-319-61630-8_3

1 Introduction

With the growing shortage of freshwater resources, the brackish water used in irrigation has become one of the important ways to alleviate irrigation water supply and demand contradiction [1]. As one of the salt tolerant crops, it is a major difficulty of how to improve the cotton seedling emergence and protect the seeding when it is cropped on saline land [2]. Studies show that when soil salinity was 2–3 g/kg, the cotton seedling emergence will not be greatly impacted; when soil salinity exceeds 4 g/kg, the cotton seedling emergence is only about 40%; the survival rate of cotton is less than 30% [3]. Yumian No. 15 seed germination limit of salt tolerance mass fraction is 0.7%. When NaCl is more than 0.5%, the seedling root length, cotyledon area and seedling dry weight are significantly lower than CK [4]. With increasing levels of soil salinity, transpiration rate of cotton functional leaves, water content and net photosynthetic rate decreases, but the leaf temperature increases [5]. Due to the different experimental conditions, there are differences in cotton seedling salt tolerance eigenvalues obtained. However, clearing the cotton seedling emergence salt tolerance threshold and improving brackish water use efficiency through improvement of the test conditions are important guiding significances to make full use of brackish water.

Many previous studies mainly focused on the impacts of salinity on the emergence and survival rate of cotton seedling under drip irrigation, while few studies examined the physiological parameters of cotton seedlings under film hole irrigation with brackish water. The objectives of this study were to investigate the changes on the morphology and physiology of cotton seedlings under salt stress, which may provide a theoretical basis for the later growth of cotton and improving cotton production.

2 Materials and Methods

2.1 Experiment Design

The experiment was conducted in the experiments field of Hebei University of Engineering in 2016. The soil is light loamy soil. The experiment used micro area which was district 1.5, 1.2 m wide. The experiment set up five salinity treatments, which were 0 g/L (CK), 2, 3, 5 and 7 g/L, repeated three times, and randomly arranged. Cotton variety was Handan cotton 646, sowed on 26 April. The seeding irrigate farming method of film mulching-seeding-irrigating was used with each micro area planting 2 rows, each row planting 4, row spacing 60 cm, spacing 40 cm. The irrigation method was border irrigation with 450 m^3/ha irrigation amount.

2.2 Observation Items and Methods

Soil moisture content and conductivity were measured by TDR (Moisture Meter HD2 ~ Mobile). After land leveling, the 75 cm depth of the tube hole was hit with the determination of depth of 70 cm, divided into 0–15, 15–30, 30–40, 40–50, 50–60, 60–70 cm in each micro area from 10 cm of the plant location by drilling machine, and then inserted the TDR observation tube. The initial soil moisture content and conductivity were measured before irrigation at different depths, and they were measured after 1, 7, 14, 21 and 28 d after irrigation.

Counting the number of cotton seedlings per hole per day for three days after 15 days after irrigation, the number of seedling emergence final count was the statistical closing date count, calculating emergence rate, and then leaving one seedling per hole.

Forty days after irrigation (at the latter period of the seedling stage), three plants were marked in each micro area, and the plant height and stem diameter were measured by ruler and vernier caliper. The leaf area was calculated by measuring the length and width of the leaf. The following leaf area formula [6]: leaf area = leaf length × leaf width × 0.84.

Forty-five days after irrigation, two cotton plants of the middle position were marked in each micro area by selecting the fully expanded leaves and then determining its photosynthetic gas exchange parameters (Net Photosynthetic Rate (Pn), Stomatal Conductance (Gs), intercellular CO_2 Concentration (Ci) and Transpiration Rate (E)) by Portable Photosynthesis Measurement System (Lcpro ~ SD, ADC BioScientific Ltd, Hoddesdon, UK). In the determination process, the light intensity was 1000 mol m^{-2} s^{-1}, and the leaf chamber temperature was 30 °C.

3 Results

3.1 Effects of Brackish Water Salinity on Soil Water Movement

As shown in Fig. 1, there was a large difference in the initial water content of the soil in each micro area, and the changing trend of soil moisture content was different. However, it still could be seen that after the same salinity of brackish water irrigation, topsoil moisture content increased significantly. With the depth of the soil increasing, the rate of increase in the soil moisture content decreased; 1–7 days after irrigation, the change of soil moisture content was mainly concentrated in the 0–40 cm soil layer, and the change was not obvious. Since the 5th day and 18th day after irrigation of rainfall about 15 mm, the 0–30 cm soil moisture content increased significantly. Deeper soil moisture content also gradually increased as time increased. Soil moisture movement occurred mainly in the 0–60 cm soil layer.

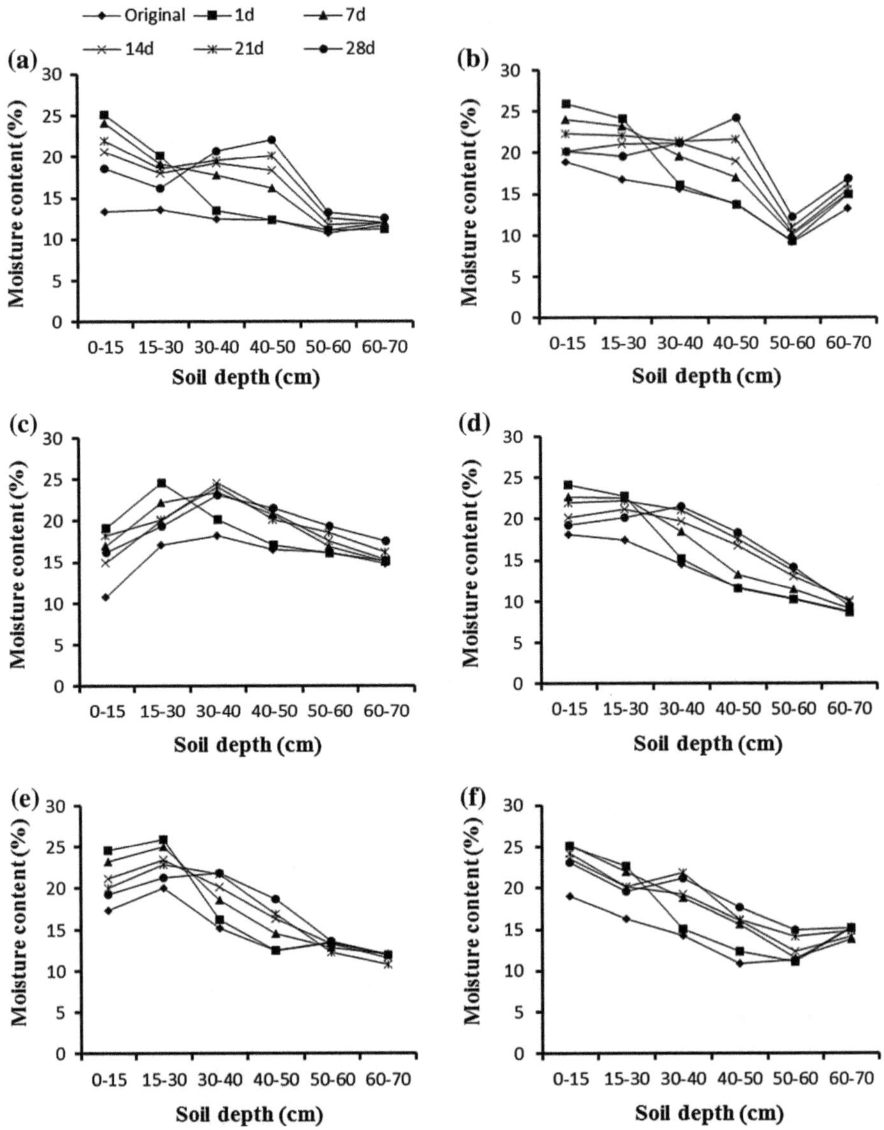

Fig. 1 Movement characteristics of soil moisture content in different salinity **a** CK(0 g/L); **b** 2 g/L; **c** 3 g/L; **d** 5 g/L; **e** 6 g/L; **f** 7 g/L

After different salinity of brackish water irrigation, soil moisture movement rates were changed. At the same time, soil moisture movement rate increased as salinity increased.

3.2 Effects of Brackish Water Salinity on Soil Salt Movement

As shown in Fig. 2, after brackish water irrigation, the soil salinity significantly increased, especially in the topsoil. Soil salt content increased in the layer of 0–60 cm soil; the salt content of 0–15 cm topsoil increased the maximum; and the amplitude of salt content decreased as the soil depth increased in each salinity treatment. Along with the prolonging of time, the topsoil salt content decreased; however, it increased in the lower soil, which showed soil salt moved down after irrigation. The salt content of the topsoil dropped the largest amount in 0–7 days and it decreased less in 7–28 days. That indicated that the soil salt moved downward faster in the early stage after irrigation, then slower.

With the increasing of brackish water salinity, soil electrical conductivity increased at the same time after irrigation. Comparing the soil moisture and conductivity with the depth at the same salinity in the cotton seedling, it can be seen that the soil salt movement had obvious synchronization with the soil moisture migration, which showed that soil salt took the soil moisture as the carrier along with the moisture movement. Two rains led to the topsoil salt content of the relative increase in each salinity treatment and soil salt moved down and accumulated at lower level in the 5th day and 18th day after irrigation.

3.3 Effects of Brackish Water Salinity on Emergence Rate and Seedling Height, Stem Diameter and Leaf Area of Cotton

As shown in Fig. 3, when brackish water salinity was 0 g/L (CK), the cotton emergence rate was 70.83%; and it had no notable change with salinity increased to 5 g/L. However, when the salinity was 6 g/L, the cotton emergence rate was 59.17%, which reduced 11.66% the salinity for 0 g/L and it changed significantly ($p < 0.05$). When continuing to increase the salinity, the cotton emergence rate decreased.

Apart from cotton emergence rate, there was an impact of brackish water salinity on cotton seedling height, stem diameter and leaf area. As Figs. 4, 5 and 6 showed that when the salinity was less than 5 g/L (including 5 g/L), there were no significant effects on cotton emergence rate and seedling plant high, stem diameter and leaf area. However, when the salinity was larger than 5 g/L, they were significantly impacted ($p < 0.05$); and with salinity increasing, the effects were more obvious. In addition, statistical results showed that the difference did not reach significant levels ($p > 0.05$) between 0 and 5 g/L treatments.

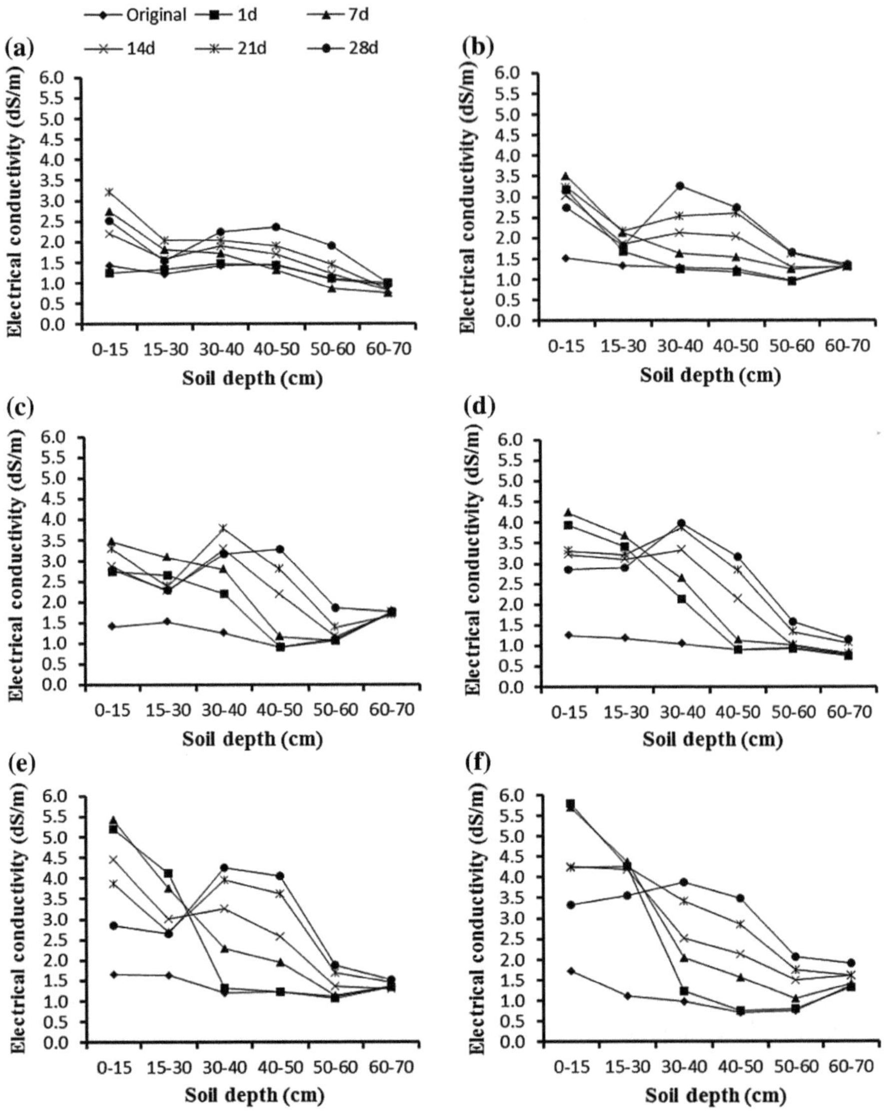

Fig. 2 Movement characteristics of soil salt content in different salinity

3.4 *Effects of Brackish Water Salinity on Gas Exchange Parameters of Cotton Seedling*

As shown in Table 1, the leaf net photosynthetic rate (Pn) decreased with the increasing of brackish water salinity. The Pn reduced by about 41.5% when the salinity changed from 7 g/L to 0 g/L, which was a significant difference ($p < 0.05$).

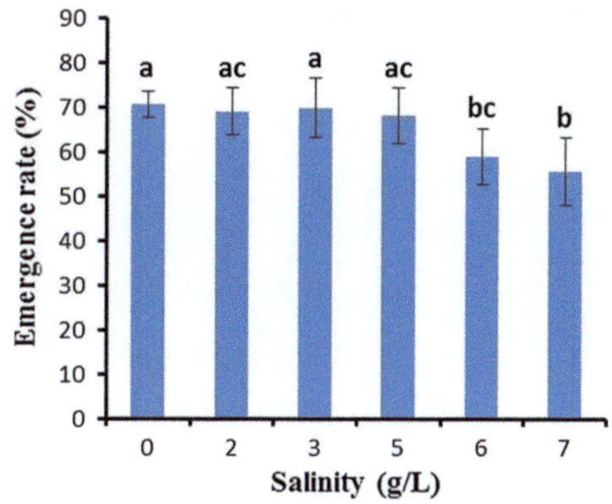

Fig. 3 Effects of salinity on emergence rate

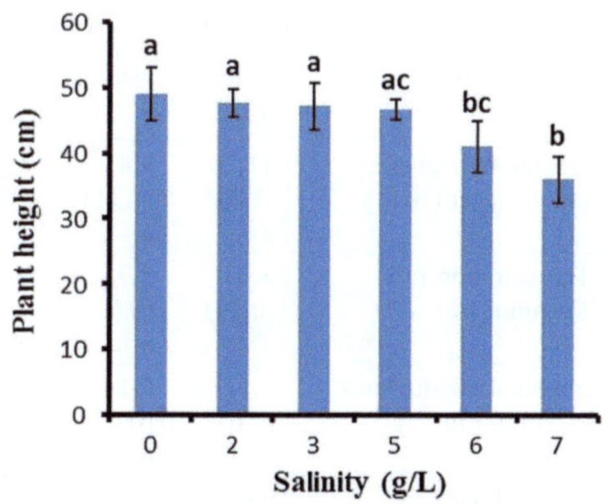

Fig. 4 Effects of salinity on height

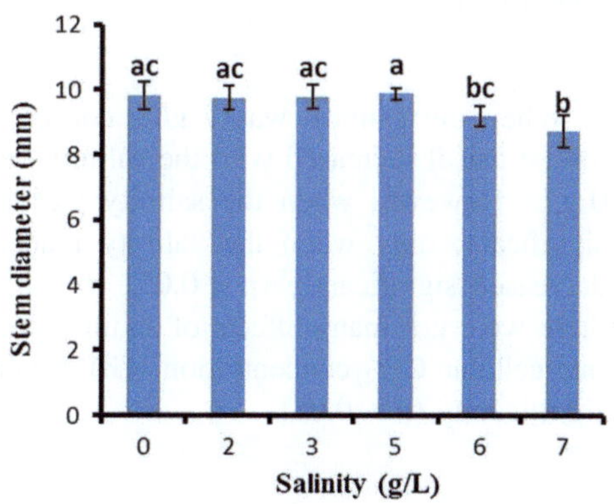

Fig. 5 Effects of salinity on stem diameter

When the salinity was less than 5 g/L, Pn did not change significantly. That indicated the brackish water did not greatly impact on leaf net photosynthetic rate in which salinity was less than 5 g/L.

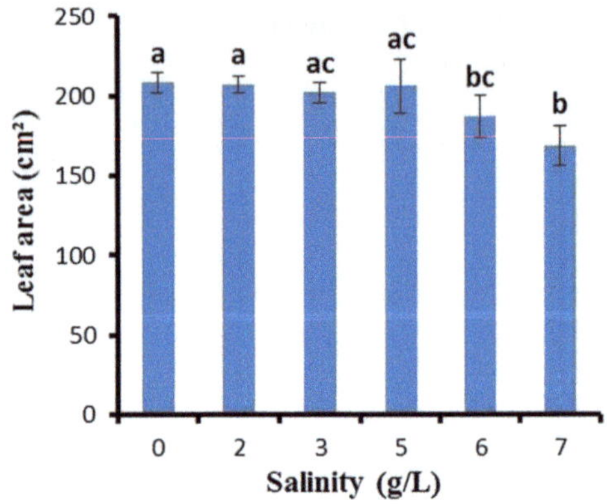

Fig. 6 Effects of salinity on leaf area

Table 1 Effects of salinity on gas exchange parameters of cotton seedling

Gas exchange parameters	Brackish water salinity (g/L)						*p*-values
	0	2	3	5	6	7	
Net photosynthetic rate P_n (μmol m^{-2} s^{-1})	11.24 (1.33) a	9.41 (4.11) a	8.54 (1.62) a	8.01 (2.97) a	7.12 (2.91) a	6.58 (3.92) b	0.323
Transpiration rate E (mmol m^{-2} s^{-1})	3.09 (0.29) a	2.75 (0.47) a	2.49 (0.64) ac	2.33 (0.73) ac	2.24 (0.89) ac	1.78 (0.67) bc	0.137
Stomatal conductance G_s (mmol m^{-2} s^{-1})	0.17 (0.03) a	0.14 (0.05) a	0.13 (0.06) a	0.11 (0.06) a	0.12 (0.09) a	0.08 (0.05) a	0.514
Intercellular CO_2 concentration C_i (mmol L^{-1})	241 (15)a	242 (28)a	252 (38)a	257 (36)a	259 (30)a	264 (31)a	0.849

When the salinity was 7 g/L, cotton seedling transpiration rate (E) was the lowest and it increased with the salinity decrease until it reached the maximum of 0 g/L. However, when the salinity was less than 5 g/L, the difference was not significant; only when the salinity reached 7 g/L, the cotton leaf transpiration decreased significantly ($p < 0.05$) with a decline of about 42.4%. In this study, there were not many effects of salinity on leaf stomatal conductance (G); the leaf intercellular CO_2 concentration (Gi) increased with increasing salinity while not significantly ($p > 0.05$).

4 Discussion

After brackish water irrigation, soil salt content increased which affected the absorption and utilization of the soil water in the cotton root and affected the growth of the cotton. The cotton emergence rate increased with the salinity of brackish water decrease; especially when the salinity was 7 g/L, the cotton emergence rate showed a significant difference. However, when the salinity was less than 5 g/L, there were no difference with emergence which showed that it did not impact cotton emergence after brackish water irrigation of 0–5 g/L. At the same time, cotton seedlings grew well after low salinity brackish water irrigation. With the increase of the salinity, cotton seedling plant height, stem diameter and leaf area decreased gradually, especially when salinity increased to 6–7 g/L, which further showed that salinity which was less than 5 g/L of brackish water can be used for irrigation of cotton seedling.

Net photosynthetic rate reflects the growth conditions of plants. The study found that salt treatment directly affected the structure and photosynthetic enzyme activity and other physiological and biochemical processes such as the photosynthesis system which led to impaired photosynthesis [7, 8]. Under salinity condition, the inhibiting nature in the plant growth process was mainly caused by a water deficit. The reason was that the massive soil salt ions made soil osmotic potential decrease and the plant root absorbed water resulting in plant physiological drought [9]. Under the related analysis between the different salinity conditions of cotton net photosynthetic rate, this paper found that the cotton leaf stomata had narrowed or closed when brackish water salinity increased and leaf net photosynthetic rate (Pn) gradually decreased; however, if the salinity was less than 5 g/L, there was not much impact on Pn. Cotton seedling transpiration rate was lowest with the 7 g/L of salinity and it increased with the decrease of salinity. The effects of brackish water salinity on leaf stomatal conduction were not obvious; the leaf intercellular CO_2 concentration increased with increasing salinity, but not significantly.

5 Conclusion

1. After brackish water irrigation, the soil moisture moved down gradually with the time increase when migration was mainly concentrated in the 0–60 cm of soil layer; the soil salt moved along with the water transport and accumulated at the edge of the water movement.
2. Brackish water had no significant effect ($p > 0.05$) on the rate of cotton seedling emergence rate and seedling growth (plant height, stem diameter, leaf area) when its salinity was less than 5 g/L.
3. Cotton seedling leaf photosynthetic rate and transpiration rate decreased with the increase of brackish water salinity; when the salinity was greater than 5 g/L, the effect was significant ($p < 0.05$).

Acknowledgements This work was supported by the Natural Science Foundation program of Hebei Province (No. E2016402169) and Hebei province science and technology research projects (No. 14227002D).

References

1. Wang, Q.J., Shan, Y.Y.: Review of research development on water and soil regulation with brackish water irrigation. J. Trans. Chin. Soc. Agric. Mach. **46**(12), 117–126 (2015). https://doi.org/10.6041/j.issn1000-1298.2015.12.017
2. Zhang, J.P., Sun, J.S., Li, K.: Effects of water salinity before sowing on cotton emergence rate under different irrigation methods. J. Irrig. Drain. **30**(03), 51–55 (2011)
3. Dong, H.Z., Xin, C.S., Li, W.J.: Characteristics of salinity and fertility in coastal saline cotton fields in Shandong and their effects on cotton emergence. J. Cotton Sci. **21**(4), 290–295 (2009)
4. Xie, D.Y., Wang, V., Wang, F.X.: Application of lactate polyacrylamide gel electrophoresis on the testing of hot pepper seed purity. J. Seed. **3**, 10–13 (2000). https://doi.org/10.3969/j.issn.1001-4705.2000.03.004
5. Zhang, L., Zhang, G.W., Meng, V.: Changes of related physiological characteristics of cotton under salinity condition and the construction of the cotton water stress index. J. Scientia Agricultura Sinica. **46**(18), 3768–3775 (2013). https://doi.org/10.3864/j.issn.0578-1752.2013.18.004
6. Hong, J.R., Fang, G.H., Chen, R.M.: Cotton Experiment Method, M. Agriculture Press, Beijing (1985)
7. Kao, W.Y., Tasi, T.T., Tasi, H.C.: Response of three glyeine species to salt stress. Environ. Exp. Bot. **56**, 120–125 (2006)
8. Yang, S.P., Wei, C.Z., Liang, Y.C.: Effects of NaCl stress on the characteristics of photosynthesis and chlorophyll fluorescence at seedlings stage in different sea island cotton genotypes. J. Scientia Agricultura Sinica. **43**(8), 1585–1593 (2010)
9. Yang, X.H., Jiang, W.J., Wei, M.: Review on plant response and resistance mechanism to salt stress. J. Shandong Agric. Univ. Nat. Sci. **37**(2), 302–305 (2006). https://doi.org/10.3969/j.issn.1000-2324.2006.02.036

Risk Assessment of Rainstorm Waterlogging in New District Based on MIKE Urban

Ying Wang, Qinghua Luan, Haichao Wang, Jiahong Liu and Jun Ma

Abstract Due to the urban rainstorm waterlogging problem which has frequently occurred in recent years, this paper establishes a rainwater pipe network system model for the newly built future science park area in Beijing based on the MIKE Urban model under the support of fundamental data in aspects of catch basin, storm sewer, land utilization, etc. Because of the lack of measured runoff data, this paper uses the runoff coefficient as the check objective to implement the experience calibration of the model parameters, verifies the model through the "6.23" storm runoff process in Beijing and establishes the regional model. On this basis, this paper simulates three designed rainfall runoff processes under different recurrence periods and obtains the simulation result of cumulated water quantity, pipe load, etc. in the corresponding catchment area of the research area. This paper concludes through analysis that the easy flooding and waterlogging points under different simulation scenarios in the research area provide the decision basis for the storm waterlogging risk management and rainfall resource utilization in this area.

Y. Wang · Q. Luan (✉) · J. Ma
College of Water Conservancy and Hydropower Engineering,
Hebei University of Engineering, Handan 056021, China
e-mail: carol97011202@163.com

Y. Wang
e-mail: 18731097102@163.com

J. Ma
e-mail: 2858278263@qq.com

Y. Wang · Q. Luan · J. Ma
Research Center for Water Ecological Civilization & Social Governance
of Hebei Province, Handan 056038, China

H. Wang
China International Engineering Consulting Corporation, Beijing 100038, China
e-mail: wanghaichao@ciecc.com.cn

J. Liu
China Institute of Water Resources and Hydropower Research, Beijing 100038, China
e-mail: liujh@iwhr.com

W. Dong et al. (eds.), *Sustainable Development of Water Resources and Hydraulic Engineering in China*, Environmental Earth Sciences,
https://doi.org/10.1007/978-3-319-61630-8_4

Keywords Stormwater drainage · Scenario simulation · Risk assessment
MIKE Urban · Rainwater utilization

1 Introduction

The reasons for frequent occurrence of urban waterlogging in recent years can be analyzed and summarized as follows: 1. rapid urbanization causes the urban impervious area to increase; changes the rainwater infiltration process and forms the hydrological effect of urbanization; 2. change in the urban thermal radiation and conduction cause urban areas to appear as the "heat island effect", which thereby leads to the occurrence of "rain island effect"; 3. urban drainage infrastructure is insufficient and seriously lags behind the urbanization process; for example: the designed storm return period for most of the storm water sewers in the original urban drainage system is one year or so, and the storm return period in some of the old cities is even lower than this lower limit; 4. with the development of urban three-dimensional transportation, the number of easy waterlogging areas increased, such as the recessed overpass, underground passage, deep groove road sections, etc.; 5. insufficient maintenance and management of urban drainage facilities as well as the blockage and siltation of drainage facilities in wet season directly influence the rainwater drainage effect.

The underlying surface condition and storm sewer system gradually becomes complicated and variable during the urbanization process. To reasonably and accurately simulate the urban hydrological process and assess the urban flood waterlogging risk, the urban rain flood simulation model method is commonly applied at home and abroad. At present, the widely applied urban hydrological models include SWMM, STORM, MOUSE, QQS and DR3M models [1]. The MIKE Urban water supply and drainage pipe network simulation software used in this paper is an urban water simulation system developed by Danish Hydraulic Institute (DHI) through integrating the ArcGIS of ESRI, drainage pipe network system CS and water supply pipe network WD. This software is based on the AO (ArcObject) framework and the documents are stored in the Geodatabase data format to facilitate its combination with GIS and utilization of powerful functions of GIS. This software is widely applied at home and abroad, for example: Søren Thorndahl et al. [2] realized the simulation of mutual action between the underground water movement and sewage pipe flow through coupling of MIKE SHE and MIKE Urban. Lee et al. [3], used MIKE Urban to simulate for reducing the flood damage of green park in flood-prone areas around South Korea's Incheon Bridge landfill. Wu et al. [4] applied MIEK Urban, MIKE 21 and MIKE FLOOD to assess the current situation of the rainwater system in Wuhan City. In addition, this software was also applied in the flood-prevention simulation of rainwater system for the Beijing Olympic central area [5]; simulation of running optimization and energy-saving and cost-reducing of the drainage pipe network system in Haining City; urban flood simulation in Lujiazui area of Shanghai City, China [6, 7]; and simulation of urban rainfall flood disaster in Jinan urban district, China, etc.

2 Regional Overview

The future science park in Beijing is in the northwest suburbs of Beijing City, and its construction was officially started in July 2009. It is in the most eastern end of the high-tech industry corridor on the North 7th Road in the Changping district and has an overall flat terrain; however, it also faces a high pressure of flood drainage and mitigation. In accordance with the Fig. 1 Distribution of Historical Waterlogging Disaster Risk, it is seen that the research area is in the region with a relatively higher occurrence frequency of waterlogging disaster risk. The inter-annual and intra-annual distribution of precipitation is uneven; the annual precipitation is 600–650 mm concentrated in June–September; and the multi-year average evaporation is 1700.5 mm. The soil type is dominated by light loam soil and sandy soil with a good permeability, and the permeability coefficient of surface soil is within 1.50×10^{-6}–9.25×10^{-5}. Main types of land use include roads among land plots, main roads, green space, water area, industrial and residential land, etc. In combination with the planning report and site investigation result, the

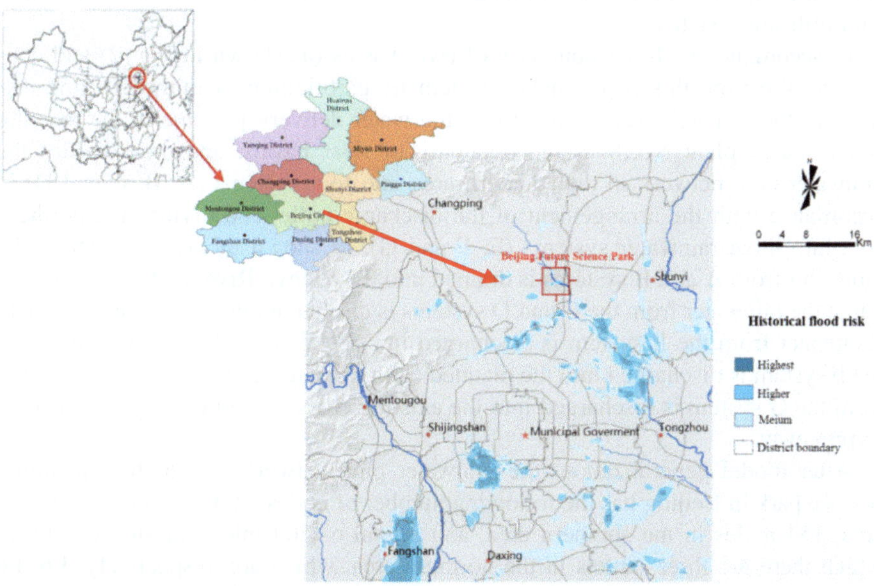

Fig. 1 Historical flood risk layout

impermeable stratum area of this area is about 45% and is mainly dominated by the area of impermeable pavement and buildings.

Scope of the research area: it reaches the western extension of Shunyu Road in the north, Beijing-Chengde Expressway and Changping district boundary in the east, Planning Road No. 28 in the south and east boundary of the central cluster of Beiqijia Town in the west, with a total planning area of 10.00 km^2. The future science park is divided into the north and south parts by using Wenyu River and Dingsi Road (Government Street) as the boundary; the north area is in the southeast of Xiaotangshan Town, with a floor area of 2.15 km^2; the south part is in the east of Beiqijia Town, with a floor area of 4.47 km^2. The area of core green land between these two parts is 3.38 km^2. This paper only researches the northern and southern area covered by the rainwater pipe network, and the green land between these two areas is the zero discharge area and is not within the scope of this research.

3 Construction of Model

3.1 System Generalization

In accordance with the Storm Sewage Discharge Planning for the Future science park (South Area) (modified) and plan for the rainwater drainage planning in the External Storm Sewage Planning; and based on the MIKE Urban model, this paper establishes the rainwater pipe network model for the future science park in Beijing under the support of fundamental data about the catch basin, storm sewer conduit, land utilization status, etc.

In accordance with the boundary of river basins (as shown in Fig. 2) and confluence direction, this paper makes systematic partitioning of rainwater pipe network in the research area by regarding the water delivery pipeline in the existing water source plant No. 9 as the boundary of river basins, and then divides the rainwater pipe network in the southern area into four systems (A, B, C and D) in accordance with the arrangement of the discharge outlet and divides the northern area into three rainwater systems (E, F and G), as shown in Fig. 2 in which the rainwater from A and B systems is drained into the Wenyu River and its old course, while the rainwater from the C and D systems is discharged into the Lutuanxi Ditch. Rainwater from the E system is discharged into the Wenyu River; rainwater from the F system is discharged into the planned earth drainage ditch, while the rainwater from the G system is discharged into the existing cross-culvert of Beijing-Chengde Expressway.

After model generalization, the rainwater pipe network system for the future science park in Beijing has the following number of nodes: 50 nodes in the northern area, 154 nodes in the southern area, and a total of 204 inlets for storm water in which there are three outlets in the northern area, which are, respectively, E6, F6 and G4. The outlets in the southern area are, respectively, A8, B14, C6 and D16, as

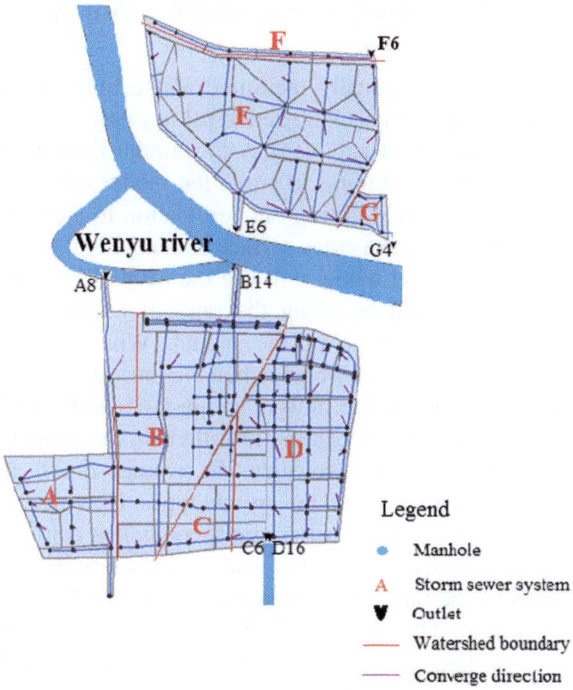

Fig. 2 Beijing Future Science Park storm sewer network

shown in Fig. 2. The length of the rainwater pipe: there are 47 sections of rainwater pipes in the northern area, with a pipe length of 13,060 m; there are 150 sections of rainwater pipes in the southern area, with the pipe length of 33,240 m. The total length of the rainwater pipeline in the research area is 46,300 m.

3.2 Division of the Catchment Area

The catchment area was divided into sub-areas in accordance with the ground elevation and small watershed line in this area in which the northern area has 32 sub-areas and the southern area has 59 sub-areas. Therefore, the total number of sub-areas is 91 in which the catchment area in the northern area is 122.67 ha and the catchment area in the southern area is 417.27 ha, so the research area is about 540 ha. Except for the area under construction and the undeveloped area, the area controlled by the rainwater pipe network is basically covered. The arrangement of confluence points in the catchment sub-area take account of the local topography and actual confluence direction and will use the adjacent nodes as the confluence points, as shown in the Fig. 2.

3.3 Model Verification and Calibration

Due to lack of actually measured and calibrated data, this paper uses the runoff coefficient as the check objective [8] to implement the experience calibration of the model parameters and verifies the model through the actually measured typical "6.23" storm in Beijing to prove the model reliability.

Model selection during the verification in the runoff model setting will include: calculation of impervious surface coverage and setting of hydrological parameters. The hydrological model will use the MOUSE Time-Area (A), while the time-area curve is selected by using the ratio of (Area-50)/Area to determine the default time-area curve type of the model. When the ratio is within 0.00–0.37, the TA-Curve3 will be selected; when the ratio is within 0.38–0.60, the TA-Curve1 will be selected; when the ratio is within 0.61–1.00, the TA-Curve2 will be selected.

In the calculation of runoff coefficient, the weighted average calculation method is usually adopted to calculate the average runoff coefficient of single confluence area in accordance with the ground types. However, for more complicated areas [9], its value of comprehensive runoff coefficient can be taken by referencing Table 1.

The actually measured typical "7.21" storm data (precipitation of 264.5 mm and rainfall duration of 26 h) in Beijing as the rainfall boundary conditions of the model is typed in. The runoff coefficient of each catchment area during this precipitation through loading the accumulative rainfall simulation result is calculated; the comprehensive runoff coefficient of the whole research area is calculated to be 0.43 through weighted average of the catchment sub-areas in the research area; the impervious surface coverage of this area is 45% and the corresponding urban comprehensive runoff coefficient takes 0.40–0.60, which match with the model result.

After multiple adjustment of parameters with the impervious surface coverage of this area is set to be 45% (fixed value) after generalized calculation; setting of hydrological parameters adopt the default value of the model, including the average ground velocity of 0.3 m/s, hydrological attenuation coefficient of 0.9 and initial loss of 0.0006 m. The calibration result of model parameter is shown in Table 2.

Table 1 The experience value of integrated runoff coefficient

Unit type	Impervious percentage (%)	The integrated runoff coefficient
Central dense habitations	>70	0.60–0.80
Dense habitations	50–70	0.50–0.70
Sparse habitations	30–50	0.40–0.60
Very sparse habitations	<30	0.3–0.50

Table 2 Parameters calibration result

Parameters calibration result	Imperviousness (%)	Mean surface velocity (m/s)	Hydrological reduction factor	Initial loss (m)
Value	45	0.3	0.9	0.0006

On this basis, the model is verified through the actually measured typical "6.23" storm process in Beijing, and this storm has a precipitation of 121 mm and duration of 10 h. In the model, this precipitation process is used as the precipitation boundary to simulate the precipitation runoff process. The simulation result shows that the integrated runoff coefficient of the whole region is 0.403, which is within the range of urban integrated runoff coefficient. According to the sub-catchment runoff hydrograph statistics, the maximum surface flow occurred at about 17:00 and the maximum surface runoff depth is 5 cm; the total flow-accumulated of the northern region is 59816.605 m^3. The amount of runoff is small and in accordance with the actual investigation result and reflections of local residents, the precipitation process on June 23, 2011 had a small influence on the future science park and the roads had no obvious water accumulation phenomenon, which matches with the model simulation result and proves the reliability of the established model.

4 Result and Analysis

4.1 Scenario Simulation

In accordance with the Hydrologic Handbook of Beijing—Rainstorm Atlas [10] and Standard of Rainstorm Runoff Calculation for Urban Stormwater Drainage System Planning and Design of Beijing (DB11/T 969-2013) [11], and assuming that the return period is, respectively, 10 years, 20 years and 50 years and the rainfall duration is 24 h, the designed precipitation process is calculated under different return periods and the MIKE Zero software was used to process the designed precipitation data and generate the "*.dfs0" file [12, 13] complying with the model input requirements. By regarding this as the precipitation boundary input model, the simulative analysis of the designed rainfall flood of the research area was implemented and then the waterlogging risk in its urban area was assessed. The designed rainfall process is shown in Fig. 3.

4.2 Simulation Result and Analysis

By regarding the above-mentioned design rainfall process as the precipitation boundary condition of the model, the parameters of runoff model used the parameters calibrated by the model and default value of the model, and the step length of simulation time is set to be the 60 s. The pipe network confluence model should use the dynamic wave equation of the Saint-Venant equation set to simulate the pipe flow.

The accumulative runoff of each catchment sub-area in accordance with the runoff simulation result was loaded, the typical catchment sub-area with a larger

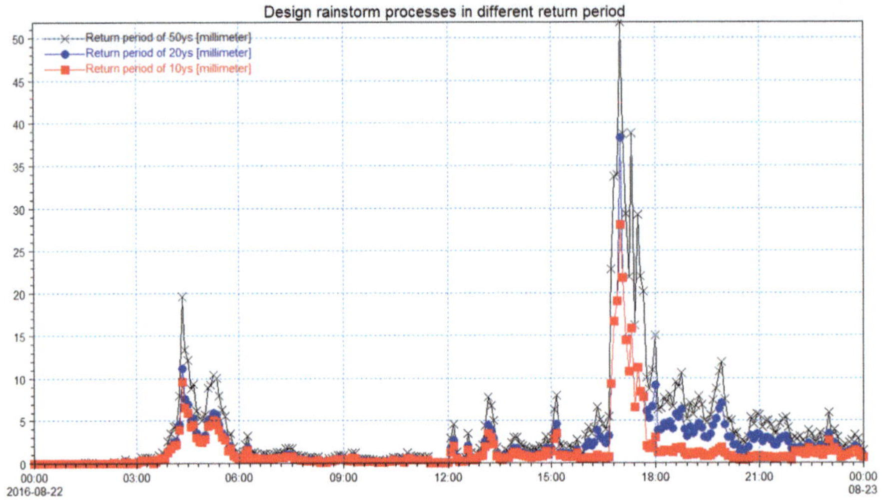

Fig. 3 Design rainstorm processes in different return period

cumulative water quantity was selected, the corresponding confluence node was determined and statistics of the maximum cumulative water volume of the typical catchment sub-area were made. The summary of simulation result of the catchment area is shown in Table 3. Statistics of the corresponding maximum ground flow in each typical catchment area were made and compared with the pipeline design standard, as shown in the following Table 4.

In accordance with the analysis of the simulation result of the typical catchment area, the catchment sub-area with a large cumulative water quantity will have the water ponding phenomenon during the whole precipitation process and there will be a high waterlogging risk, but the main problem is that the pipe network has an insufficient rainwater discharge capacity. However, not all areas will have this kind of phenomenon; for example: Catchment9 has a high pipeline design standard and can meet the drainage requirements.

Table 3 Catchment result summary

Rain event		10 yrs	20 yrs	50 yrs
Total flow-accumulated of north region (m^3)		211897.139	315867.421	565576.797
Typical catchment flow—accumulated (m^3)	Catchment50	6286.0133	31346.036	56129.144
	Catchment67	4691.4329	23392.960	41887.473
	Catchment55	3664.6886	18274.920	32723.821
	Catchment43	3049.5383	15209.545	27235.819
	Catchment30	3128.6283	15597.458	27927.571
	Catchment9	10060.982	14996.259	26850.987

Table 4 Typical catchment discharge compare

Typical catchment flow—accumulated (m^3)	Converge manhole	Pipe drainage capacity (m^3/s)	Maximum discharge		
			10 years	20 years	50 years
Catchment50	E5-6	1.247	2.428	2.708	4.865
Catchment67	E5-3	2.829	2.001	2.147	4.025
Catchment55	E5-16	1.799	1.407	1.566	2.821
Catchment43	E5-18	0.924	1.220	1.301	2.515
Catchment30	E5-14	1.022	1.442	1.555	2.843
Catchment9	E5-4	9.546	1.253	1.440	2.487

On account of the above-mentioned phenomenon, the runoff yield and confluence mode should be connected with the pipe flow model in order to further assess the waterlogging risk degree in this area. The following Fig. 4 shows the pipe pressure classification diagram under different design return period of the rainfall.

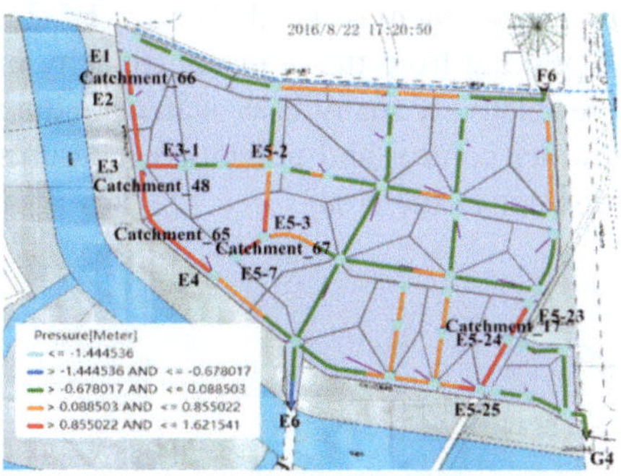

(10 yrs)

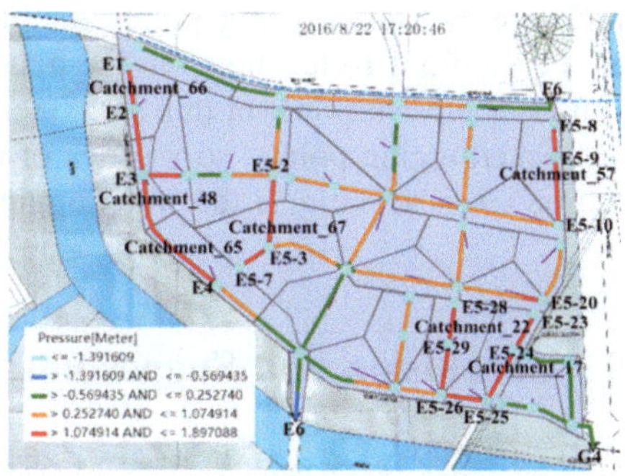

(20 yrs)

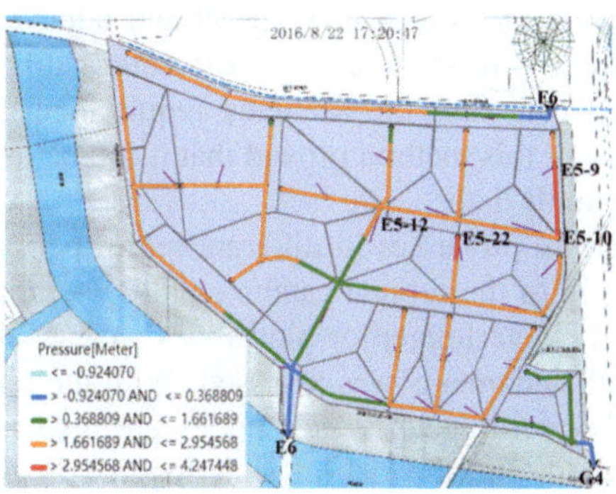

(50 yrs)

Fig. 4 The grade of pipe drainage capacity

Different colors of the pipe section represent different loads. The deeper the color, the larger the pipe load and the rainwater drainage capacity of the corresponding pipe section will be difficult to meet the demands of designed precipitation, and the waterlogging risk will be higher. The time shown in the figure is the selected observation time after taking full account of the occurrence time of the peak value of each pipe section; on this basis, the pipe operation status at the same time in the whole area can be intuitively judged so as to assess the waterlogging risk.

After integrating the simulation result of the precipitation runoff process and confluence result of the pipe network and through combined analysis on the accumulative water quantity in the typical catchment sub-areas and the diagram for classification of pipe drainage ability, it can be concluded that the sub-area Catchment50 has the highest waterlogging risk. In combination with the above cause analysis of urban waterlogging, the main reason of higher waterlogging risk in this area is that the drainage capacity of pipe network is insufficient and the pipe network arrangement is sparse. This area is mainly the scientific research and development area with an area of 2.4 ha, so it is recommended to suitably strengthen the rainwater utilization measures, which not only can accommodate the waterlogging flood, but also can realize the utilization of rainwater.

The easy waterlogging points in this area also include E5-3, E5-22, E5-21, E5-11 and E5-10, etc. shown in the figure. Speaking from the drainage capacity of rainwater pipe network in this area, there are relatively many easy water ponding and waterlogging points, so a series of rainwater control and utilization measures can be taken to mitigate the waterlogging disaster in the area.

5 Conclusion and Suggestion

In this paper, MOUSE module of MIKE Urban software is used to construct stormwater drainage system model of the new district in Beijing Future Science Park. Under the condition of no measured runoff and other related data, integrated runoff coefficient method is used for model parameter calibration, and the results obtained by simulation can meet the accuracy requirements; therefore, the availability and applicability of this method proved that it can be used for the areas in the absence of observed runoff data.

Under design rainstorm processes in different return period scenario simulation for the model, according to the typical catchments flow-simulation results and the load of pipes, the easily waterlogged or plot nodes were analyzed, and the risk of waterlogging in typical catchments was assessed. With the return period of rainfall increasing, the easily waterlogged or plot nodes are also increased; the load of network is also increased; and the risk of waterlogging in typical catchments are greater.

This area has an overall flat terrain and faces a relatively higher waterlogging discharge and mitigation pressure. It can be seen from the distribution diagram of historical flood disaster risks that the research area is located in the area with a higher occurrence frequency of flood disaster risk. Speaking from the drainage capacity of rainwater pipe network in this area, there are many easy water ponding and waterlogging points in this area. As this area is still in the planning construction stage, the rainwater control and utilization into the urban rainwater drainage system are integrated, which can enhance the drainage capacity of urban rainwater system; the regional waterlogging drainage pressure and the waterlogging disaster in this area were mitigated [14]. In addition, the future science park in Beijing is the first batch of "Wisdom City Pilot Areas" approved by the Ministry of Housing and Urban-Rural Development of the People's Republic of China and the first batch of "Green Ecological Demonstration Areas" approved in Beijing City [15]. These policy advantages can be utilized to develop the rainwater control and utilization measures to give full play to the comprehensive benefit of the rainwater control and utilization project and realize the low carbon, energy saving, harmony and ecology.

The deficient points of this paper are that it does not consider the rainwater utilization facilities under construction in this area as well as the link between the outlet of rainwater pipe network and the river water level; and the model simulation is only limited to the one-dimensional simulation. These deficient points will be further completed and discussed in the future research.

Acknowledgements The researchers would like to extend their thanks to the Chinese National Natural Science Foundation (No. 51409078 and No. 51739011) and the Natural Science Foundation of Hebei Education Authority, China (No. ZD2014020).

References

1. Haris, H., Chow, M.F., Usman, F., et al.: Urban stormwater management model and tools for designing stormwater management of green infrastructure practices. J. Inst. Phys. Publ. **32** (2016). https://doi.org/10.1088/1755-1315/32/1/012022
2. Thorndahl, S., Balling, J.D., Larsen, U.B.B.: Analysis and integrated modelling of groundwater infiltration to sewer networks. J. Hydrol. Process. **30** (2016). https://doi.org/10.1002/hyp.10847
3. Lee, K.S., Cho, W.S., Hwang, J.W., et al.: Numerical simulation for reducing the flood damage of green park using MIKE Urban. Int. J. Control Autom. **8**, 37–54 (2015). https://doi.org/10.14257/ijca.2015.8.1.04
4. Wu, S., Wang, F., Qi, T.X., et al.: Application of DHI MIKE software to assessment of typical stormwater systems for Wuhan city. China Water Wastewater **30**, 113–115 (2014)
5. Li, W.F., Chen, Q.W., Mao, J.P.: Development of 1D and 2D coupled model to simulate urban inundation: an application to Beijing Olympic Village. Chin. Sci. Bull. 1613–1621 (2009). https://doi.org/10.1007/s11434-009-0208-1
6. Huang, J., Wang, S.Z., Deng, S.Z., et al.: Numerical study on the impact of GongJi road rain pump on the waterlogging in Huinan, Pudong District. J. Geosci. Environ. Prot. **2**, 52–58 (2014). https://doi.org/10.4236/gep.2014.25008
7. Huang, J., Wang, S.Z., Deng, S.Z., et al.: Numerical research on the effect of the water level on the urban drainage network in Huinan, Pudong district, CA. In: 4th International Conference on Civil, Architectural and Hydraulic Engineering, pp. 727–730 (2016). https://doi.org/10.4236/gep.2014.25008

8. Liu, X.P.: Study on the method and application of drainage network computer simulation. Ph.D. thesis, Tongji University (2006). https://doi.org/10.7666/d.y1013667
9. Tu, C.: Application of computer simulation of urban stormwater take Seongbuk of Dayu country as an example. Ph.D. thesis, Nanchang University (2014)
10. Beijing Water Bureau: Hydrologic Handbook of Beijing City—Rainstorm Atlas (1999)
11. Beijing Academy of Urban Planning & Design: Standard of rainstorm runoff calculation for urban stormwater drainage system planning and designing of Beijing (DB11/T 969-2013) (2013)
12. Danish Hydraulic Institute: MIKE Zero User Manual (2012)
13. Danish Hydraulic Institute: MIKE Urban User Manual (2012)
14. Wang, H.C., Shang, J.S., Zhang, S.H., et al.: Analyze the important role of rainwater control and utilization in urban rainwater drainage system. J. Water Wastewater Eng. **39**, 125–129 (2013). https://doi.org/10.3969/j.issn.1002-8471.2013.05.030
15. Information on http://baike.so.com/doc/5810793-6023596.html

International Comparison of Water Resources Utilization Efficiency in China

Long Yan, Jing Ma, Lanchao He, Fei Wang and Tao Chen

Abstract To obtain knowledge of the standard of water utilization of China from an international point of view, this paper analyzes the annual variation of water resources utilization in China, and compares with other typical countries. The study shows that water resources utilization efficiency has been greatly improved in China in the last 20 years and the water use amount per ten thousand-dollar GDP has declined by 88.6%. However, there is still room for improvement of water resources utilization efficiency compared to developed countries. The comparison of water utilization and human development shows that the higher Human Development Index (HDI) the country is, the more efficient water utilization the country has. The water utilization in China corresponds to its economic and social development with a little step forward. Development is the fundamental way to improve the standard of water utilization.

Keywords Water resources · Utilization · Water use efficiency
International comparison · HDI

L. Yan (✉) · J. Ma
China Institute of Water Resources and Hydropower Research, Beijing, China
e-mail: xiangyangyl@163.com

J. Ma
e-mail: jingma@iwhr.com

L. He
Ministry of Water Resources, Beijing, China
e-mail: Helanchao@mwr.gov.cn

F. Wang
Asia Ecoenergy Development Limited, Hong Kong, China
e-mail: wangfeizf@126.com

T. Chen
Jade Investment LTD, Beijing, China
e-mail: chentao5064@qq.com

W. Dong et al. (eds.), *Sustainable Development of Water Resources and Hydraulic Engineering in China*, Environmental Earth Sciences,
https://doi.org/10.1007/978-3-319-61630-8_5

1 Introduction

China has been supporting the largest scale population and economy as well as social activities in the world based on scarce water resources and fragile ecological environment. With the rapid development of industrialization and urbanization, total water consumption will increase in the future for a long time, and water resources will face more serious challenges. China has implemented the strictest management system on water resources; it is the first time for China to incorporate total water consumption control into national strategy and establish a total water consumption index and water utilization efficiency index as restrictive indexes in the national economy and society development program. Therefore, it is imperative to solve the problem of how to achieve the goal of total water consumption control on the premise of maintaining stable social and economic growth. For this reason, China's position on water resources development and utilization level must also be determined in the rest of the countries in the world and reference made to and ideas borrowed from the course change of water utilization efficiency and total water consumption in developed countries. However, for a long time, due to the lack of effective data and failure to carry out follow-up plans in a systemic way, it is difficult to reflect the present situation of water resource development and utilization in various countries in the world in an all-around manner. An inaccurate understanding of the present situation of water resources development and utilization level in the rest of the countries, especially in developed countries in the world, makes it difficult to determine China's position on water resources development and utilization level in the world in an accurate manner. Ma et al. [1, 2] have carried out international follow-up and made comparison in terms of water resources utilization efficiency in previous researches in China. Some researches indicate that if the urbanization rate in China increases 1%, the comprehensive utilization efficiency of water resources in China will increase by 7.68% [3] and the water resource utilization efficiency in China has significantly improved because of building the water-saving society in an all-around manner in China. It is necessary to re-evaluate the current situation of water resource utilization in China; understand the present situation and development pattern of water consumption and water resource utilization efficiency in major countries in the world; keep follow-up of and make comparison with developed countries in this field so as to determine water resource development and utilization level in China from an international perspective; provide an objective basis for implementation of the strictest water resource system; and abide by total water consumption control, water utilization efficiency control, and pollution control in the water function area in China.

2 Change in Water Consumption and Water Resource Utilization Efficiency in China

2.1 Water Consumption

It is observed from the change course (Fig. 1) of water consumption more than 60 years since the establishment of the People's Republic of China (PRC), before 1980 the water consumption grew rapidly, indicating rapid growth in agricultural water consumption. With the reconstruction and expansion of irrigation area, construction of water storage project and diversion works in river channels and popularization of lifting irrigation by a pump station, agricultural water consumption in this period increased by 2.7 times. After 1980, water consumption was steady. During this period the farmland under irrigation increased by about 100 million mu, while water consumption norm decreased due to application of water saving technology. Water utilization efficiency has significantly improved, and agricultural water demand has remained at 360 billion–390 billion m^3. After 2005, with the improvement of the water supply condition and expansion of the irrigation area, China's water consumption entered a slow growth period, with a gradual increase of agricultural water and industrial water consumption and rapid increase in domestic water consumption. Total water consumption increased from 563.2 billion m^3 in 2005 to 618.3 billion m^3 in 2013. Since 1997, the per capita water consumption in China has insignificantly changed, remaining at approximately 430–460 m^3, with the exception that agricultural water consumption decreased by about 30 billion m^3 due to drought and less rainfall in south China in 2003, which resulted in the lowest per capita water consumption since 1997, being 412 m^3.

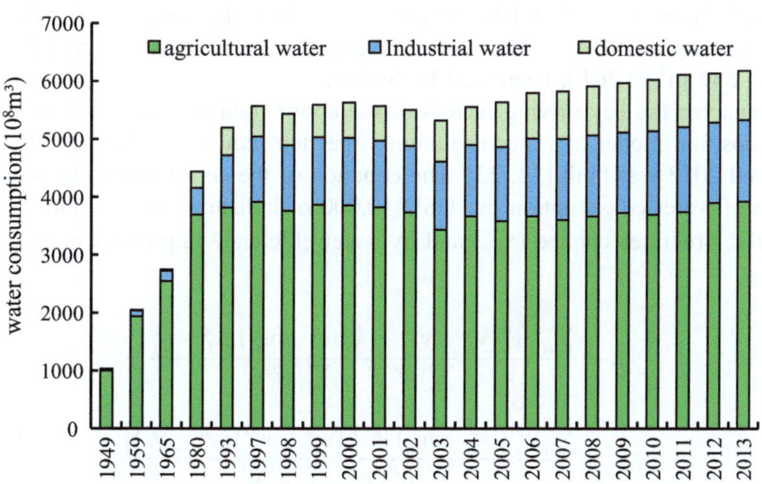

Fig. 1 Variation of water use in China from 1949 to 2013

With regard to the water consumption structure in China, in 1949, the percentage of agricultural water consumption was up to 97%; industrial water consumption was only up to 2%; and domestic water consumption was lower than 1%. Since then, with the application of water saving technology and adjustment of the industrial structure, the percentage of agricultural water consumption has been on the decrease and industrial water consumption and domestic water consumption have been on the increase. After 2005, water consumption in tertiary industries has been on the gradual increase; their water consumption structure has maintained at 63:23:14. It can be seen that with sustainable development of urbanization and industrialization in China, China's domestic water consumption and industrial water consumption will be on the increase, while agricultural water consumption will remain steady as a whole due to the constant generalization of water saving irrigation technology, making the percentage of agricultural water consumption decrease and domestic water and industrial water consumption increase to a certain extent.

2.2 Water Consumption Efficiency

Water consumption per USD 10,000 of GDP and water consumption per USD 10,000 of industrial added value are usually adopted to evaluate the comprehensive water consumption efficiency in a country in the world. This paper shows statistical analysis of comprehensive water consumption efficiency and industrial water consumption efficiency in China from 1997 to 2013 (Fig. 2), in which GDP data is based on the current price of USD from WDI database in World Bank [4]; water consumption data is from China Water Resources Bulletin since 1997 [5–21]. It is observed from the calculation that China's water consumption efficiency has greatly increased in the last 20 years, with water consumption per USD 10,000 of GDP decreased from 5843 m^3 in 1997 to 669 m^3 in 2013, decreased by 88.6%. Water consumption per USD 10,000 of industrial added value decreased from 2475 m^3 in 1997 to 347 m^3 in 2013, decreased by 86.0%.

To analyze the key driving factors of the above indicator decrease, a complete decomposition model of water use amount per ten thousand-dollar GDP between 1997 and 2013 was built [22, 23]. The equation of the contribution rate which the change of water consumption per USD 10,000 of industrial added value made by industrial structures change (γ_{str}) and by water efficiency improvement (γ_{eff}) is as follows:

$$\gamma_{str} = \frac{\sum_i \left[w_i^0 \left(y_i^n - y_i^0 \right) + \frac{1}{2} \left(w_i^n - w_i^0 \right) \left(y_i^n - y_i^0 \right) \right]}{\sum_i w_i^n y_i^n - \sum_i w_i^0 y_i^0} \tag{1}$$

$$\gamma_{eff} = \frac{\sum_i \left[y_i^0 \left(w_i^n - w_i^0 \right) + \frac{1}{2} \left(w_i^n - w_i^0 \right) \left(y_i^n - y_i^0 \right) \right]}{\sum_i w_i^n y_i^n - \sum_i w_i^0 y_i^0} \tag{2}$$

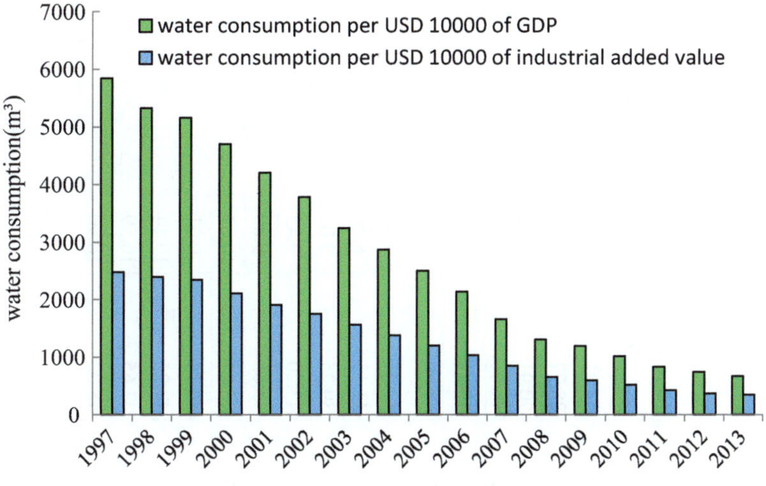

Fig. 2 Variation of water use efficiency of China from 1997 to 2013

It is observed from Table 1 that from 1997 to 2013, the improvement of industrial water consumption efficiency and structural adjustment are the key driving factors for substantial decrease in water consumption in China, with a higher rate of contribution by improvement of industrial water consumption efficiency and rate of contribution after 2006 having been higher than 70%. Contribution of improvement in industrial water use efficiency is due mainly to industry and agriculture: the rate of contribution made by industry maintained at about 18% after 2006; and the rate of contribution made by agriculture increased year after year, which is up to 47.65% in 2013. Given the worsening situation of shortage in water resources in China, improvement in water resources utilization efficiency has been upgraded to a national strategy. The total water consumption and water resource utilization efficiency have been used as a restrictive index incorporated in a national economic and social development program, embodying the position and function of water resource in national macro control. Several measures, including legislation, policy guidance and economic leverage, will be taken to improve water resource utilization efficiency. The contribution made by industrial restructuring to decrease water consumption per USD 10,000 of GDP is mainly embodied in adjustment of agricultural structure. The rate of contribution made by change of percentage of industry and other industrial structures is insignificant, the reason for which is that agricultural water consumption in China has been large in percentage. Its water consumption per USD 10,000 of GDP is much higher than water consumption of industry and other industries. From 1997 to 2013, the agricultural structure changed substantially in percentage, decreasing from 18.3% in 1997 to 10% in 2013.

Table 1 Variable effect caused by industrial structure change and water efficiency improvement

Year	γ_{str}	Rate of contribution (%)			γ_{eff}	Rate of contribution (%)		
		Industry	Agriculture	Other industrial structures		Industry	Agriculture	Other industrial structures
1997–1998	27.50	5.69	27.36	−5.54	72.50	8.94	53.91	9.64
1998–1999	48.04	5.10	49.46	−6.52	51.96	11.30	28.77	11.90
1999–2000	53.83	2.96	56.52	−5.65	46.17	15.44	20.91	9.82
2000–2001	46.91	2.98	48.93	−5.00	53.09	16.25	25.72	11.12
2001–2002	42.25	2.66	44.08	−4.48	57.75	16.38	30.63	10.73
2002–2003	38.18	1.15	40.57	−3.54	61.82	16.56	37.09	8.17
2003–2004	29.77	0.88	31.55	−2.66	70.23	17.19	43.32	9.72
2004–2005	32.50	0.06	34.88	−2.44	67.50	17.91	41.13	8.46
2005–2006	33.33	−0.20	35.79	−2.26	66.67	18.39	39.25	9.03
2006–2007	29.14	0.09	31.21	−2.16	70.86	18.21	43.20	9.45
2007–2008	25.93	0.04	27.81	−1.92	74.07	18.82	45.65	9.60
2008–2009	26.23	0.47	27.93	−2.17	73.77	18.67	45.03	10.07
2009–2010	24.93	0.27	26.64	−1.98	75.07	18.83	46.24	10.01
2010–2011	23.60	0.29	25.22	−1.90	76.40	19.01	47.25	10.14
2011–2012	22.76	0.67	24.14	−2.05	77.24	18.99	47.65	10.59
2012–2013	22.49	1.07	23.68	−2.27	77.51	18.59	48.02	10.90

3 International Comparisons

This paper presents the collections of and analyzes water consumption, water consumption structure, water consumption efficiency and industrial structure in more than 100 countries and regions in the world based on statistics, in which, global water resources quantity and water consumption data are difficult to obtain. The paper adopts the prevailing and latest data from (FAO) AQUASTAT database [24]; the database is upgraded every five years. China and the US adopt the most authoritative official data related to water resources in their countries, respectively [21, 25]. Data about added values in GDP and tertiary industries in other countries are from WDI database in the World Bank.

3.1 International Comparisons in Water Consumption

Water Consumption in Industry. In 2013, total water consumption in the world was 4 trillion m^3; the total water consumption of top 10 countries in water consumption ranking was up to 2.61 trillion m^3, accounting for 65.2% of total water consumption in the world, in which, water consumption in China was 618.34 billion m^3; next is India, which is 1.26 times that in the US. Industrial water consumption in China was 140.64 billion m^3, being the second in the world, 55.6% of that in the US. Agricultural water consumption in China was 392.15 billion m^3, being the second in the world and 0.57 times that in India and 2.46 times that in the US. Domestic water consumption in China was 85.55 billion m^3, being the first in the world.

Change of Water Consumption in Representative Countries. Economic scale and production mode determine total water consumption and water consumption efficiency. The top three countries of GDP ranking in the world in 2013 were the US, China and Japan, in which, the total water consumption in the US and Japan was up to the peak value in 1980 and 1990, respectively. Since then, their total water consumption showed negative increase or slight increase. The change course of water consumption in the US and Japan may serve as an example and reference for China to control water consumption in the future.

It can be seen from Fig. 3 that from 1950 to 1980, total water consumption in the US has grown rapidly, with total water consumption increased by 360 billion m^3 in 30 years, in which industrial water consumption increased by about 250 billion m^3, mainly due to the rapid development of high water consuming heavy industry during that period, with steel output up to all-time highs in 1973. Since 1980, low water consuming infant industry and service industry achieved rapid development and became the leading industry stimulating economic growth. With water consumption efficiency greatly improved, water consumption in industry and agriculture constantly decreased due to technological advance. Although domestic water consumption has kept increasing, total water consumption remained stable as a whole.

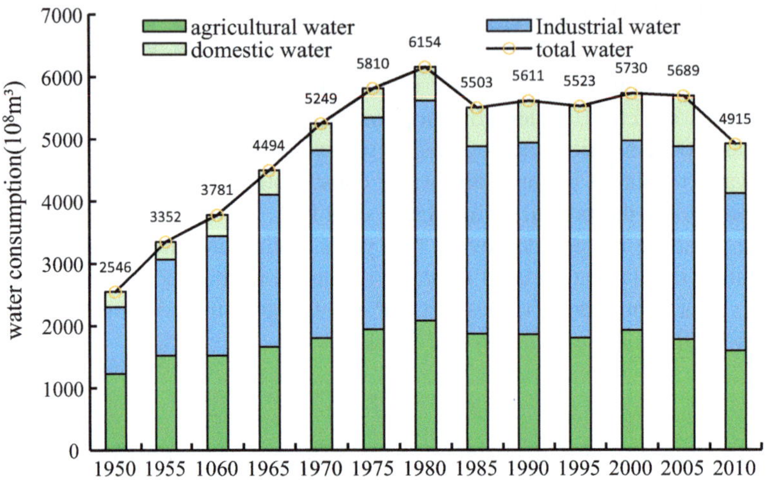

Fig. 3 Variation of water use in the US from 1950 to 2010

After 2005, the US shut down some thermal power stations to decrease coal-fuel electricity share and developed natural gas power generation with significant effort. In 2010, total water consumption in industry and agriculture decreased to a minimum level since 1970. With the economic development, relatively steady water consumption in the US can be described by a word "decoupling", proving that water consumption is not necessarily to increase based on the sustainable and steady growth of the economy and society. Therefore, the goal to stabilize water consumption is achievable with effort.

Total water consumption in Japan has not changed greatly in the past 40 years, characterized by water consumption in industry and agriculture on the decrease and domestic water consumption on slow increase (Fig. 4). In terms of agriculture, Japan popularizes water saving irrigation technology with earnest effort. They have established a waste water purification treatment and then irrigated the farmland with recycled water. In terms of industry, Japan has been striving to develop electro-mechanical device manufacture, and transportation appliance manufacture with high output and low water consumption by industrial restructuring with repeating utilization rate of industrial water significantly improved. From 1965 to 1980, its repeating utilization rate increased from 36% to 74% in these 15 years. It is also in this period that Japan's economy has developed rapidly and achieved industrialization. In terms of domestic water, the Japanese government has carried out a water saving policy with to enhance water-saving awareness, and reduce the leakage rate by renewal of water supply systems and popularization of water economizer, and established a waste water treatment and recycle system and a rainwater utilization apparatus to utilize recycled water and rain water, with domestic water consumption on a stable increase.

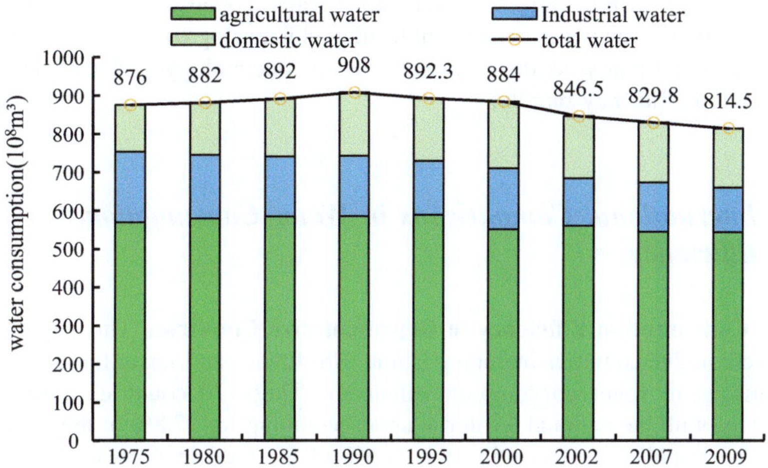

Fig. 4 Variation of water use in Japan from 1975 to 2009

The water consumption process in the US and Japan (two developed countries) shows water consumption is not always on the increase with the development of society and economy, characterized by certain stages. In the early and middle stage of industrialization, industrial water consumption is on a rapid increase, with percentage of its water consumption accounting for total water consumption which increased constantly. In the latter stage of industrialization, influenced by resource shortage, government starts to adjust the industrial structure to gradually eliminate high water consuming industries and strive to develop low water consuming process industries. In addition, with water saving technology and process popularized and industrial water consumption efficiency significantly enhanced, the economic scale in the period achieved a steady increase but the industrial water consumption remained basically stable. With the popularization and application of water saving technology, agricultural water consumption is on a slow decrease and domestic water consumption is on the increase with the increase of population and improvement of living standards. After completion of industrialization and entering the post industrialization stage, industrial structure changes, low water consuming tertiary industry begins to achieve rapid development, the water consumption in industry and agriculture remains stable with a slight decrease and total water consumption on the negative increase. It is suggested from the experience of the US and Japan that on the premise of sustainable and steady development of society and economy, water consumption is not necessarily on the increase; therefore, the goal to stabilize water consumption is achievable with great effort. China is in the middle and latter stage of industrialization, according to the experience of the US and Japan. This period is characterized by substantial industrial restructuring, with low water consuming process industry and high tech and high value-added industry coming into leading industry in industrial development and low water saving costs.

It is convenient to popularize and generalize water saving technology; therefore, during this period the water resource utilization efficiency grows rapidly. China will grasp this opportunity to develop water saving technology and enhance water consumption efficiency rapidly.

3.2 International Comparisons in Water Consumption Efficiency

Water Consumption Efficiency in Representative Countries. This paper selects 20 representative countries including China, US, France and Japan for international comparisons in water consumption efficiency. These 20 countries covering six continents, with the national territorial area accounting for 57.8% of the total in the world, the population accounting for 61.5% of the total in the world, and GDP accounting for 73.2% of the total in the world. The gap between China and developed countries in water consumption efficiency can be seen according to the difference in water consumption efficiency shown in the comparison analysis of these 20 countries.

Water consumption data per USD 10,000 of GDP in every one of these 20 countries in 2010 and 2013 are shown in Fig. 5. Water consumption data per USD 10,000 of GDP in developing countries including China, India and Egypt in 2013 obviously decreased compared with those in 2010; the water consumption per USD 10,000 of GDP in developed countries including US, France and Spain in 2013 remained unchanged or was on the slight decrease compared with those in 2010, indicating that there is room for China to enhance water consumption efficiency.

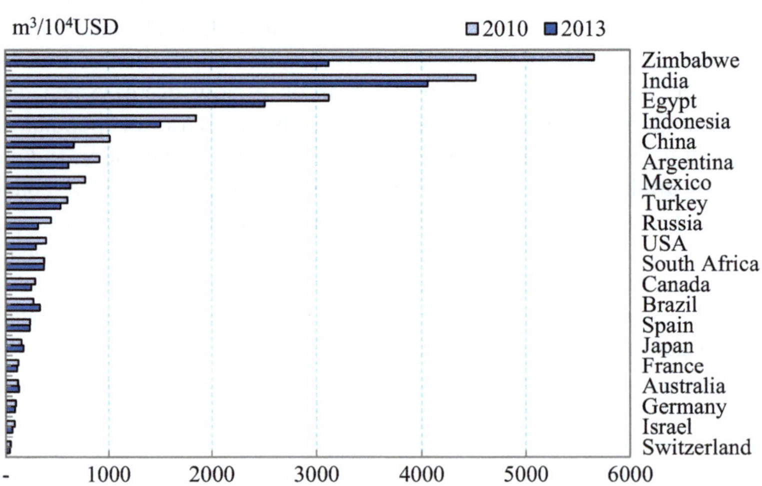

Fig. 5 Water consumption data per USD 10,000 of GDP

Water consumption per USD 10,000 of GDP in China in 2013 was 669 m^3, being 17.6 times that in Switzerland, 10.6 times that in Israel, 7.7 times that in Germany, 2.3 times that in the US, which has a wide gap compared with developed countries in this field, but the gap was narrowing compared with that in 2010. China was still in advanced level in water consumption efficiency compared with those in other developing countries, being equivalent with that in Argentina and Mexico and being far superior to that in Egypt and India.

For industrial water consumption, water consumption per USD 10,000 of industrial added value in a majority of countries in 2013 was on the decrease compared with that in 2010. The decrease in water consumption in developing countries is more than those in developed countries; water consumption in developed countries remained unchanged or was on a slight decrease. Water consumption per USD 10,000 of industrial added value in Canada, Brazil, Turkey and Spain in 2013 was higher than those in 2010, the reason for which was that the four countries had adjusted their industrial structures, with percentage of industrial added value on a slight decrease and without apparent change in industrial water consumption. Water consumption per USD 10,000 of industrial added value in China in 2013 was 347 m^3, being 12 times that in Israel, 4.2 times that Japan. Compared with 2010, China was gradually narrowing the gap between China and developed country in industrial water consumption efficiency (Fig. 6).

Water Use Efficiency in Countries at Different Development Levels. To determine China's water resource utilization level based on international perspective, Ma et al. [2] introduced Human Development Index (HDI) to analyze the difference in water resource utilization efficiency between China and other countries in different development levels. The closer the HDI nearer to 1 indicates the higher human development level. The paper divides statistical water consumption efficiency in

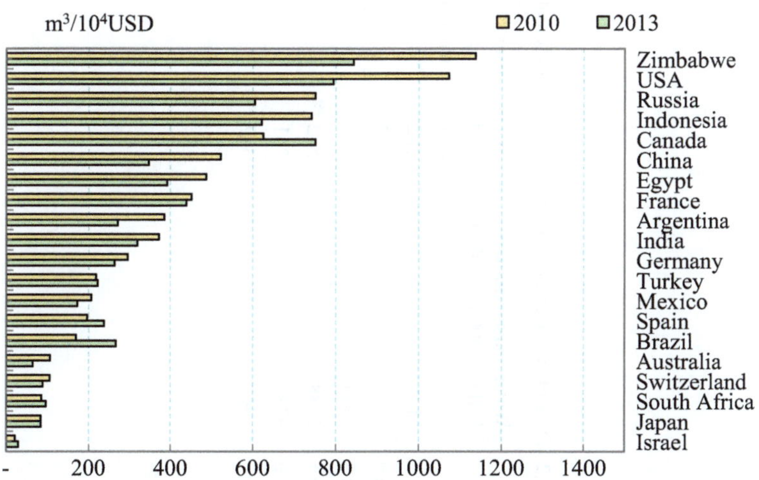

Fig. 6 Water consumption per USD 10,000 of industrial added value

Table 2 Data of water utilization efficiency with different HDI

HDI	Water consumption per USD 10,000 of GDP (m^3)	Water consumption per USD 10,000 of industrial added value (m^3)	Water consumption per capita (m^3)	Water consumption structure (%)		
				Industry	Agriculture	Other industrial structures
>0.9	145	182	778	29.7	53.5	16.8
0.8–0.9	180	182	580	53.8	27.0	19.2
0.7–0.8	627	367	502	64.5	21.0	14.5
0.6–0.7	2238	572	757	82.4	9.6	8.0
0.5–0.6	3334	261	522	90.3	2.2	7.5
China	669	347	456	63.5	22.7	13.8

more than 100 countries and regions according to HDI level and obtained water consumption efficiency of countries in different development level in 2013 (Table 2).

Development and utilization of water resource in a country is not only influenced by climate and living habit but also is closely associated with the development stage and development level of economy and society. A country with a higher development level of society and economy is superior to the country with a lower development level of society and economy, in terms of index, including water consumption structure and water consumption efficiency to evaluate water resource development and utilization level. Table 2 shows that with the increase of HDI, water consumption per USD 10,000 of GDP and water consumption per USD 10,000 of industrial added value are on the decrease and water consumption per capita is on the increase. For three major industries, lower percentage in agriculture will result in lower percentage in agricultural water consumption in a water consumption structure, in which water consumption per USD 10,000 of GDP is the most obvious. Water consumption per USD 10,000 of GDP in a country with HDI > 0.9 is 145 m^3, being 4.3% of that in a country with HDI from 0.5 to 0.6.

In 2013, China's HDI was 0.719; water consumption per USD 10,000 of GDP was 669 m^3, being equivalent to those countries with HDI from 0.7 to 0.8 and much lower than those countries with HDI from 0.6 to 0.7. Water consumption per USD 10,000 of industrial added value was 347 m^3; industrial water consumption efficiency is superior to those countries with HDI from 0.7 to 0.8. Water consumption per capita was 456 m^3 is lower than those countries with HDI from 0.7 to 0.8. Agricultural water consumption percentage was 63.5%, slightly superior to those countries with HDI from 0.7 to 0.8. In a word, water resources utilization efficiency in China is superior to that in a similar country in HDI, indicating that water resource utilization efficiency is not only suitable for but also ahead of the economic development level in China.

4 Conclusion

This paper clearly determines water resource development and utilization level based on information on water resource development and utilization in the rest of the countries in the world. In 2013, China's HDI was 0.719, but the water consumption efficiency was slightly superior to those countries with HDI from 0.7 to 0.8, indicating that there was no change in water resource utilization efficiency in China. China has significantly improved the water resource utilization efficiency in recent years, although there is a wide gap compared with developed country in this field, the gap has been narrowing constantly.

It is observed from the experience of the two developed countries of the US and Japan that on the premise of sustainable and steady development of society and economy, water consumption is not necessarily on the increase. Therefore, the goal to stabilize water consumption is achievable with great effort. The 18th CPC

National Congress put forward that GDP in 2020 should be doubled compared with that in 2010. In addition, with the constant increase in population and continuous improvement in urbanization, contradiction between supply and demand in water resource will become more and more obvious. No. 1 Central Document in 2011 indicated that the threshold for development and utilization of water resource should be established to control total water consumption and that total water consumption in 2020 should be within 670 billion m^3. China is in the middle and latter stage of industrialization, according to the experience of the US and Japan. This period is characterized by substantial industrial restructuring, with low water consuming process industry and high tech and high value-added industry becoming leading industries in industrial development and low water saving costs; and it is convenient to popularize and generalize water saving technology. Therefore, during this period water resources utilization efficiency grows rapidly. China well grasp this opportunity to develop water saving technology and enhance water consumption efficiency rapidly and fulfill the goal to control total water consumption in industrialization.

At present, global water resource shortage is becoming more and more serious, especially in those countries in lower development level, with total water demand on the increase due to economic development and increase in population. Water consumption structure and water consumption efficiency are closely associated with the development stage and development level of society and economy. The experience of developed countries shows that industrial restructuring due to national economic development is the source power of stabilizing total water consumption. Countries with higher development level of society and economy are superior to those countries with lower development level of society and economy, in terms of index, including water consumption structure and water consumption efficiency to evaluate water resource development and utilization level. Water resource utilization efficiency in China is not only suitable for but also ahead of their economic development level. The fundamental approach to improve water resource utilization efficiency and achieve the goal to control total water consumption is none other than development.

Acknowledgements The research was supported by the National Key Research and Development Program of China (No. 2016YFC0503502) and the National Science Foundation for Distinguished Young Scholars of China (Grant No. 51409271)-study on threshold depth to groundwater for vegetation stability.

References

1. Ma, J., Yang, Z.H., et al.: Tracing and comparison of international water resources utilization efficiency. China Water Resour. **5**, 13–17 (2012). (in Chinese)
2. Ma, J., Chen, T., Shen, B.F., et al.: Comparison of water resources utilization and its development at home and abroad. Adv. Sci. Technol. Water Resour. **27**(1), 1–6 (2007). (in Chinese)

3. Ma, H.L., Xu, J., Wang, P.C.: Water resources utilization and China's urbanization. Resour. Sci. **36**(2), 0334–0341 (2014). (in Chinese)
4. World Bank: World Development Indicator. http://www.worldbank.org
5. The Ministry of Water Resources of the People's Republic of China: China Water Resources Bulletin 1997. Water & Power Press, Beijing, China (1998). (in Chinese)
6. The Ministry of Water Resources of the People's Republic of China: China Water Resources Bulletin 1998. Water & Power Press, Beijing, China (1999). (in Chinese)
7. The Ministry of Water Resources of the People's Republic of China: China Water Resources Bulletin 1999. Water & Power Press, Beijing, China (2000). (in Chinese)
8. The Ministry of Water Resources of the People's Republic of China: China Water Resources Bulletin 2000. Water & Power Press, Beijing, China (2001). (in Chinese)
9. The Ministry of Water Resources of the People's Republic of China: China Water Resources Bulletin 2001. Water & Power Press, Beijing, China (2002). (in Chinese)
10. The Ministry of Water Resources of the People's Republic of China: China Water Resources Bulletin 2002. Water & Power Press, Beijing, China (2003). (in Chinese)
11. The Ministry of Water Resources of the People's Republic of China: China Water Resources Bulletin 2003. Water & Power Press, Beijing, China (2004). (in Chinese)
12. The Ministry of Water Resources of the People's Republic of China: China Water Resources Bulletin 2004. Water & Power Press, Beijing, China (2005). (in Chinese)
13. The Ministry of Water Resources of the People's Republic of China: China Water Resources Bulletin 2005. Water & Power Press, Beijing, China (2006). (in Chinese)
14. The Ministry of Water Resources of the People's Republic of China: China Water Resources Bulletin 2006. Water & Power Press, Beijing, China (2007). (in Chinese)
15. The Ministry of Water Resources of the People's Republic of China: China Water Resources Bulletin 2007. Water & Power Press, Beijing, China (2008). (in Chinese)
16. The Ministry of Water Resources of the People's Republic of China: China Water Resources Bulletin 2008. Water & Power Press, Beijing, China (2009). (in Chinese)
17. The Ministry of Water Resources of the People's Republic of China: China Water Resources Bulletin 2009. Water & Power Press, Beijing, China (2010). (in Chinese)
18. The Ministry of Water Resources of the People's Republic of China: China Water Resources Bulletin 2010. Water & Power Press, Beijing, China. (2011). (in Chinese)
19. The Ministry of Water Resources of the People's Republic of China: China Water Resources Bulletin 2011. Water & Power Press, Beijing, China (2012). (in Chinese)
20. The Ministry of Water Resources of the People's Republic of China: China Water Resources Bulletin 2012. Water & Power Press, Beijing, China (2013). (in Chinese)
21. The Ministry of Water Resources of the People's Republic of China: China Water Resources Bulletin 2013. Water & Power Press, Beijing, China (2014). (in Chinese)
22. Sun, J.W.: Changes in energy consumption and energy intensity; A complete decomposition model. Energy Econ. **20**(1), 85–100 (1998)
23. Tong, J.P., Ma, J.F., Liu, G.F.: Analysis of variation and factors of water use amount per ten thousand Yuan GDP in China based on a complete decomposition model. Resour. Sci. **33**(10), 1870–1876 (2011). (in Chinese)
24. FAO, AQUASTAT. http://www.fao.org
25. Maupin, M.A., Kenny, J.F., Hutson, S.S., et al.: Estimated Use of Water in the United States in 2010. U.S. Geological Survey Circular (2014)

Dynamic Water Environmental Capacity Calculations of Rivers Based on Hydrological Processes

Wei Deng, Jing Ma, Long Yan and Ying Zhang

Abstract The process and mechanism of the water environmental capacity, and the coupling mechanism between it and the hydrological processes were studied. Then, a dynamic approach based on the hydrologic processes was developed in which distributed hydrological and water environmental capacity models were coupled to solve the problem in a traditional way. The Tieling Section of the Liao River Basin was taken as the study area for the purpose of demonstrating the proposed method. The results indicated that the water environmental capacity was uneven during different times and spaces. The results also explained that, contrary to the dynamic methods which separately calculated the years with different frequencies, seasons, or months, the traditional method was too conservative to make full use of the water environmental capacity.

Keywords Water environment management · Dynamic water environmental capacity calculation · Distributed hydrological model · Hydrologic process

1 Introduction

Water pollution is one of the most serious water issues in China, which is essentially due to a severely overdrawn water environment carrying capacity (WECC). Therefore, it is important to regard the WECC as a rigid constraint for the future sustainable development of China's economy and society. The water environmental

W. Deng (✉) · J. Ma · L. Yan · Y. Zhang
China Institute of Water Resources and Hydropower Research, Beijing, China
e-mail: dwsylar@163.com

J. Ma
e-mail: jingma@iwhr.com

L. Yan
e-mail: xiangyangyl@163.com

Y. Zhang
e-mail: cwzhangying@iwhr.com

W. Dong et al. (eds.), *Sustainable Development of Water Resources and Hydraulic Engineering in China*, Environmental Earth Sciences,
https://doi.org/10.1007/978-3-319-61630-8_6

capacity (WEC) is usually used as a quantitative indicator of carrying capacities of the water environment. In China, the WEC is normally defined as the maximum amount of pollution the system allows within a unit of time and in given hydrological conditions, discharging patterns, and water quality targets [1]. The calculation methods include analytical formula [2, 3], model trial [4, 5], and system optimization methods [6, 7]. Regardless of the method used, the WEC is calculated under designed hydrological conditions (usually the most withered month average discharge in a 90% assurance rate, or the most withered month average discharge during the most recent ten years) without consideration being given to the uncertainty of the hydrology within a time scale according to the definition. Furthermore, the design hydrological conditions are often derived from the analysis of the hydrological station data, which also ignore the influence of the hydrological process along the rivers. Many mathematical approaches have been introduced to improve the calculation methods. These have generally included negative mathematical methods [8], a stochastic differential equation [9] and stochastic optimization models [10]. These approaches have been found to only partially solve the uncertainty problem of hydrology and water quality. The water environmental capacity is considered to be closely related to the hydrological factors, and two-dimensional or three-dimensional hydrodynamic models have been developed and applied on a small scale, such as in lakes [11] and reservoirs [12]. However, the excessive data requirement has been found to limit further applications on larger scales. A dynamic accounting method which uses the historical hydrographic data of hydrological stations has also been introduced to solve the problem [13]. However, due to the limitations of the layouts of the hydrological stations, it cannot dock well with the computing unit and the spatial changes of the hydrological elements.

The changes of the water environments are the associated phenomena of the water cycle. The WEC is a significant quantitative indicator of the water environment, and is hypersensitive to the river runoff and hydrodynamic conditions. Therefore, based on the study of the process and mechanism of the water environmental capacity, and the coupling mechanism between it and the hydrological process, a dynamic WEC calculation method was developed in which the distributed hydrological and water environmental capacity models were coupled in order to solve the hydrological uncertainty problem for which the traditional methods had previously ignored. The Tieling Section of the Liao River Basin was taken as the case study area for the purpose of demonstrating the proposed method, and to explore the feasibility of the application on a large scale. The results potentially provided technical support for both future dynamic and delicate water environmental management.

2 Methods

2.1 Study Area

Tieling City (41°59′–43°29′ N, and 123°27′–125°06′ E) is in the northeastern section of Liaoning Province, east of Jilin Province, and west of the Inner Mongolia Autonomous Region. The total area of Tieling City is 13,000 km^2. With the rapid development of the economy and society, the pressure on the water environment has increased. The total discharge of the waste water increased from 75 million tons in 2008, to 86 million tons in 2013. Also, this special geographical location has led to serious pressures from pollution which originates from the upstream. In this study, seven rivers which had drainage areas of at least 500 km^2 were chosen, including the Liao, Zhaosutai, Qing, Chai, Fan, Kou, and Erdao Rivers (Fig. 1). The basic calculation units were the 19 water function zones (including 5 types) (Fig. 1) in these seven rivers. In Fig. 1, AWCR, PDCR, BR, TR, DWSR, respectively, are the agricultural water consumption region, the pollutant discharge

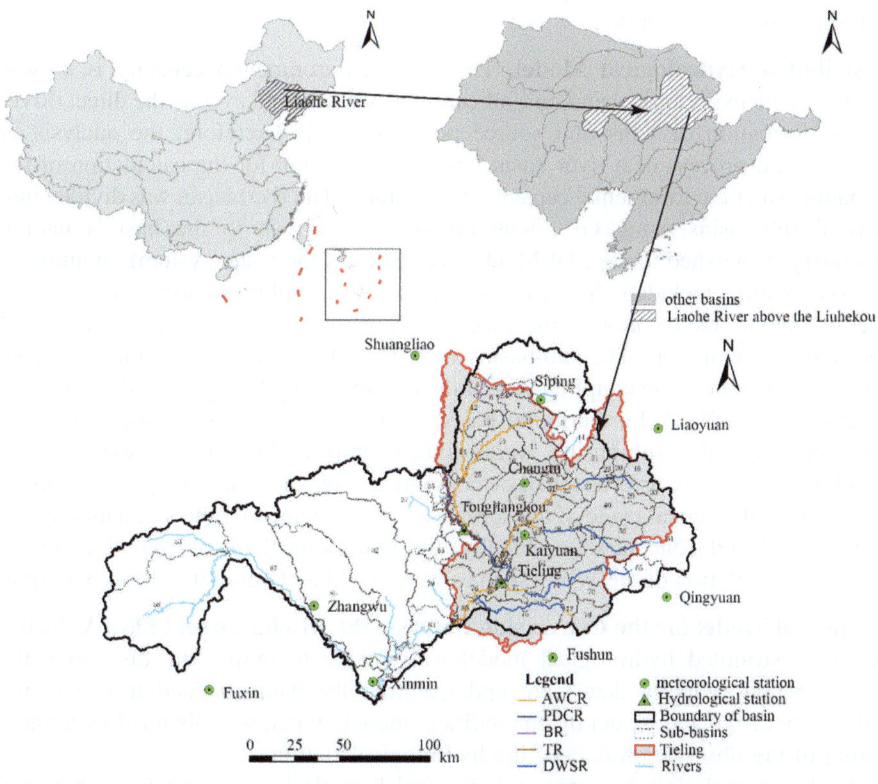

Fig. 1 Location of basin, distribution of water function zones and delineation of sub-basins

control region, the buffer region, the transitional region and the drinking water source region.

The seven rivers were mainly located in the Liaohe River Basin above Liuhekou, of which the total area was 30,000 km^2. A distributed hydrological model was established in the river basin.

The control pollutant index of the Liao River Basin was COD (chemical oxygen demand) and NH_3-N, according to the planning documents of the management department of the Liao River Basin. Also, the water environmental monitoring results for the Tieling section of the Liao River also showed that the primary pollutant was NH3-N. Therefore, this study focused on the COD and NH_3-N levels in the basin.

2.2 *Framework and Methodology*

In this study, a dynamic method was developed to calculate the WEC, which was composed of a distributed hydrological model, an empirical model for the conversion between the discharge and flow velocity, and a water environmental capacity calculation model.

Distributed Hydrological Model. The water environmental capacity is closely related to the hydrological and rainfall-runoff processes, and even is the direct driver of the formation of non-point source pollution [14]. Therefore, the analysis of hydrological process of a river basin laid the foundation for the calculation of the dynamic water environmental capacity in this study. The river basin was divided into several sub-basins using GIS, with consideration given to the basin's natural boundary (watershed terrain, DEM, characteristics of the water system); monitoring station (mainly included the hydrologic and water quality station); management factors (mainly the location of the water function zone); and the characterization of the pollutant emission. The sub-basins were nested with the water function zone, which formed the basic coupling unit of the distributed hydrological and calculation models of the WEC. Furthermore, based on the soil, land use, slope, and other characteristics, the sub-basins were divided into hydrological response units (HRUs), which were used as the basic computing units of the hydrologic simulation. Then, the hydrological cycle of the sub-basins was obtained by the simulation of the runoff yield and concentration. Finally, the hydrological factors required by the dynamic calculation of the WEC, such as water yield and outlet runoff, were output.

Empirical Model for the Conversion Between the Discharge and Flow Velocity. As the distributed hydrological model was unable to output the discharge, the dynamic flow velocity data were updated from the data obtained from the distributed hydrological model by an empirical model, which was obtained by a linear fitting of the observation data of the hydrological stations.

Under normal circumstances, an empirical formula for the conversion between the discharge and discharge flow velocity can be given by Eq. (1) as follows [15]:

$$u = \alpha Q^{\beta} \tag{1}$$

where u (m/s) is the average velocity of the river; Q is the discharge of the river section; and α and β are the parameters which can be determined by the observational data.

Then, Eq. (1) was converted to a linear form (Eq. (2)) as follows:

$$y = a + bx \tag{2}$$

where $y = \ln u$, $x = \ln Q$, $a = \ln \alpha$ and $b = \beta$; and a and b are the parameters which can be determined by the observational data.

Dynamic Water Environment Capacity Calculation Model. The basic calculation unit of the WEC is usually the water function zone (in this study, the sub-basin combination was the basic calculation unit). The water quantity and pollutant balance in every river link of the sub-basin is conceptualized in Fig. 2.

The water quantity balance equation in the river link is as follows [16]:

$$\Delta V_i = Q_{i-1} + R_i + Q_{ai} - q_i - Q_i \tag{3}$$

The water quality balance equation in the river link is as follows [16]:

$$\Delta C_i = \frac{Q_{i-1}C_{i-1} + W_P + R_i C_{Ri} + Q_{ai}C_{ai} - C_i(\Delta V_i + Q_i) - C_{qi}q_i - W_d}{V_i} \tag{4}$$

where ΔV_i is the incremental discharge of the river link i; Q_{i-1} is the inflow discharge; Q_i is the outflow discharge; q_i is the local water withdrawal; R_i is the local water yield; Q_{ai} is the inflow of the tributary into the river link; V_i is the water storage of the river link i; C_{Ri} is the pollutant concentration of the local water yield (dependent on the non-point source pollution); C_{ai} is the pollutant concentration of the tributary; C_{i-1} and C_i are the pollutant concentrations of the inflow and outflow, respectively; C_{qi} is the pollutant concentration of the local water withdrawal; W_P is

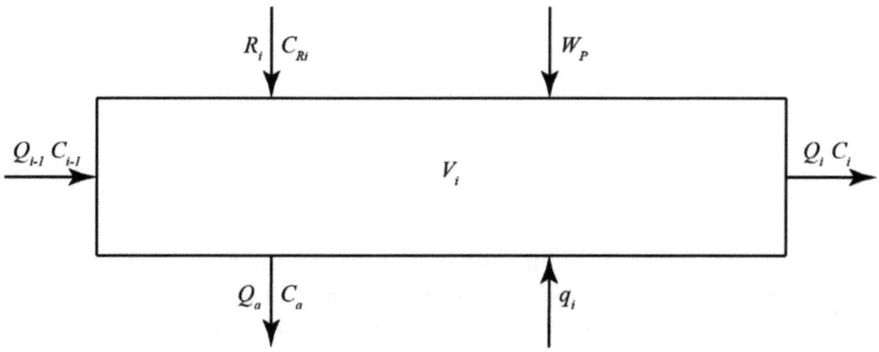

Fig. 2 Water quantity and quality balances in the river link

the total point source pollution load of the river link i; ΔC_i is the incremental pollutant concentration of the river link; and W_d is the pollution degradation amount of the river link.

The time scale management is generally months or years (the time scale for this study was months) in which the water quality can be considered stable [16]. That meant that the ΔC_i could be assumed to be 0. Then, the WEC was obtained by Eq. (5) as follows:

$$W_i = W_P + R_i C_{Ri} = C_i(Q_{i-1} + R_i + Q_{ai}) + (C_{qi} - C_i) - Q_{i-1}C_{i-1} - Q_{ai}C_{ai} + W_d \tag{5}$$

Therefore, C_i is the water quality target in the time scale management in this study. As the water quality target was often set up to consider the requirements of the water withdrawal, it could be assumed that C_{qi} was equal to C_i. Then, Eq. (5) could be converted to Eq. (6) as follows:

$$W_i = W_P + R_i C_{Ri} = C_i(Q_{i-1} + R_i + Q_{ai}) - Q_{i-1}C_{i-1} - Q_{ai}C_{ai} + W_d \tag{6}$$

where $C_i(Q_{i-1} + R_i + Q_{ai}) - Q_{i-1}C_{i-1} - Q_{ai}C_{ai}$ is the dilution capacity; and W_d is self-purification capacity which could then be divided into four parts: the pollutant degradation of the upper reaches (W_{dS}), tributary (W_{dZ}), point source pollution (W_{dP}), and non-point source pollutants (W_{dM}).

Also, the W_d could be calculated using the water quality model. A one-dimensional steady water quality model was used in this study when considering the large-scale applications. This one-dimensional steady water quality model is described by Eq. (7):

$$C = C_0 e^{-K\frac{x}{u}} \tag{7}$$

where C is the pollutant concentrations following the degradation; C_0 is the initial pollutant concentration; u is the average velocity of the river; L is the distance along the channel; and K is the degradation coefficient.

According to Eq. (7), W_{dS} and W_{dZ} can be computed by Eqs. (8) and (9), as follows:

$$W_{dS} = (1 - e^{-K\frac{L}{u}})C_{i-1}Q_{i-1} \tag{8}$$

$$W_{dZ} = \sum_{j=1}^{n} (1 - e^{-K\frac{x_j}{u}})C_{aij}Q_{aij} \tag{9}$$

where Q_{aij} is the inflow of the tributary j into the river link; C_{aij} is the pollutant concentration of the tributary j; x_j is the distance from the end of the river link to the tributary j ($j = 1, 2, \ldots, n$, n is the number of the tributary); and L is the length of the river link.

W_{dP} and W_{dM} are related to the pollutants entering the river and the route of the pollutant emission. If the pollutant emission was simplified to emission through a generalized outlet, then W_{dP} and W_{dM} could be obtained by Eq. (10), as follows:

$$W_{dg} = W_{dP} + W_{dM} = \left(1 - e^{-K\frac{x_G}{u}}\right)(W_P + R_i C_{Ri}) \tag{10}$$

where W_{dg} is the pollutant degradation of the generalized outlet discharge; and x_G is the distance from the generalized outlet to the end of the river link. The specific location of the generalized outlet is determined by the analysis of the current pollutants entering the river, and the route of the pollutant emission.

Finally, the WEC calculation model of the river link was obtained using Eq. (11) as follows:

$$W_i = e^{-K\frac{x_G}{u}} C_i(Q_{i-1} + R_i + Q_{ai}) - e^{K\frac{x_G - L}{u}} Q_{i-1} C_{i-1} - \sum_{j=1}^{n} (e^{K\frac{x_G - x_j}{u}}) C_{aij} Q_{aij} \tag{11}$$

The WEC calculation model was coupled with the empirical models for conversion between the discharge and flow velocity and distributed hydrological factors in the sub-basin, and the output of the WEC in each sub-basin. The hydrological factors in Eq. (11), such as the local water yield, the inflow discharge, inflow of the tributary and average velocity of the river were dynamic data which could be output by the distributed hydrological and empirical models.

3 Data Preparation

Distributed Hydrological Model. A SWAT (soil and water assessment tool) was selected to be used in this study. A SWAT is a basin or watershed scale model developed at the USDA-ARS [17] and has a strong physical mechanism. The model showed good applicability in large areas and had been applied and verified in many regions of the world.

The necessary data for a SWAT model include a digital elevation model (DEM), soil, land use, and weather data. However, other data, such as river discharge, reservoir information, sediment, pesticide, fertilizer, chemical, and water quality data are optional for the model development [18]. The basic data, and its sources and uses, are listed in Table 1.

Empirical Model for the Conversion between the Discharge and Flow Velocity. This model included the discharge and velocity data from the hydrological stations.

Dynamic Water Environment Capacity Calculation Model. This model included the degradation coefficient K and length data (distance from the end of the river link to the tributary and the length of the river link). The K of the COD was set

Table 1 Basic data for the research

Data type		Data type	Source	Usage
DEM	Raster graphics	Elevation, slope, slope length and slope direction of river channel	USGS, 30 m resolution	SWAT (water cycle simulation)
Land use	Raster graphics	Spatial distribution of land use type	Chinese Academy of Sciences, 1:250000	SWAT (water cycle simulation)
Soil	Raster graphics	Spatial distribution of soil types	FAO, 1:1000000	SWAT (water cycle simulation)
	Database	Physical and chemical properties of soil	FAO	SWAT (water cycle simulation)
Meteorological data		The maximum and minimum temperature, daily precipitation, relative humidity, solar radiation, wind speed	10 meteorological stations within or around the calculation area (Fig. 1)	Basic hydrologic data
Hydrological data		Monthly mean discharge and velocity	10 hydrological stations in the calculation river (Fig. 1)	Basic hydrologic data
Data of water function zone		Length data	"*The report of water environment function zones of Tieling*"	WEC calculation

between 0.05 and 0.2, and the K of the NH_3-N was set between 0.05 and 0.15, as per the "Guideline for Check and Ratification Technology of National Water Environmental Capacity" and the research results put forward by Meng [19]. The length data was determined by the water function zone of Tieling City.

4 Results and Discussion

4.1 Calibration, Validation, and Simulation of the SWAT Model

The basin was divided into 88 sub-basins according to the DEM, river water system, and distribution of the water function zone. Then, the sub-basins were subdivided into 1,624 HRUs based on the soil, land use, slope, and other characteristics of the study region. The sub-basins were nested with the water function

Table 2 Results of the hydrological simulation

	Hydrological station	NSE	R^2
Calibration (1991–2000)	Tongjiangkou	0.71	0.72
	Tieling	0.83	0.84
Validation (2008–2010)	Tongjiangkou	0.65	0.85
	Tieling	0.76	0.86

zone, which formed the basic coupling unit of the distributed hydrological and calculation models of the WEC. The basic corresponding relationship between the water function zone and sub-basins is shown in Fig. 1.

The observational flow rate data of two typical hydrological stations (Tieling and Tongjiangkou) were used for the calibration and validation. The model calibration was performed using a monthly time scale. The calibration period was from 1991 to 2000, and the validation period was from 2008 to 2010. Two quantitative statistics, the Nash-Sutcliffe efficiency (*NSE*) and the coefficient of determination (R^2), were used to evaluate the performance of the model in this study. The results of the hydrological simulation were shown in Table 2.

Generally speaking, a score of the R^2 which was above 0.5 was considered acceptable. Regarding the *NSE*, the simulation results were "excellent" when the $NSE > 0.75$ in both the calibration and validation periods, and the simulation results were "good" when the $0.65 < NSE < 0.75$ in both the calibration and validation periods [20]. Therefore, the simulation results of the Tieling Station were determined to be "excellent", and the simulation results of the Tongjiangkou Station were determined to be "good" in this study. This implied that the results of the model were reliable, and could thereby be used to calculate the WEC.

4.2 Fitting the Results of the Discharge and Velocity

A least squares linear fitting was applied; and the main parameters of the empirical model for the conversion between the discharge and velocity of each station, as well as the linear regression coefficients, were obtained (Table 3). The results showed that the fitting effect was relatively ideal when r was above 0.8 in all the stations except the Songshu Station ($r = 0.79$). These results confirmed that the empirical model was reliable for the computation of the WEC.

4.3 Dynamic Calculation Result of the WEC

In this study, the hydrological process between 1991 and 2010 was simulated; and the related discharge data were obtained by the calibrated SWAT. Then, the

Table 3 Results of least squares linear fitting

Rivers	Hydrological station	α	β	r
Liao River	Fudedian	0.12	0.44	0.93
	Tongjiankou	0.04	0.42	0.93
Chai River	Chaihe	0.15	0.43	0.95
Qing River	Kaiyuan	0.29	0.23	0.90
	Bakeshu	0.26	0.31	0.95
Fan River	Guanliangjiao	0.10	0.44	0.81
Zhaosutai River	Wangbaoqing	0.19	0.41	0.91
Kou River	Songshu	0.42	0.23	0.79
Erdao River	Baoli	0.29	0.39	0.97

discharge data were transformed to velocity data using the empirical model. Finally, the WEC of the Tieling Section of the Liao River Basin was calculated. The location of the sewage outfall was difficult to generalize when the lack of data regarding the two extreme cases ($x_G = L$ and $x_G = 0$) was considered in this study. The results are shown in Table 4 and Figs. 3 and 4.

The annual average WEC of COD and NH_3-N were determined to be 30,620 ton/year and 1,504 ton/year, respectively, when $x_G = L$; and 25,764 ton/year and 1,243 ton/year, respectively, when $x_G = 0$. The WEC of the COD and NH_3-N were found to be 10,551 ton/year and 513 ton/year, respectively, in the most unfavorable conditions (when the frequency was 90%, and $x_G = 0$). For the most optimistic conditions (when the frequency was 10%, and $x_G = L$), the WEC of the COD and NH_3-N were determined to be 55,640 ton/year and 2,418 ton/year, respectively.

There were found to be great differences among the WEC during years with different frequencies. For example, the WEC in especially dry years was less than half that in the normal years, or even a quarter of that in the especially wet years.

Table 4 The WEC (ton/year) in flood and non-flood season with different frequencies

Frequency (%)	$x_G = L$				$x_G = 0$			
	Flood season		Non-flood season		Flood season		Non-flood season	
	COD	NH_3-N	COD	NH_3-N	COD	NH_3-N	COD	NH_3-N
10	37,382	1,928	18,258	904	35,462	1,844	16,165	817
25	32,336	1,642	11,817	579	30,320	1,560	9,633	489
50	17,676	867	9,311	444	16,292	809	7,695	379
75	9,676	475	6,848	321	8,648	431	5,440	265
90	7,386	348	5,538	251	6,296	305	4,255	208
Annual average	20,418	1,013	10,202	491	17,129	826	8,636	417

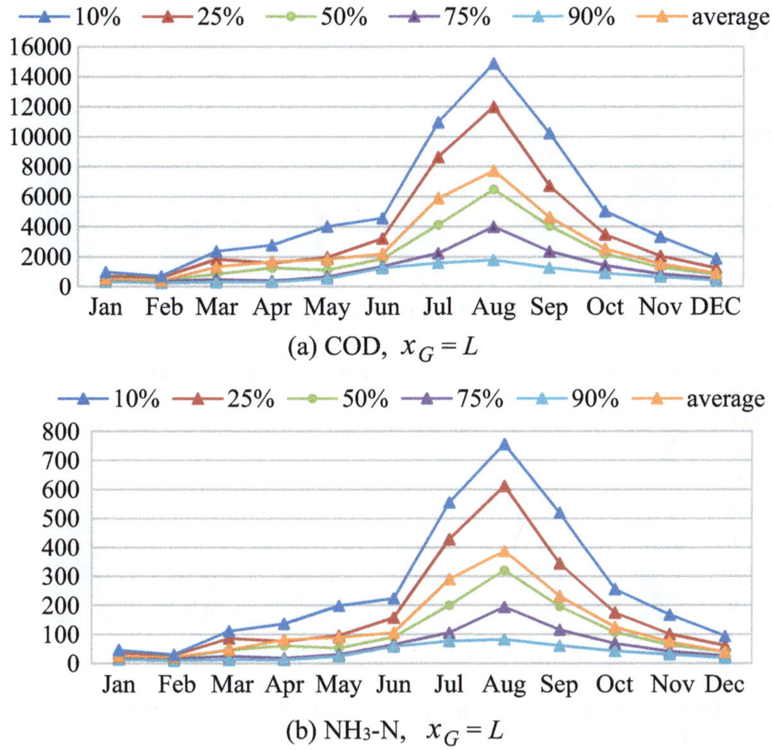

Fig. 3 The WEC (ton/year) in different months during years with different frequencies **a** COD, $x_G = L$ **b** NH_3-N, $x_G = L$

The WEC in the flood (from June to September) and non-flood seasons was also found to be uneven, and the capacity during the flood seasons was above 60%. The differences of the WEC among the months were more obvious; and August was determined to account for 20% of the entire year, while February accounted for only 1–2%. These findings revealed that the traditional method, which did not consider the annual and inter-annual variation of the hydrology, tended to be conservative. However, the dynamical method was able to make a greater use of the WEC.

In this research study, specifically related to these seven rivers, the WEC ranked consistently in different frequencies, from top to bottom, as follows: the Liao, Zhaosutai, Kou, Erdao, Qing, Chai, and Fan Rivers. The capacity of the Liao River made up a significant share (more than half) of the entire region, with the annual average WEC of the COD and NH_3-N determined to be 15,246.13 ton/year and 762.31 ton/year, respectively.

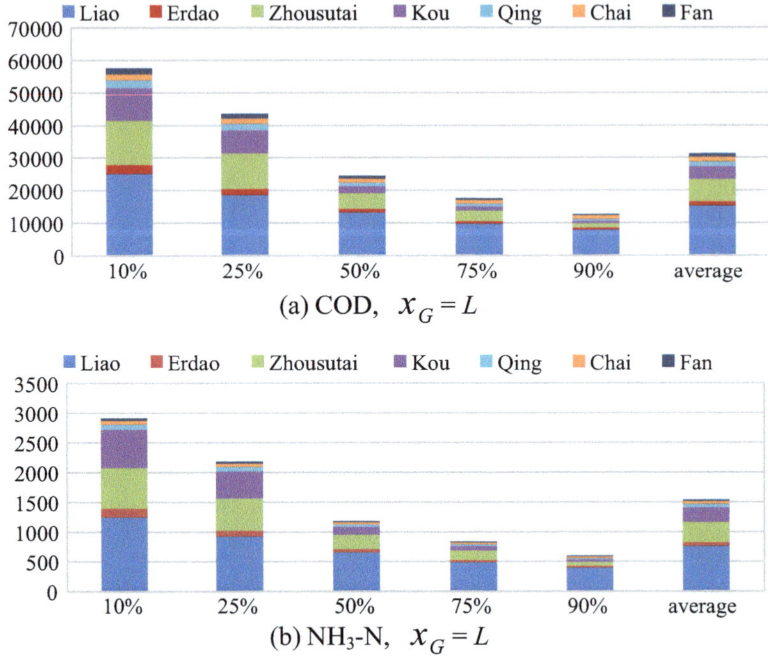

Fig. 4 The WEC (ton/year) in rivers during years with different frequencies **a** COD, $x_G = L$ **b** NH_3-N, $x_G = L$

5 Conclusions

(A) In this research study, by using the analysis of the hydrological factors which led to the changes of the WEC, along the coupling mechanism between the hydrological factors and the WEC, a dynamic calculation method of the WEC based on the hydrologic process was developed. In this method, distributed hydrological and water environmental capacity models were coupled.

(B) The Tieling Section of the Liao River Basin was taken as the case study area, for demonstrating this method. There were great differences among the WEC observed in both time and space which suggested that the dynamic method was able to consider the annual and inter-annual variations of the hydrology in a more scientific manner. The calculation results could potentially be used as a scientific basis for future pollutant emission reductions.

(C) In this study, the sewage outfall was generalized as two types of extreme cases (the beginning and end of the river link) due to the limited observational data. That meant that the calculation results of the two extreme cases were the upper and lower thresholds. Moreover, a scientific calculation of the pollutant discharges, as well as the pollutants discharged into the river, should be conducted to analysis the environmental assets and liabilities. These calculations could ultimately provide scientific support for future water environmental management processes.

Acknowledgements The research was supported by the National Key Research and Development Program of China (No.2016YFC0503502) and the National Science Foundation for Distinguished Young Scholars of China (Grant No.51409271)-study on threshold depth to groundwater for vegetation stability.

References

1. Chinese Academy for Environmental Planning: Guideline for Check and Ratification Technology of National Water Environmental Capacity. Chinese Academy for Environmental Planning, Beijing, China (2003). (in Chinese)
2. Li, Y., Qiu, R., Yang, Z., Li, C., Yu, J.: Parameter determination to calculate water environmental capacity in Zhangweinan Canal sub-basin in China. J. Environ. Sci. **22**(6), 904–907 (2010). https://doi.org/10.1016/S1001-0742(09)60196-0
3. Chen, Q., Wang, Q., Li, Z., Li, R.: Uncertainty analyses on the calculation of water environmental capacity by an innovative holistic method and its application to the Dongjiang River. J. Environ. Sci. **26**(9), 1783–1790 (2014). https://doi.org/10.1016/j.jes.2014.06.025
4. Zeng, S., Xu, Y., Zhang, T.: Application of unsteady water quality model for looping river network to water pollution control planning. Adv. Water Sci. **02**, 193–196 (2004). https://doi.org/10.14042/j.cnki.32.1309.2004.02.012. (in Chinese)
5. Li, S., Li, H., Xia, J.: Dapeng Bay water environment capacity analysis on the base of Delft 3D model. Res. Environ. Sci. **05**, 91–95 (2005). https://doi.org/10.13198/j.res.2005.05.93.lisw.023. (in Chinese)
6. Deng, Y., Zheng, B., Fu, G., Lei, K., Li, Z.: Study on the total water pollutant load allocation in the Changjiang (Yangtze River) Estuary and adjacent seawater area. Estuar. Coast. Shelf Sci. **86**, 331–336 (2010). https://doi.org/10.1016/j.ecss.2009.10.024
7. Zhou, G., Lei, K., Fu, G., Mao, G.: Calculation method of river water environmental capacity. J. Hydraul. Eng. **02**, 227–234 (2014). https://doi.org/10.13243/j.cnki.slxb.2014.02.013. (in Chinese)
8. Li, R., Wang, J., Wang, C., Qian, J.: Calculation of river water environmental capacity under unascertained information. Adv. Water Sci. **04**, 359–363 (2003). https://doi.org/10.14042/j.cnki.32.1309.2003.04.013. (in Chinese)
9. Hu, B.: Using probabilistic dilution model to calculate the permissible pollutant capacity. Res. Environ. Sci. **5**(5), 21–25 (1992). https://doi.org/10.13198/j.res.1992.05.23.hubq.004. (in Chinese)
10. Li, S., Morioka, T.: Optimal allocation of waste loads in a river with probabilistic tributary flow under transverse mixing. Water Environ. Res. **71**(2), 156–162 (1999). https://doi.org/10.2175/106143099X121472
11. Cheng, S., Qian, Y., Zhang, H.: Estimation and application of macroscopic water environmental capacity of total phosphorus and nitrogen for Taihu lake. Acta Sci. Circumst. **10**, 2848–2855 (2013). https://doi.org/10.13671/j.hjkxxb.2013.10.032. (in Chinese)
12. Huang, Z., Li, Y., Li, J., Chen, Y.: Water environmental capacity for the reservoir of Three Gorges Project. J. Hydraul. Eng. **03**, 7–14 (2004). https://doi.org/10.3321/j.issn:0559-9350.2004.03.002. (in Chinese)

13. Jiang, X., Xu, S., Lian, J., Meng, Q.: Analysis and calculation of dynamic water environmental capacity of rivers in North China. J. Ecol. Rural Environ. **29**(4), 409–414 (2013). https://doi.org/10.3969/j.issn.1673-4831.2013.04.001. (in Chinese)
14. Niu, C., Jia, Y., Wang, H., Gao, H.: Integrated simulation and evaluation on water quantity and quality of the Yellow River Basin. Yellow River **11**, 58–60 (2007). https://doi.org/10.3969/j.issn.1000-1379.2007.11.027. (in Chinese)
15. Yang, B., Li, Z.: Preliminary study on simulating average flux and velocity at hydrological section in dry seasons with experience equation. Hai River Hydrol. **6**, 18–19 (2002). https://doi.org/10.3969/j.issn.1004–7328.2002.06.007. (in Chinese)
16. Xia, J., Wang, M., Wang, Z., Niu, C., Yan, D.: An integrated assessment method of water quality & quantity applied to evaluation of available water resources. J. Nat. Resour. **05**, 752–760 (2005). https://doi.org/10.3321/j.issn:1000-3037.2005.05.015. (in Chinese)
17. Arnold, J.G., Srinivasan, R., Muttiah, R.S., et al.: Large area hydrologic modeling and assessment part I: model development 1. J. Am. Water Resour. Assoc. **34**, 73–89 (1998). https://doi.org/10.1111/j.1752-1688.1998.tb05961.x
18. Saha, P.P., Zeleke, K., Hafeez, M.: Streamflow modeling in a fluctuant climate using SWAT: Yass River catchment in south eastern Australia. Environ. Earth Sci. **71**(12), 5241–5254 (2014). https://doi.org/10.1007/s12665-013-2926-6
19. Meng, W.: Technique of Total Amount Control for Water Pollutants in Watershed and its Application. Chinese Environmental Science Press, Beijing, China (2008). (in Chinese)
20. Moriasi, D.N., Van Liew, M.W., et al.: Model evaluation guidelines for systematic quantification of accuracy in watershed simulations. Trans. ASABE **50**(3), 885–900 (2007). https://doi.org/10.13031/2013.23153

Hourly Campus Water Demand Forecasting Using a Hybrid EEMD-Elman Neural Network Model

Xiao Deng, Shuai Hou, Wen-zhu Li and Xin Liu

Abstract Accurate and reliable water demand forecasting is important for effective and sustainable planning and use of water supply infrastructures. In this paper, a hybrid EEMD-Elman neural network model for hourly campus water demand forecast is proposed, aiming at improving the accuracy and reliability of water demand forecast. The proposed method combines the Elman neural network, EEMD method, and phase space reconstruction method providing favorable dynamic forecast characteristics and improving the forecasting accuracy and reliability. Simulation results show that the proposed model provides a better performance of hourly campus water demand forecast by using the real data of water usage of our campus.

Keywords Elman neural networks · EEMD · Phase space reconstruction
Water demand forecasting

1 Introduction

In recent years, with the development of the economy and improvement of people's living standards, residential water consumption has been increased, which makes great pressure on the water supply system and challenges the reasonable scheduling of water resources. The better water resource scheduling approaches should be to

X. Deng · S. Hou · W. Li · X. Liu (✉)
School of Information and Electric Engineering,
Hebei University of Engineering, Handan 056038, Hebei, China
e-mail: xin.liu@hebeu.edu.cn

X. Deng
e-mail: dengxiaozm@163.com

S. Hou
e-mail: houshuai20072@163.com

W. Li
e-mail: liwenzhu@hebeu.edu.cn

W. Dong et al. (eds.), *Sustainable Development of Water Resources and Hydraulic Engineering in China*, Environmental Earth Sciences,
https://doi.org/10.1007/978-3-319-61630-8_7

reduce the cost of water supply and demand as far as possible, so as to save energy and water. Hence, an accurate and reliable forecasting model will provide an important data basis for the scheduling in the water supply enterprises.

Traditional methods for water demand forecasting include the quota method, the end-use model, time series analysis, and linear regression analysis. The annual water demand of Baiyangdian was predicted considering the economic factors by the quota method [1]. However, the influence factors of annual water demand method are determined by the experience of researchers which deteriorate the prediction accuracy and application range of prediction model. The end use model was established by a large number of historical data of demographic data and end use water data [2, 3]. Time series analysis usually is used if the climatic and socioeconomic data is unavailable, with forecasting future water demand only to the past values without considering other factors [4]. Regression analysis forecasts the future values not only based on the past data but also those determinants that cause a change in the water demand [5]. Some static prediction models are often used, including autoregressive moving average (ARMA) model [6], autoregressive integrated moving average (ARIMA) model [7] and the seasonal autoregressive integrated moving average (SARIMA) model [8, 9]. However, the prediction results of these methods are not ideal when the fluctuation of the water use data is large.

With the rapid development of computer technology, several intelligent algorithms with high prediction accuracy, such as the artificial neural network (ANN), support vector machine (SVM) and other machine learning algorithms, were introduced to predict the urban water consumption. Vijayalaksmi set up an adaptive neural fuzzy inference system (ANFIS) to predict urban water demand, and obtain a good prediction effect [10]. Candelieri presented a reliable urban water demand forecasting for the entire day based on Support Vector Machine (SVM) regression method [11]. Adamowski compared the results of three water demand forecasting models which are established by multiple linear regression, time series analysis and artificial neural network, and pointed out that the latter has a better performance in the city's daily water demand [12] Bennett used three conventional ANNs, which are two feed-forward back propagation networks and one radial basis function network, to build the more accurate forecasting models [13].

These methods mentioned above all were used to forecast urban water demand; however, aiming to achieve more accurate water supply scheduling in a city, a smaller target area as a community or a town was predicted. Forecasting water demand for the small area can reduce the cost of urban water supply facilities and save more water. In this paper, a hybrid ensemble empirical mode decomposition (EEMD)-Elman neural network (ENN) model for hourly campus water demand forecast is proposed by using the real data of water usage of the campus of the Hebei University of Engineering.

As a feedback neural network, ENN is widely used in different fields. ENN is optimized based on back-propagation (BP) neural network [14], which adds a context layer to the hidden layer of feed-forward network as the time delay operator so as to memorize and, therefore, make the system have the characteristic of dealing with the dynamic information and enhance its capacity of nonlinear fitting.

In consideration of the nonlinearity and randomness of water data, ENN will be an effective method of water demand forecasting. ENN has better performance than the BP neural network [15], but there are still shortcomings. On the one hand, non-stationarity of the real data of water usage will reduce the accuracy of water demand forecasting by the neural network method. On the other hand, it is difficult to determine the structure of Elman neural network, limiting its applicability. Therefore, this paper proposes an associated forecasting method based on Elman neural network. First, for the non-stationarity of data, the EEMD is used to decompose the original data into a series of subsequences as the input of the ENN, that are relatively simple and show a regular pattern. Then, aiming at the difficult problem of ENN's structure, this paper presents the use phase space reconstruction method to optimize the structure of ENN. Finally, the simulation results based on the real data show that the hybrid forecasting model proposed in this paper has higher accuracy than other methods.

2 Data Preprocess Based on EEMD

Empirical mode decomposition (EMD) is widely used for pre-process non-stationary signal, so as to make the complex series of time smooth, and obtain several intrinsic mode functions (IMF) components that are independent of each other. However, when the time scale of the signal dramatic shift has occurred, the EMD is not competent to separate effectively the different modal components according to features of time scale. In other words, there will be the phenomena of mode mixing among IMF components after the EMD decomposition, therefore, improving the applicability of the method of water demand prediction. In this paper, the EEMD improved from EMD method to pre-process water data was used. The method will add white noise sequences to original series and then the EMD decomposition is carried out repeatedly, and the mean value of the multi-group IMF is obtained as the real component, thereby avoiding the mode mixing. Specific steps are as follows:

Step 1: Initialization procedure. Initialize 'm' which is the execution time of EMD and 'k' that is amplitude of noise.

Step 2: Add a white noise series to the original signal and decompose the signal with added white noise into IMFs using EMD.

Step 3: Repeat steps 1 and 2 again and again but with different white noise series each time (repeated 15 times in this work).

Step 4: Obtain the corresponding IMF components of the decompositions.

Step 5: Adopt the means of ensemble corresponding to the residue components of the decompositions as the final result.

The white noise sequence added in the decomposition satisfies the normal distribution and is independent to each other. According to the literature [16], it is known that in the case: m = 100 and k = 0.01, the result is better.

3 Forecast Model Based on the Improved ENN

3.1 *Elman Neural Network*

The type of Elman neural network consists of the input layer, the middle layer (hidden layer), the context layer and the output layer, as shown in Fig. 1 [17]. In this sense, it is similar to a three-layer feed-forward neural network; the neurons of the input layer only play the role of signal transmission; the output layer neurons have a linear weighting function; the transfer function of hidden layer can use a linear or nonlinear function. The context layer feeds back the hidden layer outputs in the previous time steps. The neurons are contained in each layer used to propagate information from one layer to another.

The Elman neural network has a character that the context neurons receive input from the hidden layers and pass their output to the hidden layers. The context layers always store the output from the hidden layer and relay this information in the next iteration. This behavior allows them to form a sort of short-term memory. Adding to internal feedback network increases the ability of dealing with dynamic information, so as to achieve the purpose of dynamic modeling [18].

As Fig. 1 shows, the nonlinear state space expression of Elman neural network is follows:

$$y(k) = g(w^3 x(k)) \tag{1}$$

$$x(k) = f(w^1 x_c(k) + w^2(u(k-1))) \tag{2}$$

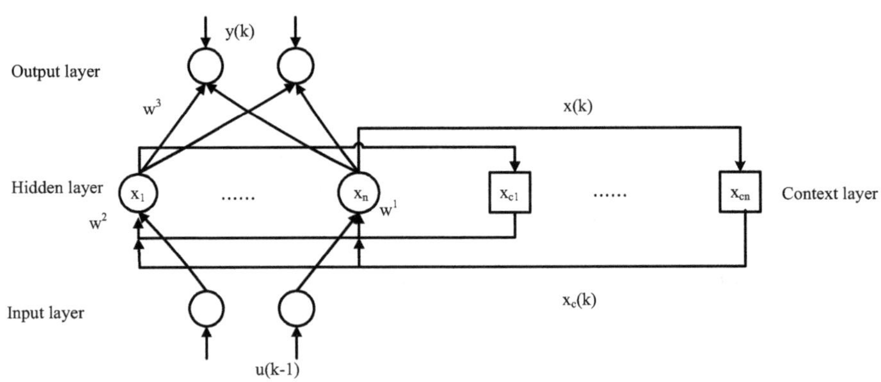

Fig. 1 Elman neural network architecture

$$x_c(k) = x(k-1) \tag{3}$$

where y is the node vector of m dimension output layer, x is the node vector of n dimension hidden layer, u is the input vector of r dimension and x_c is the feedback state vector of n dimension. w^3, w^2, w^1 denote the corresponding weights of the hidden layer to the output layer, the input layer to the hidden layer, and the context layer to the hidden layer, respectively. $g(\cdot)$ is the transfer function of output neuron, which is a linear combination of the middle layer output. $g(\cdot)$ is the transfer function of the hidden layer neurons, often using sigmoid function.

3.2 Improved ENN Based on Phase Space Reconstruction

In the process of forecasting, the previous and real value is usually used as input to predict the water consumption of the next time, m, the number of nodes in the input layer, is determined by the common test method and empirical formula. To avoid the arbitrariness and loss of information in choosing the number of nodes in the input layer,in this paper, the phase space reconstruction method is proposed, which can reconstruct the nonlinear water data time series, and the phase space is obtained to reflect the internal law of the sequence.

Its specific operation is to choose a suitable embedding dimension 'm' and delay time 'τ', for the time series of water consumption $\{P(t)\}$, $t = 1, 2, \ldots, N$; then, the new state space as the formula (4) can be obtained that is the reconstructed phase space of the time series.

$$P(t+1) = \{P(t), P(t+\tau), \ldots, P(t+(m-1)\tau)\} \tag{4}$$

There have been many mature algorithms for time delay and embedding dimension of the reconstructed phase space. In this paper, the C-C method [19] is used to select the saturation embedding dimension as the number of input nodes of neural network.

The number of hidden layer nodes of Elman neural network greatly influence on prediction accuracy. If the number is too small, neural networks are barely able to learn well; on the other hand,huge training data make the training time of Elman neural network remarkably increased and the network becomes over fitted. Kolmogorov theorem proved that the three-layer forward network can be satisfactory reproduction of any continuous function, if the reasonable structure and appropriate weight conditions are selected. When the number of nodes in the input layer of the network is m, the number of nodes in the hidden layer is about $2m+1$. After determining the theoretical value of the number of nodes in the hidden layer, the optimal node number is selected by the method of minimum training error. The specific steps are as follows:

(1) Use phase space reconstruction method to reconstruct the water consumption sequence, and obtain the input data and node number m.
(2) Range the number of hidden layer nodes to be optimized as $N \in [2M - 4, 2M + 6]$
(3) Select cyclically the number of nodes of the hidden layer to train the Elman neural network, and the node with the minimum training error will be the optimal node.

4 EEMD-Improved ENN

In this paper, the general idea of the method of water demand prediction: firstly, decompose water data step by step according to the different frequency by EEMD, so as to get different IMF components to reduce the non-stationary characteristics. Then the phase space is reconstructed for each component, and respectively the forecasting model of Elman neural network that is improved is established to reduce the mutual interference between components. Finally, the final forecasting results are superposed by prediction data of each component. The flow chart is shown in Fig. 2.

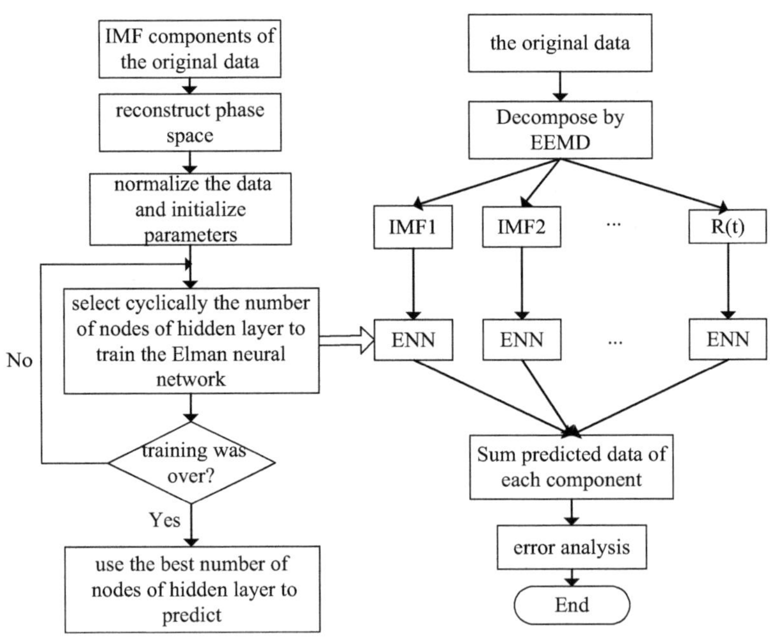

Fig. 2 Forecast flow chart of the EEMD—improved ENN

5 Simulation

In this paper, the water data recorded the water consumption hourly for 31 days on the campus of Hebei University of Engineering. The first 30 days of data are for training; the thirty-first day of water consumption data is selected to validate the EEMD-ENN algorithm. The water consumption data curve is shown in Fig. 3.

The EEMD method is used to decompose the original water data, resulting in 8 IMF components and 1 residue, as shown in Fig. 4.

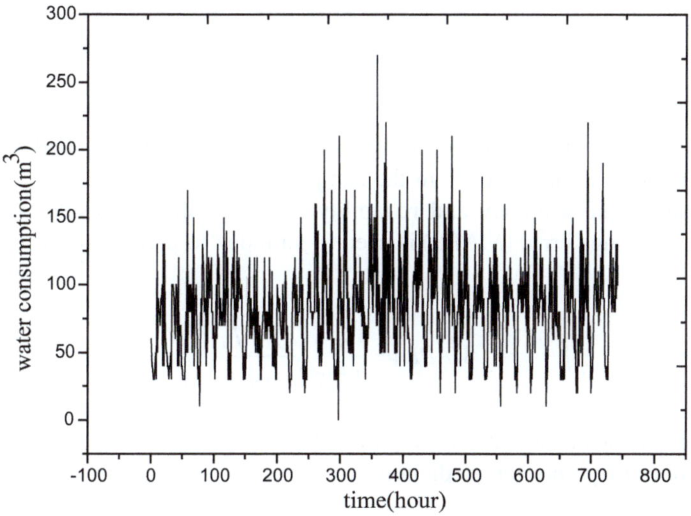

Fig. 3 Water consumption data curve

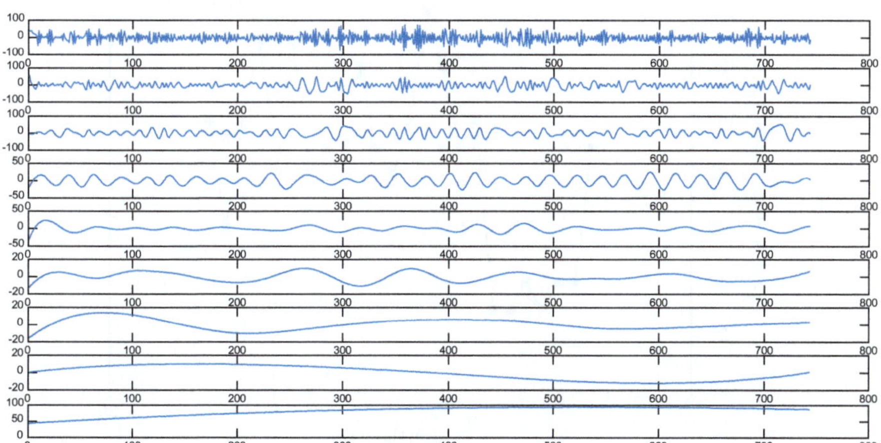

Fig. 4 Diagram of each wind power component through EEMD decomposing

As the sub-sequences shown in Fig. 3, the frequency of components IMF1 and MF2 is higher with their periodicity not obvious, and their volatility is still very wide. The frequency of IMF3 to IMF7 is low, and their periodicity is more obvious. The residue R(T) is a curve with higher amplitude and monotonically increasing. Therefore, the EEMD method can effectively avoid the mode mixing phenomenon of decomposition process.

After the water data is decomposed by EEMD, the value of each component is larger; therefore, it is necessary to normalize the data before forecasting. It is normalized and reduced, such as the formula (5) and (6):

$$y_i = \frac{x_i - x_{\min}}{x_{\max} - x_{\min}} \tag{5}$$

$$x_i = (x_{\max} - x_{\min}) * y_i + x_{\min} \tag{6}$$

where: y_i is the results after the normalization of a sample, $x_{\max}$ and $x_{\min}$ are the maximum and minimum values in the sample.

To evaluate the accuracy and reliability of the prediction results, the average absolute percentage error (MAPE) is used as an index to predict the accuracy of the prediction.

$$MAPE = \frac{1}{n}\sum_{i=1}^{n}\left|\frac{A_i - P_i}{A_i}\right| \tag{7}$$

where A_i is the actual water consumption, P_i is the predicted water consumption; n is the number of points for prediction.

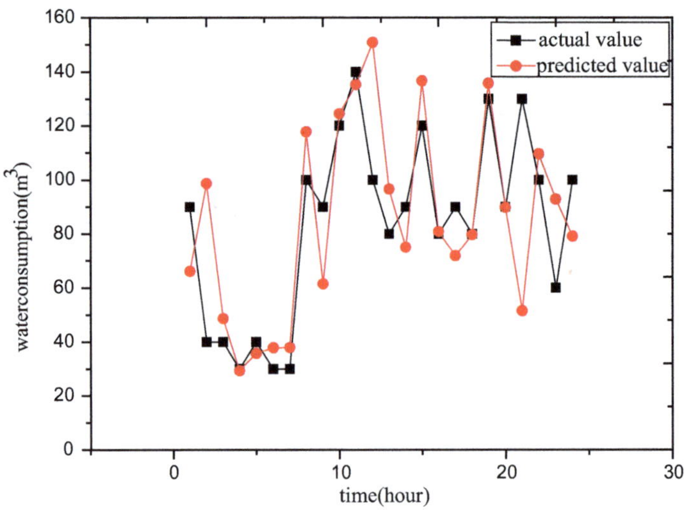

Fig. 5 Water demand forecasting results based on EEMD-improved ENN

Table 1 Model performance index

Model	MAPE
BP	0.3342
Elman	0.3269
EEMD-ENN	0.2458

Figure 5 shows the comparison of the predicted results of EEMD-improved ENN model with the actual water consumption. From the graph we can see that the trend of the change of forecast value and the actual value is roughly analogical, which has a high fitting accuracy, and shows the validity of the proposed model.

At the same time, it is compared with the single BP and Elman neural network, and the prediction error is shown in Table 1. From the results we can know that the prediction accuracy of the combined forecasting model is higher than that of the single prediction method.

6 Conclusion

Water demand forecasting is an essential component in effective water resources planning and management because it can help to identify appropriate alternatives to ensure the balance between water demand and supply. In this study, the EEMD-improved ENN models for hourly campus water demand forecasting is investigated. The original data decomposed to a series of relatively simple sub-components by EEMD can effectively reduce the non-stationarity of the data, so as to lay the foundation to the prediction of residential water demand. Elman neural network has strong nonlinear fitting ability, and the parameters of model structure can be obtained according to the method of phase space reconstruction and minimum training error, which can improve the prediction accuracy. Finally, the results indicate that the combined model have a good performance in forecasting hourly water demand.

However, the combination model is bound to increase the complexity of the algorithm and reduce the computational efficiency. Therefore, it should be possible to reduce the complexity of the algorithm to improve the prediction accuracy.

References

1. Qin, T.L., et al.: Water demand forecast in the Baiyangdian basin with the extensive and low-carbon economic modes. J. Appl. Math. **2014**, 1–10 (2014). https://doi.org/10.1155/2014/6734
2. Beal, C.D.: SEQ residential end use study. J. Aust. Water Assoc. **38**(1), 80–84 (2011). https://www.researchgate.net/publication/279662589)

3. Fielding, K.S., et al.: Using individual householder survey responses to predict household environmental outcomes: the cases of recycling and water conservation. Resour. Conserv. Recycl. **106**, 90–97 (2016). https://doi.org/10.1016/j.resconrec.2015.11.009
4. Shen, J.-C., et al.: Real-time correction of water stage forecast using combination of forecasted errors by time series models and Kalman filter method. Stoch. Env. Res. Risk Assess. **29**(7), 1903–1920 (2015). https://doi.org/10.1007/s00477-015-1074-9
5. Luo, X., Jiaqi Yang, J.: Study on the imbalance of shipping demand and supply of inland water transportation of Yangtze River. ICTIS 2013: Improving Multimodal Transportation Systems-Information, Safety, and Integration, p. 2211–2221 (2013). https://doi.org/10.1061/9780784413036.297
6. Adamowski, J., Chan, H.-F., Prasher, S.O.: Comparison of multiple linear and nonlinear regression, autoregressive integrated moving average, artificial neural network, and wavelet artificial neural network methods for urban water demand forecasting in Montreal, Canada. Water Res. Res. **48**(1) (2012). https://doi.org/10.1029/2010wr009945
7. Mombeni, H.A., et al.: Estimation of water demand in Iran based on SARIMA models. Environ. Model. Assess. **18**(5), 559–565 (2013). https://doi.org/10.1007/s10666-013-9364-4
8. Braun, M., et al.: 24-hours demand forecasting based on SARIMA and support vector machines. Procedia Eng. **89**, 926–933 (2014). https://doi.org/10.1016/j.proeng.2014.11.526
9. Mombeni, H.A., et al.: Reducing water consumption after targeted subsidy plan in Iran. Water Resour. **42**(3), 389–396 (2015). https://doi.org/10.1134/S0097807815030100
10. Vijayalaksmi, D.P., Jinesh Babu, K.S.: Water supply system demand forecasting using adaptive neuro-fuzzy inference system. Aquat. Procedia **4**, 950–956 (2015). https://doi.org/10.1016/j.aqpro.2015.02.119
11. Candelieri, A., Archetti, F.: Identifying typical urban water demand patterns for a reliable short-term forecasting—the icewater project approach. Procedia Eng. **89**, 1004–1012 (2014). https://doi.org/10.1016/j.proeng.2014.11.218
12. Adamowski, J.F.: Peak daily water demand forecast modeling using artificial neural networks. J. Water Res. Plann. Manage. **134**(2), 119–128 (2008). https://doi.org/10.1061/(ASCE)0733-9496(2008)134:2(119)
13. Bennett, C., Stewart, R.A., Beal, C.D.: ANN-based residential water end-use demand forecasting model. Expert Syst. Appl. **40**(4), 1014–1023 (2013). https://doi.org/10.1016/j.eswa.2012.08.012
14. Rumelhart, D.E., Hinton, G.E., Williams, R.J.: Learning representations by back-propagating errors. Nature **323**(6088), 533–536 (1986). https://doi.org/10.1038/323533a0
15. Jia, W., et al.: Study on optimized Elman neural network classification algorithm based on PLS and CA. Comput. Intell. Neurosci. **2014**, 724317 (2014). https://doi.org/10.1155/2014/724317
16. Wu, Z.H., Huang, N.E.: Ensemble empirical mode decomposition: a noise-assisted data analysis method. Adv. Adapt. Data Anal. **1**(1), 1–41 (2009). https://doi.org/10.1142/S1793536909000047
17. Wang, J., Zhang, W., Li, Y., Wang, J., Dang, Z.: Forecasting wind speed using empirical mode decomposition and Elman neural network. Appl. Soft Comput. **23**, 452–459 (2014). https://doi.org/10.1016/j.asoc.2014.06.027
18. Lin, C.-M., Boldbaatar, E.-A.: Autolanding control using recurrent wavelet Elman neural network. IEEE Trans. Syst. Man Cybern. Syst. **45**(9), 1281–1291 (2015). https://doi.org/10.1109/tsmc.2015.2389752
19. Kim, H.S., Eykholt, R., Salas, J.D.: Nonlinear dynamics, delay times and embedding windows. Phys. D **127**, 48–60 (1999). https://doi.org/10.1016/s0167-2789(98)00240-1

The Study on Groundwater Recharge and Evolution in Northwestern China

Yunpu Zheng, Lili Guo, Liang Liu, Lihua Hao, Qiang Ma, Jingui Wang, Fei Li, Lishu Wang and Ming Xu

Abstract The recharge and evolution of groundwater in the Wuwei Basin were investigated using chemical indicators, stable isotopes, and radiocarbon data. The results showed that the concentrations of Na^+ and K^+ in the groundwater were controlled by the dissolution of halite and sylvite from fine-grained sediments, whereas the increase of Na^+ and Cl^- was not in accordance with a ratio of 1:1, indicating that the dissolution of halite and sylvite barely affected the concentrations of Na^+ and K^+ in groundwater. Meanwhile, HCO_3^- was the dominant ion with a decreased ratio in the groundwater. The SO_4^{2-}/Cl^- ratio decreased with the sample profile from Southwest to Northeast due mainly to the increases of Cl^- concentration. The Cl^- was enriched in the hydrodynamic sluggish belt, and thus the Ca^{2+}/Cl^- ratio decreased with the enhancement of Cl^-. In addition, the $\delta^{18}O$ and δ^2H

Y. Zheng · L. Guo · L. Liu · L. Hao · J. Wang · F. Li · L. Wang (✉)
School of Water Conservancy and Hydropower, Hebei University of Engineering, 178 S. Zhonghua Street, Handan 056021, Hebei, China
e-mail: fc8232@126.com

Y. Zheng
e-mail: zhengyunpu_000@sina.com

L. Guo
e-mail: 898367221@qq.com

L. Liu
e-mail: liuliang4048086@163.com

L. Hao
e-mail: haolihua_000@sina.com

J. Wang
e-mail: wangjingui@126.com

F. Li
e-mail: 2365580744@qq.com

Q. Ma
Yahoo! Inc., 701 1st Ave, Sunnyvale, CA 94089, USA
e-mail: qiang@yahoo-Inc.com

W. Dong et al. (eds.), *Sustainable Development of Water Resources and Hydraulic Engineering in China*, Environmental Earth Sciences,
https://doi.org/10.1007/978-3-319-61630-8_8

values of groundwater gradually increased from Southwest to Northeast along the flow path. Compared with the isotopic values of precipitation, the heavy isotopic values were strongly depleted in the groundwater samples, suggesting that the recharge of groundwater in the plain region was very limited from precipitation. Moreover, the groundwater in the phreatic aquifer was younger water with ^{3}H values from 47 to 71 a.BP, while the groundwater age in the confined aquifer was 1000–5800 BP evidenced by the ^{14}C values between 48 and 88 pmc. These results suggested that the recharge duration of the groundwater was from the late Pleistocene or early Holocene. These results might have important significance for inter-basin water allocation and groundwater management of the Wuwei Basin.

Keywords Environmental isotopes · Hydrochemistry · Groundwater circulation Iron

1 Introduction

Water demand for agricultural, household and environmental uses is rapidly rising due to the continuously increasing population, especially in the developing world; and thus, many areas throughout the world are expected to experience imbalance on the supply and demand for water in the near future [1]. Meanwhile, the climate change, population growth, and economic development may also affect the availability of water resources in the arid and semi-arid regions. One of the most prominent issues in these arid areas is climate change which may have impacts on the quality and quantity of the water resources, which are supplied to the growing population. Previous studies have reported that the increasing population density, economic activity, and unsustainable water management practices have result in over-exploitation of many more easily accessible freshwater resources [2]. Some cities in the drier parts of the world are likely to have exhausted their groundwater reserves in a little more than a decade. Water scarcity has brought into focus the urgent need for planned action to manage water resources effectively as it is widely

M. Xu (✉)
Key Laboratory of Ecosystem Network Observation and Modeling,
Institute of Geographic Sciences and Natural Resources Research,
Chinese Academy of Sciences, Beijing 100101, China
e-mail: mingxu@crssa.rutgers.edu

M. Xu
School of Resources and Environment, University of Chinese Academy
of Sciences, Beijing 100039, China

M. Xu
Department of Ecology, Evolution and Natural Resources, Rutgers University,
Center for Remote Sensing and Spatial Analysis, 14 College Farm Road,
New Brunswick, NJ 08901, USA

acknowledged that water is a major limiting factor in the socio-economic development of a world with a rapidly expanding population. Sustainable management of water resources to meet human and ecosystem needs will require accurate estimates of groundwater recharge, as the surface water resources are generally scarce and highly unstable in semi-arid and arid regions with the result that groundwater is the primary source of water in these regions [3, 4].

Situated deep in the hinterland of Eurasia, China's northwestern inland zone has a very arid climate. The landscape in these regions is fragile due to the low and irregular precipitation, high temperatures and evaporation, and notable drought periods [5]. In such arid and semi-arid environments, groundwater is not only an important source of public water supply, but is also important to the regional ecology. Since the 1950s, human activities, and particularly the exploitation of land and water resources associated with dramatic population growth, have led to significant changes in the water regime and have created serious environmental consequences, including declines in the regional groundwater levels, desertification, and drying of rivers and lakes in the lower reaches of river basins [6–10]. For example, the groundwater level has dropped widely by as much as 35 m in the Minqin Basin since the 1960s [5].

The "Green Corridor" and the highway along the Lower Tarim River are endangered by degrading riparian vegetation and desertification [10]. Desertification in the Shule River Basin is also a very severe problem. The sand dune along the Shule River has evolved from immobilization to mobile dunes because of the degeneration of the vegetation cover [11]. If this situation continues, further deterioration of the environment and ecosystems of this vast area is unavoidable [12]. Careful studies of the characteristics of the groundwater and its evolution under natural water circulation processes help to provide scientific guidelines for sustainable exploitation of the region's water resources and prevention of further degradation of the regional environment.

Many studies have examined the characteristics and utilization of the groundwater resources in the arid regions of northwestern China. Most of the research has taken place in the eastern and middle parts of the Hexi Corridor (i.e., the Zhangye, Ejina, and Minqin Basins), with important results that provide guidance for regional water management. For example, the deep groundwater of the Minqin Basin was recharged under cool and wet conditions during the late Pleistocene to Holocene periods based on an analysis of isotopic, noble gas, and chemical indicators [5, 12, 13]. Some authors have studied the hydrogeochemical evolution and residence time of water along the groundwater flow path in the Zhangye and Ejina Basins [6, 8]. The exchanges between groundwater and surface water in the Zhangye Basin have also been examined, using either a 3D groundwater flow simulation model or ^{222}Ra analyses [14, 15]. As of the present, similar studies have not been performed in the western part of the Hexi Corridor, Wuwei Basin is a type locality of this area, and Wuwei was an important city on the ancient Silk Road. It is certain that increasing surface water in the basin will increase the groundwater recharge.

The main objective of this study is to present the results from a wide selection of geochemical and isotopic indicators revealing the main characteristics of the groundwater in the Wuwei Basin. The specific goals included: (1) using environmental isotopes (^{18}O, ^{2}H, ^{3}H and ^{14}C) to determine the evolution and age of the groundwater under natural conditions; (2) using the chemistry of major ions to determine the dominant geochemical processes that take place along the groundwater flow paths. These results may not only improve the understanding of the groundwater system in the whole Hexi Corridor, but also provide essential information and theoretical basis for the design of effective water resources management in the Wuwei Basin.

2 Study Area

2.1 General Setting

The Shiyang River Basin, located in the eastern part of the Hexi Corridor in the middle of Gansu province, Northwestern China, is considered a typical arid and semi-arid area. As a result of the long-term water resources development in the Basin, the Shiyang River Basin has become one of the most over-exploited inland basins. Water resources shortage is an increasingly serious issue in China, especially in the arid and semi-arid areas of the Northwest. The Shiyang River is an eco-environmentally fragile area, which is characterized by low and irregular rainfall, high temperatures and evaporation, and notable drought periods. In such arid and semi-arid environment, groundwater is an important part of the total water resource, and plays a vital role as a water supply for drinking and irrigation.

Research interests have been focused on aspects such as utilization survey, evolutionary prediction, reserve estimation and systematic assessment of natural water resources. Some reasonable allocation and utilization plans have been proposed. The effective development and management of the valuable groundwater resources in the area require a better understanding of hydrochemical characteristics of the groundwater and its evolution under natural water circulation processes. During the last 30 years, environmental isotopic techniques have been commonly and largely used in the overall domain of water resources development and management.

The application of these relatively new techniques in the case of arid and semi-arid zones has proven to be an attractive tool for the quantitative evaluation of groundwater systems. However, comprehensive approaches to groundwater understanding using chemical and isotopic indicators are relatively few, although there are several investigations on the interaction and mechanism among surface water, groundwater and rocks in recent years.

The systematic analysis of the hydrochemical types and features of the whole river basin were combined using computer simulation and isotopic geochemical methods in this study. The objective was to elucidate the controls of groundwater quality evolution, recharge, circulation and mixing processes and, furthermore, to explain the mechanism of these processes from the standpoint of environmental geochemistry. The results of the study may provide a theoretical basis for the government to make utilization strategies for water resources and development policies in northwest China in the new century. The purpose of this paper is, thus, to make a new assessment of the availability of groundwater resources in the Wuwei Basin as the typical area of arid northwestern China based on a better understanding of hydrogeological features.

2.2 Geology and Hydrogeology

The Shiyang River Basin, located in the eastern part of Hexi Corridor in the middle of Gansu province, is a typical arid and semi-arid area (Fig. 1). The Shiyang River originates from Lenglongling Glacier of the Qilian Mountains, and ends gradually in the Tenggeli Desert. The Qilian Mountains form the southern part of the basin. Many modern glaciers are distributed above 4,500 m. The middle and northern part of the basin is a plain that can be divided into two sub-basins (i.e., Wuwei Basin and Mingqing Basin).

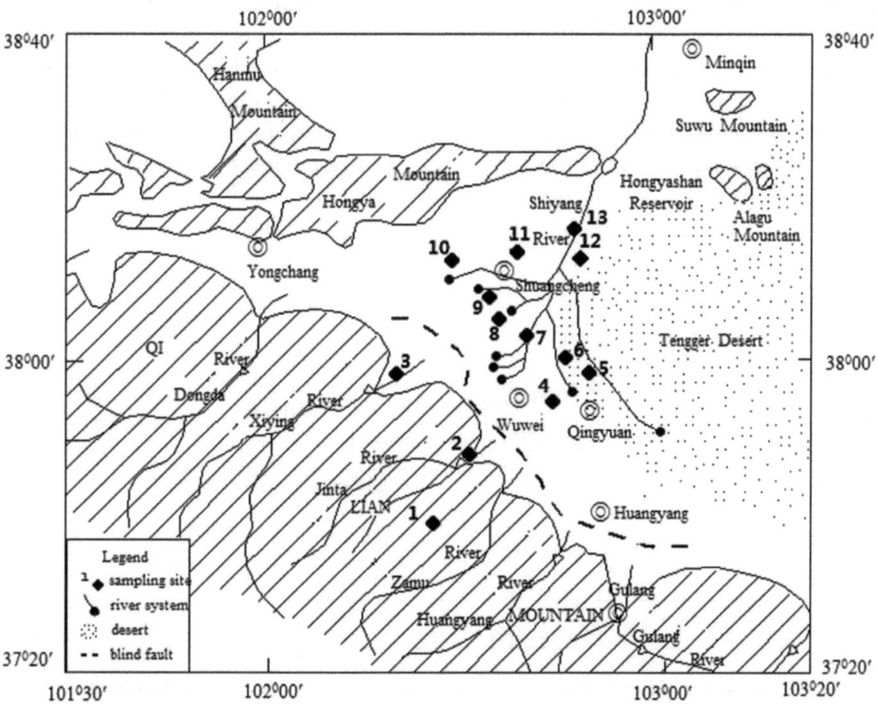

Fig. 1 Geographic location of the study area and water quality sampling point

Most of the basin belongs to the arid climate zone. Solar radiation is strong and sunny. Summer is hot and short, while winter is cold and long. The temperature difference is extremely significant. Less rainfall and intense evaporation lead to dry air. The annual precipitation is 200–800 mm, while the evaporation capacity is 2000–3000 mm.

Consisting of Quaternary alluvial and pluvial sediments, the Wuwei Basin lies in a depression at the foot of the Qilian Mountains with an elevation ranging from 1,400–2,500 m. The loose Quaternary sediments filling the basin are of great thickness, which provide an ideal place for groundwater preservation. In terms of the hydrogeological characteristics, two geomorphologic units are recognized in the Wuwei Basin, a faulting terrace zone and its sub-basin. The former is in the proximity of Qilian Mountains and lies on the head of a pluvial fan. The aquifer there is composed of gravel and sand. The well-connected porosity of the aquifer provides a good conduit for groundwater flow, which is the basin's recharge region. The sub-basin is the enrichment zone of the groundwater. As the grain size of the aquifer gradually decreases from south to north and the flow rate becomes slower, the hydraulic gradient becomes smaller. Along with the uplifting of the Quaternary basement, the burial depth of the aquifer becomes shallower, resulting in the springs of groundwater from the middle part of the Wuwei basin (Fig. 2).

In the southern part of the Wuwei Basin, the diluvial aquifer is formed from highly permeable cobble and gravel deposits that are between 200 and 300 m thick. The aquifer specific capacity is about 3–30 $l\ s^{-1}\ m^{-1}$, and aquifer permeability is up to 100–600 $m\ d^{-1}$. This allows a large amount of surface runoff in the piedmont fan to seep down and recharge the aquifer. From the northern edge of this diluvial fan, the aquifer, comprising interbedded cobble gravel, fine sand and clay, becomes confined or semi-confined with piezometric levels less than 5 m depth. In many places the groundwater overflows as springs and re-emerges as streams.

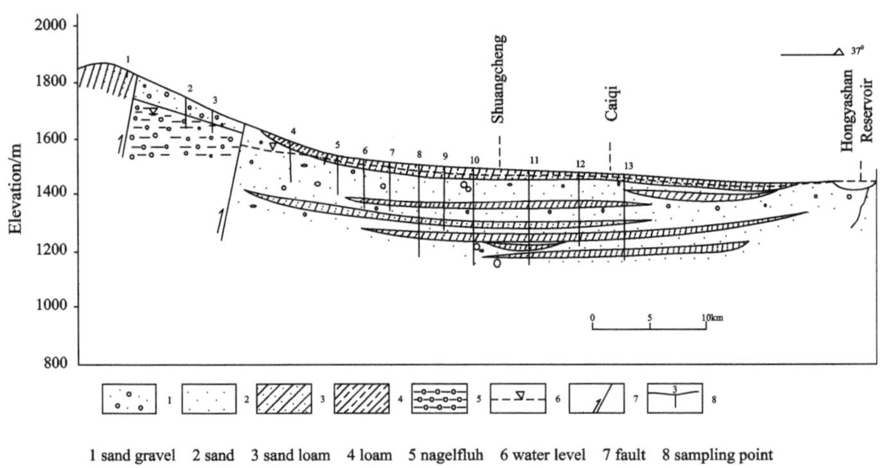

Fig. 2 Geologic section of the Wuwei Basin

The limited precipitation in the plains of the river basin accounts for less than 7% of the total groundwater recharge. Hence, as found in other inland river basins in arid northwest China, the aquifer in Shiyang River Basin to a great extent is recharged by surface water via river infiltration, canal system seepage and farmland irrigation water seepage. In other words, surface water and groundwater are integrated in the middle and lower basins; and the upper, middle and lower reaches are united in using water resources. The water resources in the different parts of the basin are inseparably linked: the groundwater resources in the plain area are recharged from precipitation in the Qilian Mountains. From the Qilian Mountain reach to the Hongyashan section, groundwater and surface water are closely linked due to the gradient and the uniquely permeable nature of the sediments near the mountains. To a large extent, the groundwater is transferred back to surface water before flowing out of the section. Surface water in the Hongyashan reservoir, therefore, mainly comes from surface water and spring water originally present as short residence time groundwater within the Wuwei Basin.

3 Materials and Methods

According to the topography, geomorphology and hydrogeology, 13 groups of sampling locations from Qilian Mountain to Hongyashan Reservoir were chosen (Fig. 1). The positions of the sample points were positioned using a GPS locator. The well depth, latitude, longitude, groundwater depth, water temperature, and pH value were measured on the scene, respectively. Sample bottles were washed three times with water sample before collecting water samples. All water samples were packed in plastic bottles, which were then sealed with wax. No air bubbles were allowed in the bottles to avoid exchanging with carbon dioxide in the air.

Precipitation materials were collected from the global precipitation isotope which was established by International Atomic Energy Agency (IAEA) and the World Meteorological Organization (WMO). Ca^{2+}, Na^{+}, K^{+}, Mg^{2+}, SO_4^{2-}, Cl^{-} and NO_3^{-} were measured using Ion Chromatograph. Deuterium (D) and oxygen-18 (^{18}O) were measured by MAT-252 mass spectrometry. Hydrogen samples were prepared by the zinc reduction method under temperatures of up to 7000 °C, and ^{18}O samples by a H_2O-CO_2 equilibrium method. Here, SMOW (standard mean ocean water) was adopted as the criterion for measuring H and O isotopic compositions. Duplicate analyses of δD and $\delta^{18}O$ were within ±1‰ and ±0.1‰, respectively.

4 Results and Discussion

4.1 Groundwater Geochemistry Evolution Along the Flow Path

The total soluble solids (TDS) characteristics of the groundwater and changes along transects from the southwestern part of the basin to the northeastern part of the basin are shown in Fig. 3. The TDS values were between 131 and 1750 mg/L, gradually increasing along the flow direction. It was generally low in the piedmont area, with a mean concentration of 288 mg L^{-1} and a range from 131 to 444 mg L^{-1}. In such arid regions, groundwater with TDS below 1000 mg L^{-1} is thought to have good quality. The pH of the groundwater ranged between 6.0 and 7.0, indicating an acidic nature. In the Wuwei oasis, the salinity of the groundwater gradually increased along transects. The TDS increased from 500 to 1000 mg L^{-1} in the unconfined groundwater, with no obvious trend along transects.

From the groundwater TDS contour map (Fig. 3), it can be seen that TDS has better zoning characteristics from the piedmont to the middle basin. Groundwater salinity values along the Shiyang river runoff were on the low side, but the TDS value was higher on both sides of the river irrigation district. The TDS values were respectively 1480 and 1750 mg/L in sample Gaogou of Qingyuan irrigation and sample Qijiahu of Yongchang irrigation, which is because the two are in the

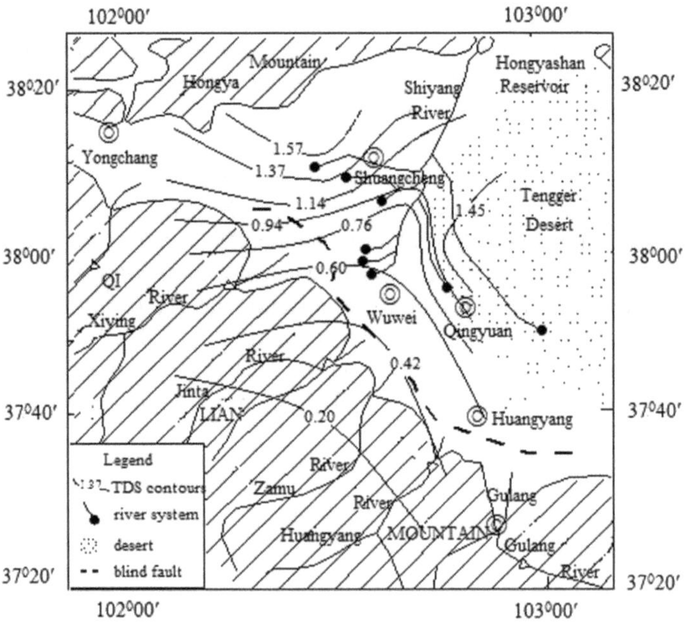

Fig. 3 TDS contours of groundwater in the study area

irrigation district center, where groundwater exploitation is frequent, and they belong to the salty water area. Groundwater TDS value is higher on the eastern desert edge, which was 1063 mg/L. The reason is that with the increase of buried depth of groundwater the supply is less, and the salinity is increasing.

Chlorine ion can be used as the first tracer in the research of groundwater hydrochemistry (Fig. 4). The chloride ion concentration from the piedmont plain increased from 1.5 to 198 mg L^{-1}. The dissolution of halite and sylvite from fine-grained sediments controls concentrations of Na^+ and K^+ in the groundwater, but Na^+:Cl^- molar ratios shows no growth in accordance with the 1:1 in most samples, which are also indicative that the dissolving effect is not significant. HCO_3^- content is dominant, gradually reducing the proportion along the way. With the increase of Cl^-, SO_4^{2-}/Cl^- ratio decreases, but the SO_4^{2-} is increased along the runoff. Cl^- enrichment in the hydrodynamic sluggish belt, Ca^{2+}/Cl^- reduce with the increase of the content of Cl^-.

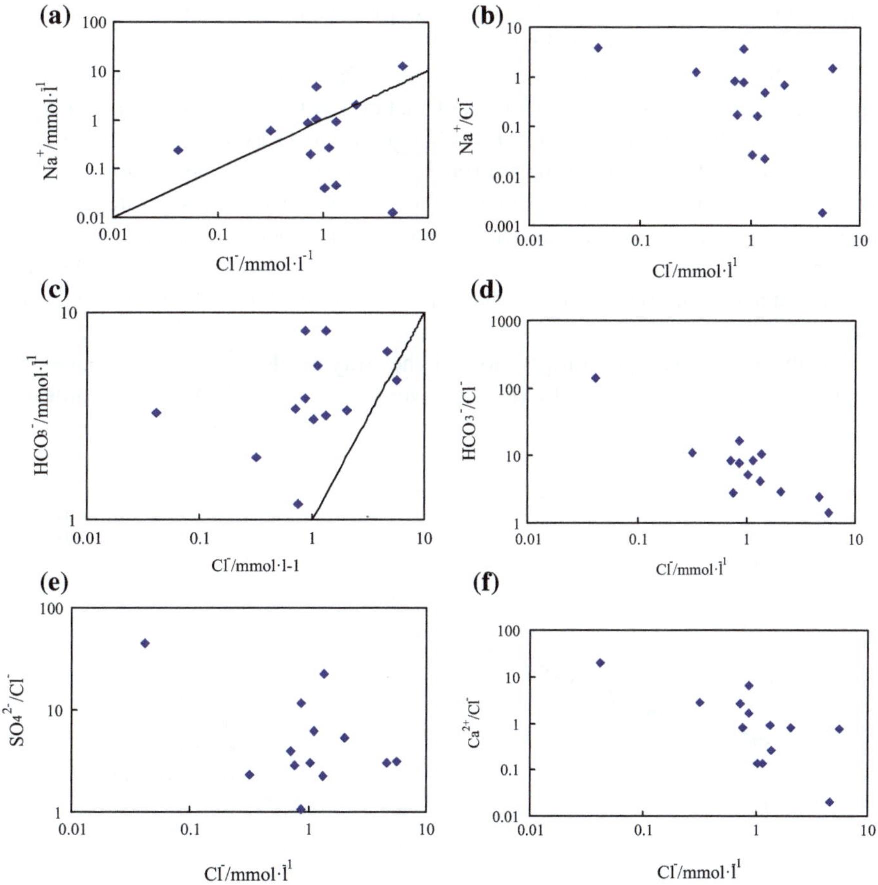

Fig. 4 Plots of molar ratios of major ions versus Cl^-

4.2 Groundwater Recharge Based on Stable Isotope Ratios

The results of stable isotope analysis both for water samples in the basin and precipitation collected in the IAEA were plotted in Fig. 5. Values of $\delta^{18}O$ and δ^2H in precipitation varied from −20.6 to 3.8‰ and from −158 to 33.1‰, respectively, but they are linearly similar to the global meteoric water line (GMWL) with an equation of $\delta^2H = 7.8\ \delta^{18}O + 5$ ($r^2 = 0.92$), which is defined as local meteoric water line (LMWL). These values were slightly enriched and indicated modification due to evaporation in the monsoon rains producing departure from the GMWL. Weighted mean rainfall values lay around $\delta^{18}O$ −7.3‰ and δD −46.9‰. These values were most likely to be the representative of the present local waters.

Some deviation of H and O isotopic compositions from the meteoric water line was shown. This was mainly because of the effect of local variations in rainfall. In addition, the poor coverage of vegetation in the Shiyang River basin was another influencing factor. when the meteoric water recharged into groundwater, drastic evaporation can produce a much larger isotopic fractionation and cause the deviation from the meteoric water line.

Near Qijiahu, the ^{18}O value was lighter than that of other areas, which may be caused by the distribution of granite and metamorphic rocks. When groundwater contacts with those rocks, water can exchange H and O isotopes with minerals in rocks. Such a water-rock equilibrium fractionation can reduce the content of ^{18}O in water. The isotopic fractionation intensity and the water-rock system depended on the temperature, quantity ratio of water and rock, and reaction time. The water-rock equilibrium fractionation was very small during groundwater leakage. The slow groundwater circulation rate enabled a full isotopic exchange between water and rock.

For the stable isotope composition of the Shiyang River running course, the mean values of $\delta^{18}O$ increased along the river course. The $\delta^{18}O$ and δD values for

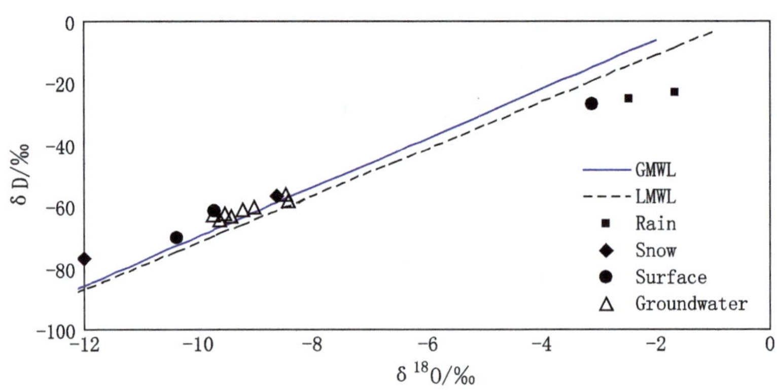

Fig. 5 Plot of $\delta^{18}O$ and δD for groundwater from the Quaternary aquifer system of the Shiyang River with plots of the GMWL and LWML

these samples ranged from −11.99 to −8.4‰ and from −76.7 to −56‰, respectively. Figure 5 illustrated that the samples from the Shiyang River running course had the similar stable isotope compositions to those from the wells drilling into the unconfined aquifers near the river bank. This indicated that evaporation probably occurred in channels, reservoirs and fields in the study area.

The majority of the isotope compositions of the groundwater samples plot were close to the GMWL (Fig. 5). This indicated that the groundwater recharge resource mainly originated from precipitation, and was weakly affected by evaporation. These more depleted signatures were considered to be from the late Pleistocene and Holocene periods, during which the climate was wetter and colder than that of the present day.

4.3 Residence Time Based on ^{14}C Data

This study of groundwater ^{14}C age in the studied area was calculated using carbon-13 correction model (Table 1). Meanwhile, Fig. 6 shows the distribution of ^{14}C value in the research area. ^{14}C value ranged between 48 and 88 pmc and decreased gradually from mountain front radial directions. ^{14}C value is bigger in piedmont, in 77–88 pmc. ^{14}C content change is relatively stable in fine soil plain, which slowly fell to 77 from 67 pmc. ^{14}C is between 56 and 67 pmc in rump region.

The age range of the deep groundwater (>100 m) in the study area was between 1000–5800 a (Fig. 7), which showed an increased trend from piedmont to rump. The piedmont groundwater was mainly from the recharge of piedmont infiltrated water, and thus the deep groundwater is younger than the low layer groundwater. The groundwater age in the central basin featured a small variation ranging from 2920 to 3400 a, which may mainly be due to the mixture with ooze water from the confined aquifer during exploring mining groundwater for irrigation. In the rump area, the deep groundwater age is more than 4000 a with a slow water

Table 1 Sample point ^{14}C values

Sample no.	Site	pH	^{14}C (pmc)	Age (BP)
4	Shiyangzhan	7.0	81.4996	1645
5	Dawan	6.0	66.9869	3220
6	Gaogou	6.0	74.7928	2335
7	Yujiawan	6.0	72.7173	2560
8	Hongshuihe	6.5	48.5654	5800
9	Baiyun	5.5	88.2733	1000
10	Qijiahu	6.0	62.7277	3745
11	Xigou	6.5	54.2718	4910
12	Liuquan	6.5	56.5280	4580
13	Canqi	6.0	86.0702	1205

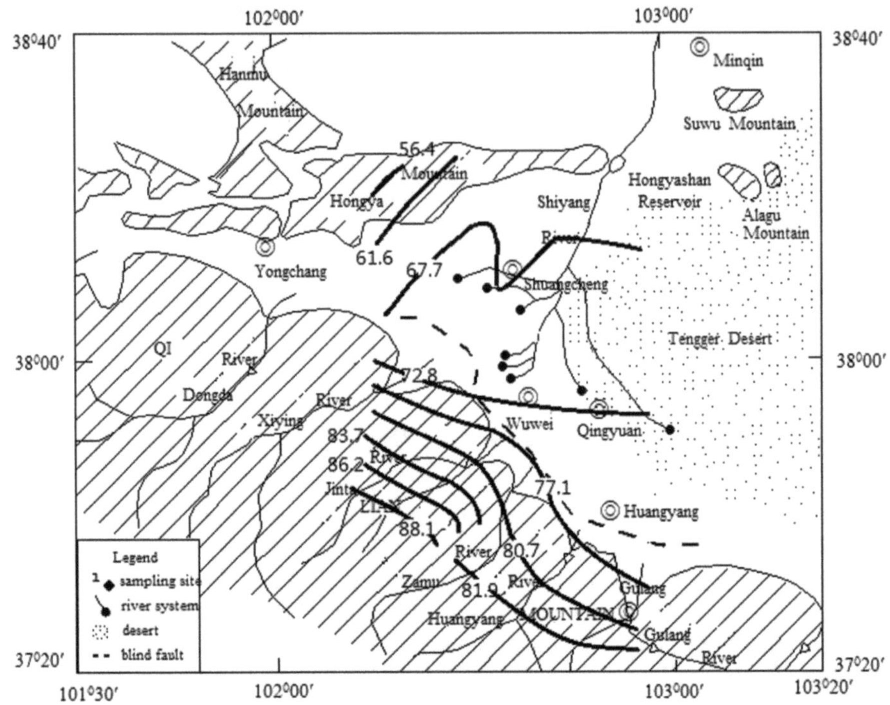

Fig. 6 The ^{14}C simulation in the study area

cycle and a limited water quantity for groundwater recharge, and thus the groundwater age of the rump area is older.

4.4 Conceptual Model for Groundwater Recharge

A conceptual model for groundwater recharge is proposed based on the isotopic data (Fig. 7). The rainfall was more abundant in the region during the late Pleistocene and the aquifer was probably replenished by rainfall through mountain-front recharge and direct diffusive recharge. However, modern direct infiltration provides negligible recharge to the shallow unconfined aquifer. Most recharge originates from the Qilian Mountains where there is higher precipitation and melt water. The natural rivers in the oasis area have mostly been changed to artificial channels, the seepage from which plus irrigation return water, are the main recharge sources for the shallow phreatic aquifer. The confined aquifer of the piedmont has been recharged through a number of mechanisms, including lateral flow from the piedmont and modern water through directly vertical infiltration from

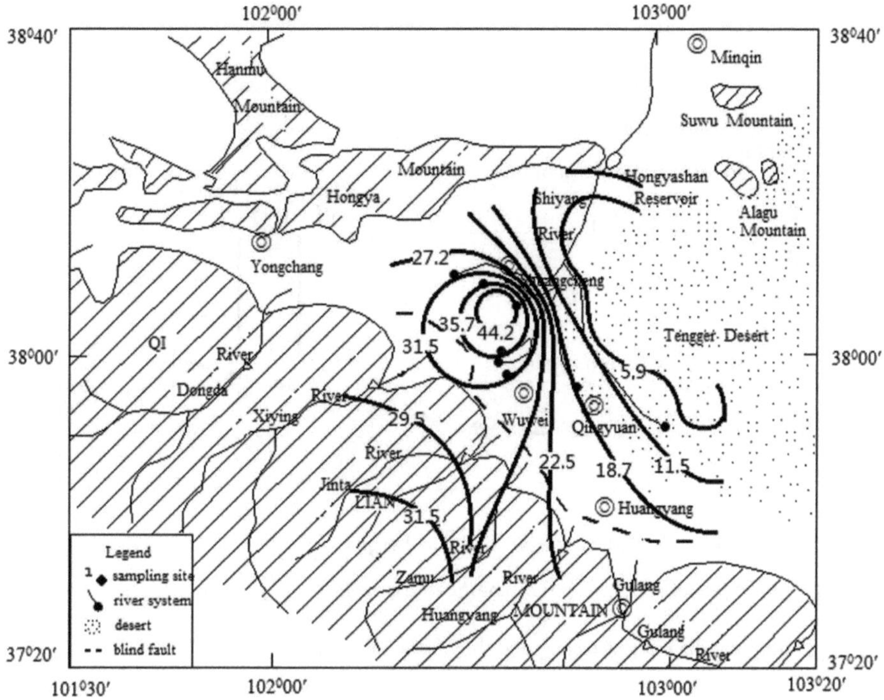

Fig. 7 Confined water of ^{14}C age distribution in the study area

river channels. Groundwater in the confined aquifer flows towards the discharge zones to the NE.

The groundwater flow velocity can be estimated using the difference in dates and positions between the two confined groundwater samples: sample 8 is about 31 km further along the flow path from sample 4, indicating an approximate flow velocity of 3.04×10^8 m^3 a^{-1}. This result is coincident with that calculated based on Darcy's law. This indicated that groundwater flow is relatively slow and that the water resources sustained by this flow are essentially non-renewable in the confined aquifer. In summary, the confined groundwater dates back to the late Pleistocene and has continued throughout the Holocene, as has been observed in other basins in northwestern China [8, 16–18]. The aquifer system retains this ancient groundwater, but also reveals modern recharge (particularly in unconfined aquifers) as well as mixing of water from different periods (Fig. 8).

Isotopic data and hydrogeochemical tracers were used to understand the recharge and geochemical evolution of groundwater in the Quaternary aquifers beneath the Wuwei Basin. The Quaternary aquifers are all closely connected with streams originating in the Qilian Mountains. The mean values of δ^{18}O and δ^2H in the unconfined groundwater tended to be enriched along the overall groundwater flow path from the SW to the NE, increasing from −9.71‰ to −8.61‰ for δ^{18}O and

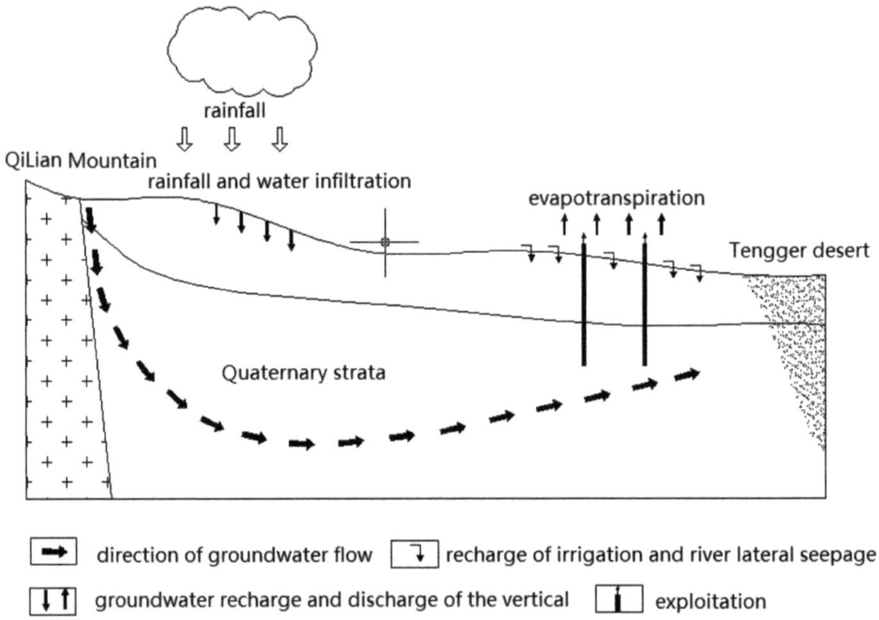

Fig. 8 Conceptual diagram of recharge to the Quaternary aquifer beneath

from −64.5‰ to −56‰ for δ^2H, respectively. The values in surface runoff and melt water from the Qinghai-Tibetan Plateau south of the study area suggest that melt water from the Qilian Mountains strongly determines the isotopic composition of water in the Shiyang River, and contributes significantly to the steady long-term runoff of this river. However, the stable isotope data, which are markedly depleted in heavy isotopes compared to modern rainfall in the plains area, clearly indicate that there is no direct relationship between the confined groundwater and modern direct recharge. The groundwater exhibited a range of radiocarbon activities from 48 to 88 pmc, with younger water in the upper reaches of the Shiyang River and older water in the lower reaches. The unconfined water was generally young with 0.9–50.78 TU. At the northern edge of the basin, confined groundwater was much older, with a radiocarbon content of 48 pmc, indicating an age of ~6 ka, which corresponds to the humid climatic phases of the late Pleistocene.

The strong and significant linear relationships between Na^+ and Cl^- and between K^+ and Cl^- indicate that the dissolution of halite and sylvite strongly control the concentrations of these three ions. Groundwater was far below saturation for halite, so halite minerals in the fine-grained sediments can easily enter the groundwater. The Na^+/Cl^- molar ratio values were all >1 in the groundwater, indicating the presence of excess Na^+, possibly due to the weathering of feldspar. Sulfate, HCO_3^-, Mg^{2+}, and Ca^{2+} were not derived from the simple dissolution of calcite, dolomite and gypsum.

The results have important implications for management of the inter-basin water allocation program and groundwater resources in the Wuwei Basin, one of several water-stressed basins in northwestern China that lies in a region destined for rapid development under China's West Development Strategy [19]. As most of the deep groundwater resources derived from recharge during a period with a colder and wetter climate, exploitation of this groundwater is considered to be "mining" if it cannot be replenished. The glaciers of the Qilian Mountains have exhibited large absolute losses and the altitude of the equilibrium line has generally increased because the ablation rate has exceeded the mass accumulation rate as a result of global warming since 1950 [20–24]. This has significantly influenced the sustainable source of Shiyang River and recharge of the unconfined groundwater. However, the groundwater has seen over-exploitation due to a lack of knowledge of recharge and evolution mechanism. The results of this study will help the local government to protect this limited and precious water resource [22, 25, 26]. A rational land-use plan for agriculture, forestry and animal husbandry to regulate water allocation on a whole-basin basis should be set up.

Acknowledgements This study was supported by the Natural Science Foundation of China (No. 31400418), the Natural Science Foundation of Hebei Province (No. E2015402128, C2016402088), the Young Outstanding Innovative Talents of Hebei Province (BJ2016012), and the China Postdoctoral Science Foundation funded project (2014M561044 and 2016T90128).

References

1. Vairavamoorthy, K., Gorantiwar, S.D., Pathirana, A.: Managing urban water supplies in developing countries-climate change and water scarcity scenarios. Phys. Chem. Earth. Parts A/B/C **33**, 330–339 (2008). https://doi.org/10.1016/j.pce.2008.02.008
2. Vörösmarty, C.J., McIntyre, P.B., Gessner, M.O., Dudgeon, D., Prusevich, A., Green, P., Glidden, S., Bunn, S.E., Sullivan, C.A., Liermann, R.C., Davies, P.M.: Global threats to human water security and river biodiversity. Nature **467**, 555–561 (2010). https://doi.org/10.1038/nature09440
3. Kinzelbach, W., Bauer, P., Siegfried, T., Brunner, P.: Sustainable groundwater management problems and scientific tools. Episodes **26**, 279–284 (2003)
4. Scanlon, B.R., Keese, K.E., Flint, A.L., Flint, L.E., Gaye, C.B., Edmunds, W.M., Simmers, I.: Global synthesis of groundwater recharge in semiarid and arid regions. Hydrol. Process. **20**, 3335–3370 (2006). https://doi.org/10.1002/hyp.6335
5. Ma, J.Z., Wang, X.S., Edmunds, W.M.: The characteristics of ground-water resources and their changes under the impacts of human activity in the arid northwest China—a case study of the Shiyang River Basin. J. Arid Environ. **61**, 277–295 (2005). https://doi.org/10.1016/j.jaridenv.2004.07.014
6. Feng, Q., Wei, L., Su, Y.H.: Distribution and evolution of water chemistry in Heihe River Basin. Environ. Geol. **45**, 947–956 (2005). https://doi.org/10.1007/s00254-003-0950-7
7. Chen, Y.N., Chen, Y.P., Xu, C.C., Ye, Z.X., Li, Z.Q., Zhu, C.G., Ma, X.D.: Effects of ecological water conveyance on groundwater dynamics and riparian vegetation in the lower reaches of Tarim River, China. Hydrol. Process. **24**, 170–177 (2010). https://doi.org/10.1002/hyp.7429

8. Chen, Z.Y., Nie, Z.L., Zhang, G.H., Wan, L., Shen, J.M.: Environmental isotopic study on the recharge and residence time of groundwater in the Heihe River Basin, northwestern China. Hydrogeol. J. **14**, 1635–1651 (2006). https://doi.org/10.1007/s10040-006-0075-7
9. Li, J.: Water shortages loom as northern China's aquifers are sucked dry. Nature **328**, 1462–1463 (2010). https://doi.org/10.1126/science.328.5985.1462-a
10. Huang, T.M., Pang, Z.H.: Changes in groundwater induced by water diversion in the lower Tarim River, Xinjiang Uygur, NW China: evidence from environmental isotopes and water chemistry. J. Hydrol. **387**, 188–201 (2010). https://doi.org/10.1016/j.jhydrol.2010.04.007
11. Ding, H.W., Zhao, C., Huang, X.H.: The ecological environment and desertification in the Shule River basin. Arid Zone Res. **18**, 5–10 (2001). https://doi.org/10.13866/j.azr.2001.02.002
12. Zhu, G.F., Su, Y.H., Feng, Q.: The hydrochemical characteristics and evolution of groundwater and surface water in the Heihe River Basin, northwest China. Hydrogeol. J. **16**, 167–182 (2008). https://doi.org/10.1007/s10040-007-0216-7
13. Edmunds, W.M., Ma, J., Aeschbach-Hertig, W.: Groundwater recharge history and hydrogeochemical evolution in the Minqin Basin, North West China. Appl. Geochem. **21**, 2148–2170 (2006). https://doi.org/10.1016/j.apgeochem.2006.07.016
14. Wu, Y.Q., Wen, X., Zhang, Y.: Analysis of the exchange of groundwater and river water by using Radon-222 in the middle Heihe Basin of northwestern China. Environ. Geol. **45**, 647–653 (2004). https://doi.org/10.1007/s00254-003-0914-y
15. Hu, L.T., Chen, C.X., Jiao, J.J., Wang, Z.J.: Simulated groundwater interaction with rivers and springs in the Heihe river basin. Hydrol. Process. **21**, 2794–2806 (2007). https://doi.org/10.1002/hyp.6497
16. Ma, J.Z., Gao, Q.Z.: Groundwater vulnerability and its assessing method in the arid land of NW China. Arid Land Geogr. **26**, 44–49 (2003). https://doi.org/10.13826/j.cnki.cn65-1103/x.2003.01.008
17. Ma, J., Ding, Z., Gates, J.B., Su, Y.: Chloride and the environmental isotopes as the indicators of the groundwater recharge in the Gobi desert, northwest China. Environ. Geol. **55**, 1407–1419 (2008). https://doi.org/10.1007/s00254-007-1091-1
18. Ma, J., Pan, F., Chen, L.H., Edmunds, W.M., Ding, Z.Y., He, J.H., Zhou, K.P., Huang, T.M.: Isotopic and geochemical evidence of recharge sources and water quality in the Quaternary aquifer beneath Jinchang city, NW China. Appl. Geochem. **25**, 996–1007 (2010). https://doi.org/10.1016/j.apgeochem.2010.04.006
19. Ding, Y.J., Liu, S.Y., Li, J., Shangguan, D.H.: The retreat of glaciers in response to recent climate warming in western China. Ann. Glaciol. **43**, 97–105 (2006). https://doi.org/10.3189/172756406781812005
20. Edmunds, W.M., Carrillo-Rivera, J.J., Cardona, A.: Geochemical evolution of groundwater beneath Mexico City. J. Hydrol. **258**, 1–24 (2002). https://doi.org/10.1016/s0022-1694(01)00461-9
21. He, J.H., Ma, J.Z., Zhang, P., Tian, L.M., Zhu, G.F., Edmunds, W.M., Zhang, Q.H.: Groundwater recharge environments and hydrogeochemical evolution in the Jiuquan Basin, northwest China. Appl. Geochem. **27**, 866–878 (2012). https://doi.org/10.1016/j.apgeochem.2012.01.014
22. Schoeller, H.: Geochemistry of groundwater. In: Brown, R.H. (ed.) Groundwater Studies–An International Guide for Research and Practice, pp. 1–18. UNESCO, Paris (1977)
23. Wang, N.L., He, J.Q., Pu, J.C., Jiang, X., Jing, Z.F.: Variations in equilibrium line altitude of the Qiyi Glacier, Qilian Mountains, over the past 50 years. Chin. Sci. Bull. **55**, 3810–3817 (2010). https://doi.org/10.1007/s11434-010-4167-3
24. Wang, P.Y., Li, Z.Q., Gao, W.Y.: Rapid shrinking of glaciers in the Middle Qilian Mountain region of northwest China during the last 50 years. J. Earth Sci. **22**, 539–548 (2011). https://doi.org/10.1007/s12583-011-0195-4

25. Zhang, Y.H., Wu, Y., Su, J., Wen, X., Liu, F.: Groundwater replenishment analysis by using natural isotopes in Ejina Basin, northwestern China. Environ. Geol. **48**, 6–14 (2005). https://doi.org/10.1007/s00254-004-1214-x
26. Zhang, H.S.: A brief introduction of the changes in groundwater resources in the Hexi Corridor. Hydrogeol. Eng. Geol. **32**, 81–84 (2005). https://doi.org/10.16030/j.cnki.issn.1000-3665.2005.04.020

Evaluation System Building for Groundwater Overexploitation Zone Management

Siyu Liu, Genfa Chen and Hongzhen Ni

Abstract China's demand for groundwater is growing along with its economic development. Such problems as continued declining of water table due to groundwater overexploitation, ground subsidence and water quality deterioration as well as other environmental issues have become increasingly prominent. Therefore, China has carried out a series of groundwater management measures in the overexploitation zones. However, management measures and experience cannot be evaluated and summarized systematically as a management evaluation system has not been put in place in a timely manner. Against this background, this paper aims to establish a relatively scientific and rational groundwater overexploitation management and evaluation system by combining with the generally applicable management experience in groundwater overexploitation zones oriented in management results, control measures, monitoring system and regional characteristics. The system is also applied to evaluate two typical overexploitation zones of Hebei Province and Jiangsu Province from 2012 to 2014 with their final scores of 70 and 86, respectively. The evaluation results are analyzed to prove the rationality of the system as a basis for reference to fully promote the management of groundwater overexploitation zones.

Keywords Groundwater · Overexploitation zone · Management
Evaluation

S. Liu (✉) · G. Chen · H. Ni
China Institute of Water Resources and Hydropower Research,
Beijing 100038, China
e-mail: liusy0130@163.com

G. Chen
e-mail: chengf02@163.com

H. Ni
e-mail: nihongzhen@163.com

W. Dong et al. (eds.), *Sustainable Development of Water Resources and Hydraulic Engineering in China*, Environmental Earth Sciences,
https://doi.org/10.1007/978-3-319-61630-8_9

1 Introduction

Groundwater is an important element in the eco-environment and plays a decisive role in maintaining socio-economic development, human water use security as well as a sound eco-environment. Since the 1970s, groundwater has been widely exploited and utilized along with China's accelerated industrialization and urbanization process, especially in those regions without abundant surface water resources, leading to their increasingly growing reliance on groundwater. According to statistics, overexploited groundwater in China had grown from about 20 billion m^3 in 1972 to 110.7 billion m^3 in 2010. Under the strained conflicts between supply and demand of water resources, coupled with continued declining water table, ground subsidence, and water quality deterioration [1, 2], it has become a major task for China's water resources work to strengthen its management of groundwater overexploitation zones.

Although China's management work has been conducted in groundwater overexploitation zones for many years, management measures and experience have not been systematically evaluated and summarized as a management evaluation system has not been put in place in a timely manner, which is not conducive to pushing forward the all-round management of the overexploitation zones. In light of this situation and on the basis of analyzing management approaches of groundwater overexploitation, this paper aims to establish a relatively scientific and easily operational groundwater overexploitation management evaluation system for conducting a quantitative appraisal and evaluation of groundwater resources management level and the determination of the weak links so as to put forward rational suggestions. It will be of great significance to strengthen groundwater management of overexploitation zones and to implement the most stringent water management system.

2 Evaluation Methods

According to scientific, comprehensive, representative and practical principles, the evaluation system adopts two types of index in this paper, namely national general index and regional characteristic index. National general index is mainly set up by combining with the generally applicable management experience in groundwater overexploitation zones oriented in three levels of management results, control measures, monitoring system general management experience; while regional index mainly considers regional characteristics for evaluation. Firstly, a series of indices which can reflect the management level most should be identified before their weights are determined and they are scored with corresponding graded standards. Then relevant data before and after the implementation of management measures in groundwater overexploitation zones are acquired. Finally, a comprehensive conclusion for management of groundwater overexploitation zones is summed up according to a quantitative evaluation mode to score assignments by the expert judgment methods.

3 Evaluation System Building for Groundwater Overexploitation Zone Management

3.1 National General Index

3.1.1 Evaluation Index of Management Effects

(1) Groundwater Table Variation. Water table is the core of restricting groundwater overexploitation, so the change of groundwater table is an important index of groundwater zone management effects. However, the underground water table variation is subject to the dual effects of natural precipitation and management of overexploitation zones; therefore, the impact of natural precipitation variation should be deducted in the evaluation. The relationship between groundwater table variation and natural precipitation [3] can be expressed as:

$$\sum \Delta h = \mathrm{a} \times \mathrm{P}/\mathrm{u} \tag{1}$$

where Δh is the groundwater changing range due to all the rainfalls within a year, a is precipitation infiltration recharge coefficient, u is specific water yield in aquifer, and P is precipitation.

(2) Variation Rate of Groundwater Exploitation Amount. Groundwater exploitation variation is a fundamental factor to reflect management enforcement and effects of overexploitation zones. As the groundwater exploitation amount cannot be directly compared in different regions, the variation rate of groundwater exploitation amount is adopted to reflect the effects of groundwater management, and the calculation formula is:

$$\mathrm{Vea} = \frac{\mathrm{EA}' - \mathrm{EA}}{\mathrm{EA}} \times 100\% \tag{2}$$

where Vea is variation rate of groundwater exploitation amount, EA is groundwater exploitation amount in the baseline year, EA′ is groundwater exploitation amount in the evaluation year.

(3) Groundwater Quality. Groundwater is not only an important source for water supply, but plays a prominent role in supporting ecosystem and maintaining a virtuous cycle of water system. Once the groundwater is contaminated, damage

to the ecology and human health will be more profound. Therefore, the focus must be on groundwater quality. In this paper, water quality of monitoring results is scored according to the National Groundwater Quality Standard (GB/T 14848-93).

3.1.2 Evaluation Index of Management Measures

(1) Laws, Regulations and Institutional System. Laws, regulations and institutional system, are the “superstructure” of groundwater management, including groundwater development and utilization system, groundwater protection system, reward and punishment system, etc. The index can be evaluated qualitatively according to the fact that whether it is perfect and effectively enforced.
(2) Economic Measures. In a water management approach proposed by Chinese President Xi Jinping, namely “top priority given to water-saving, systematic management, spatial balance and emphasis on both the government and market”, the roles of both the administration and market are especially highlighted. The role of the market is mainly reflected in economic measures, which can be evaluated through whether economic measures are adopted to encourage everyone to save water, improve water use efficiency and reduce groundwater exploitation.
(3) Technical Measures. Science and technology is an important force in promoting the development of modern productive forces. It is also of significance in groundwater management of overexploitation zones, which are embodied in the research on water-saving technology in industry and agriculture, groundwater monitoring technology, artificial back irrigation technology, ground subsidence prevention and control technology.

3.1.3 Evaluation Index of Monitoring System

(1) Groundwater Monitoring Well Density. Monitoring Wells are important facilities to learn the groundwater table and dynamic change of water quality in overexploitation zones. The common method to indicate groundwater monitoring well density in China is the number of wells per square kilometer on average. This method facilitates the deployment and analysis of the exploration work. The Netherlands tops the world’s groundwater monitoring station network density with one station per one square kilometer on average. Therefore, the goal should be to strive to keep abreast with developed countries in terms of monitoring stations for the scoring criteria.
(2) Effective Data Monitoring Rate. Although there are many places with network coverage, it is difficult or simply impossible to obtain some data in the process of actual monitoring and operation. The effectiveness of monitoring data collection is an important guarantee of monitoring results, so effective data monitoring rate is required for the evaluation. The index should also have high

standards, or the established monitoring wells will not fully give its role to play, then the efficiency and accuracy will be greatly affected.

3.2 Regional Characteristic Index

In this paper, three typical areas, namely North China, the Yangtze River delta region, northwestern region are selected due to their differences in natural conditions and economic bases. They are representatives in terms of groundwater overexploitation with outstanding problems, which can provide references to other regions for building the evaluation system of groundwater management.

Regional Characteristic Index in North China. North China, as one of the "big four" groundwater overexploited zones as well as the biggest "funnel zone" in the world, is seriously overexploited. In addition, some cities in this region have been plagued with seawater intrusion; therefore, the funnel area variation and seawater intrusion constitute its regional characteristic index.

(1) Funnel Area Variation Rate. An underground funnel is the most severe in North China, so the funnel area variation rate can most directly reflect management results in overexploitation zones in this region, and the calculation formula is as follows:

$$\mathrm{Vfa} = \frac{\mathrm{FA}' - \mathrm{FA}}{\mathrm{FA}} \times 100\% \tag{3}$$

where Vfa is funnel area variation rate, FA is funnel area in the baseline year, FA′ is funnel area in the evaluation year.

(2) Seawater Intrusion. Seawater intrusion is mainly caused by groundwater overexploitation. As the water table sharply subsides, the original dynamic equilibrium between seawater and fresh water is destroyed, resulting in their interface moving toward the land direction. According to *China Maritime Disaster Bulletin*, severe seawater intrusion distance is defined as 10–30 km away from the coast, and in this paper severity will be scored according to seawater intrusion distance.

Regional Characteristic Index in China's Yangtze River Delta Region. The Yangtze River delta region is the first big economic zone in China with its water demand increasingly growing due to its rapid economic development. Moreover, its groundwater overexploitation zone is facing serious geological problems, such as ground subsidence and ground fissures, which will be important indices for evaluating groundwater management in the Yangtze River delta.

(1) Ground Subsidence Rate. Ground subsidence, one of the main geological disasters in the Yangtze River delta, is a geological environmental problem in groundwater overexploitation zones due to excessive exploitation of groundwater. Ground subsidence rate can indirectly reflect the effects of the geological environment of groundwater overexploitation zone management, so scores will be given according to the scale and level of ground subsidence.

(2) Accumulated Length of Ground Fissure. Ground fissures are mainly formed by uneven subsidence due to variations of deep groundwater exploitation in different areas. The scale of ground fissures mainly depends on the accumulated length of one ground fissure activity. According to statistics, there are nearly 6000 large ground fissures in China [4], which are generally about 50–500 m long, and the longest is up to 8 km.

Regional Characteristic Index in China's Northwestern Region. Northwestern China features low precipitation, sparse vegetation, serious soil erosion and land desertification, and extremely vulnerable ecological system; therefore, the land desertification area variation rate and rate for phreatic to support vegetation root system are selected as the characteristic indices.

(1) Ground Desertification Area Variation Rate. Groundwater shortages cause plant communities to wither, oasis to fade and desertification area to expand unceasingly. Therefore, the area variation rate of land desertification is one of the characteristic indices in northwestern China with the calculation formula as:

$$\text{Vda} = \frac{\text{DA}' - \text{DA}}{\text{DA}} \times 100\% \tag{4}$$

where Vda is ground desertification area variation rate, DA is ground desertification area in the baseline year, DA′ is ground desertification area in the evaluation year.

(2) Rate for Phreatic Water to Support Vegetation Root System. Northwestern China's vegetation ecological stability is closely related with groundwater, as groundwater transports upward through the phreatic evaporation to supply water for vegetation growth. Therefore, the key lies in the relationship between groundwater and the critical depth of vegetation to maintain the stability of the vegetation ecosystem, and that is the phreatic water to support the vegetation root system. When the groundwater table is below the embedded depth, the vegetation root system cannot absorb groundwater; so, the growth of plants is affected. Therefore, the rate for phreatic water to support the vegetation root system can be used to evaluate groundwater in northwestern China whether it is capable of guaranteeing the stability of the vegetation ecology. Specific critical depth of groundwater to recharge vegetation can be referred to in the literature available [5].

3.3 Determination of Index Weight and Scoring Criteria

Index Weight. Index weight is a comprehensive measurement for subjective evaluation and objective reflection of its important degree. Considering the fact that some geological problems do not stem from only one factor of groundwater, it is inappropriate for regional characteristic index weight to be set above 20% with the national general index weight as 80%. Then the weight relationship among management effectiveness, management measures and monitoring system is further determined. By referring to the appraisal index system of the most stringent water resources management system [6], and the appraisal index system of national water ecological civilization city [7], and other related evaluation systems, the management effectiveness is generally weighted as 40–50%. For the convenience of calculation in this paper, management effectiveness index is weighted 45%, management measures index 15%, and monitoring system index 20%. Weights of all indices are as shown in Table 1.

Scoring Criteria. Scoring criteria is the basis of the evaluation system. According to related standards formulated by China, with references to regional actual situations at the same time, and based on a summary of the academic research results, scoring criteria rules for the evaluation system of groundwater overexploitation zone management are ultimately determined as in Table 1. The evaluation results are classified into five grades as "excellent" (above 90 points), "good" (80–90 points), "medium" (70–80 points), "pass" (6–70), and "bad" (below 60 points).

Unavailable Data Treatment. In the actual operation process, it is common that evaluation of relevant indices cannot be conducted because of certain data unavailability. In this paper, if the data is not enough to support a certain index evaluation, this index will not be evaluated, and its corresponding weight will be given to other similar types of index.

4 Application of Groundwater Overexploitation Zone Management Evaluation System

4.1 Basis for Selecting Typical Areas

In this paper, Hebei Province and Jiangsu Province from 2012 to 2014 are selected as application examples for the groundwater management evaluation system.

Hebei Province is short of water resources; therefore, groundwater constitutes an important water source for people's lives and production, accounting for 80% of the total water supply. Its groundwater overexploitation problem stands as very outstanding with the average annual overexploitation amount at more than 5 billion m^3, or 1/3 of national total groundwater overexploited area. While Jiangsu Province

Table 1 Evaluation index weight and scoring criteria

Index	Scoring criteria			
	Variation amplitude	Scores	Variation amplitude	Scores
Groundwater table variation (m) (weight 15%)	>1.5	100	−0.5 to 0	50
	1–1.5	90	−1 to −0.5	30
	0.5–1	80	−1.5 to −1	10
	0–0.5	70	<1.5	0
	0	60		
Variation rate of groundwater exploitation amount (%) (weight 15%)	<−40%	100	0–10%	50
	40 to −30%	90	10–20%	40
	−30 to −20%	80	20–30%	30
	−20 to −10%	70	30–40%	20
	−10 to 0	60	>40%	0
Groundwater quality (weight 15%)	Grade III and above	100		
	Grade IV	80		
	Grade V	60		
	Worse than Grade V	30		
Laws, regulations and institutional system (weight 5%)		Qualitative evaluation		
Economic measures (weight 5%)		Qualitative evaluation		
Technical measures (weight 5%)		Qualitative evaluation		
Groundwater monitoring well density (No./100 km^2) (weight 10%)	10 and above		100	
	10–5		80	
	5–1		60	
	1–0.5		30	
	0.5–0		0	
Effective data monitoring rate (%) (weight 10%)	Effective data monitoring rate of monitoring wells × 100			
Funnel area variation rate (%) (weight 10%)	< −40%	100	0–10%	50
	40 to −30%	90	10–20%	40
	−30 to −20%	80	20–30%	30
	−20 to −10%	70	30–40%	20
	−10 to 0	60	>40%	0

(continued)

Table 1 (continued)

Index	Scoring criteria			
	Variation amplitude	Scores	Variation amplitude	Scores
Seawater intrusion distance (km) (weight 10%)	No seawater intrusion		100	
	0–15		80	
	15–30		60	
	30–50		30	
	>50		0	
Ground subsidence rate (mm/a) (weight 10%)	<5		100	
	5–10		80	
	10–30		60	
	30–50		30	
	>50		0	
Accumulated length of ground fissure (m) (weight 10%)	0		100	
	0–100		80	
	100–500		60	
	500–1000		30	
	>1000		0	
Ground desertification area variation rate (%) (weight 10%)	< −40%	100	0–10%	50
	40 to −30%	90	10 to 20%	40
	−30 to −20%	80	20 to 30%	30
	−20 to −10%	70	30 to 40%	20
	−10% to 0	60	>40%	0
Rate for phreatic water to support vegetation root system (%) (weight 10%)	Rate for phreatic water to support vegetation root system × 100			

has abundant but generally heavily polluted surface water resources due to its rapid economic development, excessive groundwater overexploitation in many areas has produced geological problems such as ground fissures and ground subsidence. These two typical areas constitute strong representatives in their treatment of groundwater overexploitation. The system will be applied to evaluate the groundwater management in Hebei Province and Jiangsu Province in order to provide technical support for the management level in groundwater overexploitation zones through the analysis of the results to verify the rationality of the system.

4.2 Evaluation for Groundwater Overexploitation Zone Management in Hebei Province

According to the actual situation in Hebei Province in 2013–2014 and with reference to related data, including *Hebei Provincial Water Resources Bulletin, Hebei Provincial Environmental Status Bulletin, Groundwater Bulletin, Hebei Provincial Water Quality Monthly Report* and *China Maritime Disaster Bulletin*, groundwater overexploitation zone management in Hebei Province is evaluated with specific situation in the following Table 2.

4.3 Evaluation for Groundwater Overexploitation Zone Management in Jiangsu Province

According to the actual situation in Jiangsu Province in 2012–2014 and with reference to Jiangsu Provincial Water Resources Bulletins, Annual Reports of Groundwater Monitoring, website of Water Resources Department of Jiangsu Province and other related literature data [9–11], groundwater management in Jiangsu Province is evaluated with its groundwater overexploitation zone management situation summed up in the following Table 3.

4.4 Evaluation Result Analysis

According to the calculation by evaluation system, the overexploitation zone management in Hebei Province from 2012 to 2014 scores 70 points at the level of "medium", while Jiangsu Province scores 86 points at the level of "good".

Differences in results, to some extent, also depend on the local water resources situation. Confronted with an extreme shortage of water resources, Hebei Province has to ensure its socio-economic development at the cost of continual overexploitation of groundwater, while Jiangsu Province, with abundant surface water resources, can have alternative water sources to guarantee an appropriate water amount. From its specific scoring, overexploitation zone management in Hebei Province was in the initial stage from 2012 to 2014. Despite great efforts in laws, regulations, and policies, water-saving measures, which have played a significant role in the recovery of the groundwater table and water resources fee collection, it is still not optimistic in terms of the current status of seawater intrusion and funnel. Therefore, intensified efforts are still needed to manage geological disasters. Thanks for its strengthened efforts in groundwater exploitation management since 1996, the effectiveness evaluation results in Jiangsu Province are still better than that of Hebei Province in spite that management measures of Jiangsu Province register low scores. In particular, after Suzhou, Wuxi and Changzhou cities began to seal the

Table 2 Evaluation system table in Hebei Province

Area of overexploitation zone (ten thousand km^2)	6.7		Baseline year	2012
			Evaluation year	2014
Quantitative evaluation				
Index	2012	2014	Calculation formula	Scores
Groundwater table (m)	16.1	17.11	(17.1 − 16.1 + 0.9429)/3 = 0.65	80 × 25% = 20
Groundwater exploitation amount (m^3)				
Groundwater quality	Grade III	Grade III		100 × 20% = 20
Groundwater monitoring well number		2784	2784/670 = 4/100 km^2	60 × 10% = 6
Monitoring rate for effective data of monitoring wells (%)				40 × 10% = 4
Funnel area (km^2)	8815	44000	(44000 − 8815)/8815 = 399%	0
Seawater intrusion distance from coast (m)	8.4–24.9	9.75–30		60 × 10% = 6
Qualitative evaluation				
Laws, regulations and institutional system	① *Implementation plan of survey and evaluation for basic groundwater environment of pilot areas in Hebei Province* ② *Appraisal measures for implementing the most stringent water resources management system in Hebei Province* ③ *Plan for comprehensive management of groundwater overexploitation in pilot areas of Hebei Province (the year of 2014)*			100 × 5% = 5

(continued)

Table 2 (continued)

Area of overexploitation zone (ten thousand km^2)	6.7	Baseline year	2012
		Evaluation year	2014
	④ *Fund plan for comprehensive management of groundwater overexploitation in pilot areas of Hebei Province in 2014* ⑤ *Regulations of groundwater management in Hebei Province*		
Economic measures	① Increase water resources fees from 0.9 yuan/m^3 to 1.5 yuan/m^3 ② Adopt a mechanism of "one compensation for one fee increase". One increase refers to increase in water fees with different degrees according to importance of water sources; one compensation refers to government subsidy, that is the fund coming from the increase in water fees and used for average subsidy for water use units		80 × 5% = 4
Technical measures	① Popularize mixed casting technology of brackish water and fresh water. The saltwater that cannot be used originally is converted into available resources to reduce the deep underground fresh water exploitation. One cubic meter of deep underground water can be saved for per cubic meter of shallow brackish water ② Improve irrigation efficiency through a three-adjustment approach featuring "long into short, wide to narrow, big into small" to change the size of strips ③ Promote the greenhouse plus (beneath film) drip irrigation water saving cultivation in the vegetable growing areas; develop rainwater collection combined with micro-spray and drip irrigation technology in mountainous areas; develop rainwater collecting and back penetration technology to achieve in situ efficient use of urban sewage and rainwater		100 × 5% = 5
Other explanation	① According to the analysis by Meng [8], with Cangzhou city as a representative, its regional precipitation infiltration coefficient is 0.26, water yield 0.053, precipitation in Hebei Province in 2014 406 mm and average precipitation in 2012 598.2 mm. Variation amplitude of groundwater table due to natural rainfalls can be calculated in the following formula: $\Delta h \geq a \times \frac{P_{Evaluation} - P_{Baseline}}{u} = 0.26 \times \frac{(406-598.2)}{0.053} = -942.9\,mm$		

(continued)

Table 2 (continued)

Area of overexploitation zone (ten thousand km^2)	6.7	Baseline year	2012
		Evaluation year	2014
	② According to the method for unavailable data treatment, for those data of groundwater exploitation amount in overexploitation zones, which fail to be obtained, its index will not be evaluated and its weight is assigned to water table variation and water quality with respective increase as 25 and 20% ③ Although specific data of effective monitoring rate of groundwater cannot be obtained, scores are given according to the current fact that there are few dedicated monitoring wells and professional monitoring holes, and most of the monitoring methods are artificial		
Total scores 70			

Table 3 Evaluation system table in Jiangsu Province

Area of overexploitation zone (ten thousand km^2)	1.66		Baseline year	2012
			Evaluation year	2014
Quantitative evaluation				
Index	2012	2014	Calculation formula	Scores
Embedded depth of water table (m)	27.23	23.5	(27.23 − 23.5 − 0.4918)/3 = 1.1	90 × 15% = 13.5
Groundwater exploitation amount (m^3)	9 × 10^8	8.47 × 10^8	(8.47 × 10^8 − 9 × 10^8)/ 9 × 108 = −6%	60 × 15% = 9
Groundwater quality	Grade III	Grade III		100 × 15% = 15
Groundwater monitoring well number	1167		1167/166 = 7/100 km^2	80 × 10% = 8
Effective monitoring rate (%)		100%	100% * 100 = 100	100 × 10% = 10
Ground subsidence (mm/a)	<5	<5		100 × 20% = 20
Ground fissure (mm)				
Qualitative evaluation				
Laws, regulations and institutional system building	① *Circular on further strengthening groundwater resources management in Jiangsu Province* ② *Plan for groundwater overexploitation zoning in Jiangsu Province* ③ Implement management system of so-called "four ones", namely "one plate, meter, certificate and card for each well"			80 × 5% = 4
Economic measures	① Increase water resources fees from 0.2–3.2 yuan/m^3 to 0.40–10 yuan/m^3 ② Levy 2–5 times more of water fees for water withdrawal beyond the plan in addition to deduct water use plan for the next year ③ Implement the system that water resources fees firstly being collected before returning to groundwater monitoring, protection and management and reward for water conservation			80 × 5% = 4
Technical measures	① Achieve groundwater dynamic balance between exploitation and recharge by artificial recharge techniques			50 × 5% = 2.5
Other explanation	① In Jiangsu Province, the precipitation infiltration coefficient is 0.19, specific water yield 0.035, precipitation in 2014 1044.5 mm, average precipitation in 2012 953.9 mm. Variation amplitude of groundwater table due to natural rainfalls can be calculated in the following formula: $\Delta h \geq a \times \frac{P_{Evaluation} - P_{Baseline}}{u} = 0.19 \times \frac{(1044.5-953.9)}{0.035} = 491.83\,\text{mm}$ ② According to the method for unavailable data treatment, for those data of ground fissure in overexploitation zones in Jiangsu Province, which fail to be obtained, its index will not be evaluated and its weight is assigned to ground subsidence with increase as 20%			
Total scores 86				

wells, the water table in most areas stabilize and rise, fully demonstrating the management effects. However, data monitoring should still be reinforced.

Analyses indicate that the evaluation system and the results can reveal the difference in the management level in different regions and reflect the main advantages and disadvantages of each region, so the rationality of the results is relatively strong.

5 Conclusions and Suggestions

(1) This paper aims to establish a relatively scientific and rational groundwater overexploitation management and evaluation system by combining with the generally applicable management experience of groundwater overexploitation zones oriented in management results, control measures, monitoring system and regional characteristics. Fourteen indices have been selected through analysis as to evaluate groundwater overexploitation zone management and provide weights of indices and specific scoring criteria.

(2) The evaluation system is applied to evaluate two typical overexploitation zones of Hebei Province and Jiangsu Province with their final scores of 70 and 86, respectively. The evaluation results are analyzed to prove that they are able to reveal management level differences of groundwater overexploitation zone management in different regions and to reflect the drawbacks and directions for improvement, thus verifying the rationality of the system as a reference basis to comprehensively promote management in overexploitation zones.

(3) It is the most basic and important link to collect materials and data in the process of evaluation. In current evaluation of groundwater management in Hebei Province and Jiangsu Province, the accuracy of the evaluation results has been, to a certain extent, affected due to the unavailability of the detailed data of exploitation amount and ground fissure. Therefore, it is suggested to solidify the collection and sorting of basic data to ensure the accuracy so that the evaluation results can be rational and play their due part.

(4) Criteria need to be adjusted and updated. With the progress of the work, scoring criteria should be adjusted in time to constantly adapt to the current situation, and reflect the dynamics and operability of evaluation criteria so as to finally establish scientific and rational evaluation criteria.

(5) It is suggested to put in place the appraisal system, reward and punishment system for managerial personnel in the overexploitation zone. For those who fulfill their duties well with sound performance should be granted rewards to mobilize their enthusiasm so as to speed up the management process and improve the building of laws, regulations and systems in the overexploitation zone management.

References

1. Li, X., Wang, Y.: Views on development and utilization of underground water. Res. Ind. **13** (2), 116–119 (2011). https://doi.org/10.3969/j.issn.1673-2464.2011.02.024
2. Ma, R.: The current situation and existing problems of groundwater resources development in China. Agric. Technol. **32**(4), 13–14 (2012). https://doi.org/10.3969/j.issn.1671-962x.2012.04.014
3. Xu, K.: Analysis on relationship between the precipitation & groundwater replenishment. Ground Water **04**, 272–274 (2004). https://doi.org/10.3969/j.issn.1004-1184.2004.04.013
4. Wang, J., Liu, K.: Progress in ground fissures and its hazard research. Adv. Earth Sci. **16**(3), 303–313 (2001). https://doi.org/10.3321/j.issn:1001-8166.2001.03.003
5. Zhang, L.: Modeling on relation between major plants growth and groundwater depth in arid area. J. Desert Res. (1), 110–113 (2004). https://doi.org/10.3321/j.issn:1000-694x.2004.01.019
6. Guo, W., Zuo, Q.: Performance evaluation system and application of the strictest water resources management in Zhengzhou. S N. Water Trans. Water Sci. Technol. https://doi.org/10.13476/j.cnki.nsbdqk.2014.04.019
7. Tang, K.: Discussion on concept and assessment system of aquatic ecological civilization. Water Res. Prot. **04**, 1–4 (2013). https://doi.org/10.3969/j.issn.1004-6933.2013.04.001
8. Meng, S.: Research on spatial and temporal distribution of the precipitation infiltration amount over the past 50 years in North China plain. Adv. Earth Sci. **28**(8), 923–929 (2013). https://doi.org/10.11867/j.issn.1001-8166.2013.08.0923
9. Le, F., Fang, R.: Strengthen management of groundwater area in our province's situation analysis and countermeasures. Jiangsu Water Res. **05**, 42–43 (2014). https://doi.org/10.3969/j.issn.1007-7839.2014.05.018
10. Guan, F.: Research on the method and application of groundwater management evaluation. China Water Res. (11), 20–22 (2011). https://doi.org/10.3969/j.issn.1000-1123.2011.11.009
11. Ma, Q., Lu, X.: Analysis on the characteristics and sustainable utilization of water resources in Jiangsu Province. ZhiHuai (7), 9–11 (2008). https://doi.org/10.3969/j.issn.1001-9243.2008.07.005

Runoff Simulation Using SWAT Model in the Middle Reaches of the Dagu River Basin

Fu-hui Du, Li Tao, Xin-mei Chen and Huai-xian Yao

Abstract This article uses Dagu River basin that is located in Qingdao of Shandong Province as the research object, and sets up the hydrological model using SWAT hydrological model and the software of ARCGIS 9.3 which was combined with the basin DEM figure, soil, land use, the observed meteorological data and runoff data. It then established a simulation for regional hydrology on the basis of the runoff data in 1986–2000 and adjusts sensitive parameters by SWAT CUP-2012. The results showed that the determination coefficient (R2) was higher than 0.8 and Nash efficiency coefficient (NS) was higher than 0.7. Therefore, the simulation results can meet the requirements. In addition, it sets up five different land use scenarios and 25 kinds of assumptions of the weather situations to analyze the surface runoff variation of Dagu River basin with two different scenarios, and then sums up the impact of surface runoff influenced by land use change and climate change in Dagu River. The simulation results have an important reference value and practical significance for the sustainable development of this basin in the future and a reasonable allocation of water resources.

Keywords Runoff simulation · SWAT model · Land use change
Dagu River basin

F. Du (✉) · L. Tao
School of Hydraulics and Electric Power, Hebei University of Engineering,
Handan 056000, Hebei, China
e-mail: dufuhui@hebeu.edu.cn

L. Tao
e-mail: 18440409@qq.com

X. Chen
Water Resources Bureau of Handan, Hebei, Handan 056000, Hebei, China
e-mail: cxm1978155@163.com

H. Yao
Water Resources Bureau of Guantao, Guantao 056000, Hebei, China
e-mail: yaohuaixian@163.com

W. Dong et al. (eds.), *Sustainable Development of Water Resources and Hydraulic Engineering in China*, Environmental Earth Sciences,
https://doi.org/10.1007/978-3-319-61630-8_10

1 Introduction

The hydrological cycle process has become one of the hot spots in current research to deal with global warming and other environmental problems, make better use of water resources and realize the sustainable development of water resources to achieve the optimal allocation of water resources, improve the ecological environment, and improve the quality of human's life [1–7]. In order to make the simulation of the whole hydrological process more accurate, scientists have developed and applied a hydrological model [8–15], which is widely used in various hydrological processes. SWAT model is a typical distributed hydrological model, which is widely used in the scientific research of hydrology [16–19].

Qingdao is the regional economic center of Shandong Province and it is also a tourist city that enjoys a good reputation overseas. With the increased population density and the development of the economy, the water resources will increasingly become the key problem for the development of Qingdao. It is of great significance to meet the demand of agricultural, industrial and domestic water use under the premise of river basin ecological security through insight into the water resources structure in Dagu River basin. In recent years, many scholars discussed hydrological processes in Dong Jiang watershed and the development of the SWAT model.

In this paper, by using the SWAT model to simulate the Dagu River basin hydrological processes based on the spatio-temporal data, such as meteorology, land use, soil, etc., and also analyze the change of the runoff simulation under the land use and climate change, the aim is to explore the applicability of the hydrological simulation in China's eastern coastal plain area and to provide a scientific basis for river basin water resources management and ecological protection.

2 The General Situation in the Study Area

The Dagu River is located in Qingdao of Shandong Province and in east longitude 120° 03′–120° 25′, north latitude 36° 10′–37° 12′. The birthplace of Dagu River is on the west side of Zhaoyuan mound hills in Yantai; it imports the Jiaozhou Bay at Pier village in Jiaozhou. The total length is 180 km; the watershed topography is high in the south and low in the north. On the north side of the birthplace there are mountains and hills; on the south side are plains and depressions. The relative elevation of mountains is above 200–300 m while the hills are between 50 and 200 s; and the plains are under 50 m. The region is located in the warm temperate and monsoon climate zones; therefore, it is hot and rainy in summer, cold and dry in winter, dry and little rain in spring, but moderate in autumn. One of the remarkable characteristics in the meteorology is that the realized-niche breadths gap is larger, the precipitation period is unbalanced, and the low water period is longer. The annual frost-free period is about 200 days. On the other hand, the multi-year average precipitation is 707.4 mm; the temperature is between −19 and 40 °C; and the general average temperature difference in range of 40–50 °C.

Vegetation in the basin mainly includes deciduous broad-leaved forest, coniferous forest, bamboo forest, meadows, saline vegetation, thickets and irrigation grass. Soil types mainly include five categories with brown soil and sand ginger black soil being the most widespread. In addition, there are brown soil, chao soil and sandy soil. About 75% of the land in this region is used for farmland; the rest is mainly used for forest land, grassland, water, and urban construction.

3 Basic Data

The basic data needed to build the SWAT hydrological model mainly include: watershed digital elevation model (DEM), soil type data, land use type data and hydrological data. All input spatial data need to have the same geographic coordinates and the projection coordinate system. This paper combines the model structure characteristics using WGS_1984 geographical coordinates and Albers projection coordinate system (Table 1).

4 Construction of SWAT Hydrological Model of 5 Dagu River Basin

The SWAT model was built to complete a variety of operations, respectively: loading digital elevation map (DEM), generate river network, calculate sub-basin and division, the load of land use and vegetation cover file, loading soil data file, loading the meteorological data, the input required file editing, operation model. The parameters were sensitivity analysis; the sensitivity parameters for the calibration and verification of the sensitivity parameters were obtained.

Table 1 The data table of the SWAT model

Data	Project data	Precision	Format
Spatial data	DEM	30 m	GRID
	Land use map	1:100000	GRID
	Soil cover map	1:1000000	SHAPE
Attribute data	Soil properties		DBF
Meteorological data	Temperature, precipitation, wind speed, etc.	DAY	DBF
Runoff data	River runoff	DAY	DBF

4.1 Sub-basin Division

In this paper, the Haikou River is selected as the outlet of the river basin in Dagu. SWAT model was based on the DEM map and SHP format drainage map was used to generate the digital river network; and the Dagu River basin is divided into 39 sub-basins.

4.2 Land Use Data

In the SWAT model, land use data is also very important, and the final convergence of the land is affected by the data. The land use data in this model mainly includes the corresponding distribution map. In addition to the index table, for the former, the attribute data must have the type of land use information, and this type and a record in the database has a mapping relationship. Finally, a land cover/plant SWAT database was established. For the treatment of the land type map, first of all, it is transformed into SHP format, and then it is compatible with the land use type map. Then, a landuse control table file can be built. Then it carries on the re-classification, and finally obtains this model corresponding to the land use type chart as well as the classification statistics. The specific land use chart (Fig. 1) and the table (Table 2) are as follows:

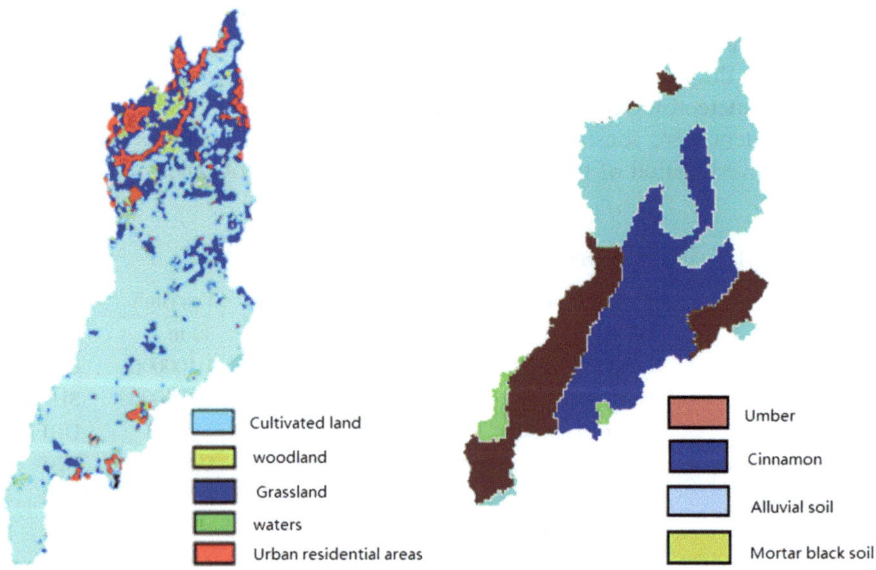

Fig. 1 Land use and soil map

Table 2 Land use classification

Number	First class type	Code	Second class type	Model code
1	Woodland	FRSD	Deciduous broad leaved forest	FODB
			other forest land	CRGR
2	Grassland	PAST	Shrub	SHRB
			Common grassland	GRAS
3	Waters	WATR		
4	Residential land		Medium density residential area	URMD
			Low density residential area	URLD
5	Agricultural land	ARGL	Cultivated land	

4.3 Soil Data

Soil data is also an important parameter in the SWAT model. The quality level of the data will have a great impact on the final simulation results. At present, the soil data mainly include the type distribution map, the index table, and the physical property files. Generally speaking, the water and air movement in the soil profile is mainly determined by the physical properties of the soil, and it has a great influence on the water cycle of HRU. So the physical property of soil becomes one of the important data to construct the model (Table 3).

In the research of this paper, the physical properties of soil are classified according to the following methods: first, according to the relevant literature and information to obtain; second, according to the parameters obtained from the relevant operation; and third, the use of SWAT comes with some of the parameters.

5 Simulation Results

The basic data DEM map, soil data, land use data and meteorological data are input into the SWAT hydrological model, and run.

5.1 Sensitivity Analysis of Parameters

In this study, the SWAT model comes with lh-oat sensitivity analysis, using this analysis method to the model affects the runoff sensitivity parameter sensitivity analysis, and selects the most sensitive parameters into the original SWAT hydrological model. The results of sensitivity analysis in this study are shown in Table 4.

Table 3 Soil attribute table

MUID	SEQN	SNAM	HYDGRP	SOL_ZMX	ANION_EXCL	SOL_CRK	SOL_Z	CLAY	SILT	SAND
Umber	3090	ZONGTU	D	550	0.5	0.5	300	23	37	40
Cinna-mon	4286	HETU	C	1000	0.5	0.5	300	22	24	54
Alluvial soil	4291	CHAOTU	C	1000	0.5	0.5	300	23	38	39
Mortar black soil	4336	SHAJIANGHEITU	D	1000	0.5	0.5	300	31	40	29

Table 4 The results in sensitivity analysis

Parameter	Format
CN2	.mgt
ALPHA_BF	.gw
GW_DELAY	.gw
GWQMN	.gw
GW_REVAP	.gw
ESCO	.hru
CH_N2	.rte
CH_K2	.rte
SOL_AWC	.sol
SOL_K	.sol
REVAPMN	.gw
CANMX	.hru

Drawn from the table, SWAT model influences on the study of the sensitive parameters are normal wet vegetation coverage value, base flow partition coefficients, available soil water, groundwater evaporation coefficient, river Manning coefficient, etc.

5.2 Parameter Setting

The parameters of the model are determined using the SWAT-CUP2012 software. The Operation SWAT-CUP2012 program results are shown in Fig. 2 in which the Nancun station model simulation index R2 is 0.85; NS coefficient is 0.79; Zhang hospital station model to simulate index R2 0.80; NS coefficient is 0.73; and the simulation results are good.

The fitting results of the model are better, model fitting results are shown in Fig. 3, so the parameters used in the final simulation of SAWT-CUP2012 are selected, and the results are shown in Table 5.

5.2.1 Simulation Results

The model can accurately simulate the runoff flow of the main rivers in the whole river basin of the Dagu River in Qingdao after the calibration of the parameters and the validation of the model. The decision coefficient (R2) and Nash coefficient (NS) are selected as the model efficiency coefficients, after the calibration of the parameters the R2 of this model is greater than 0.8 and its NS is greater than 0.7; thus, this study of SWAT model simulation result is credible.

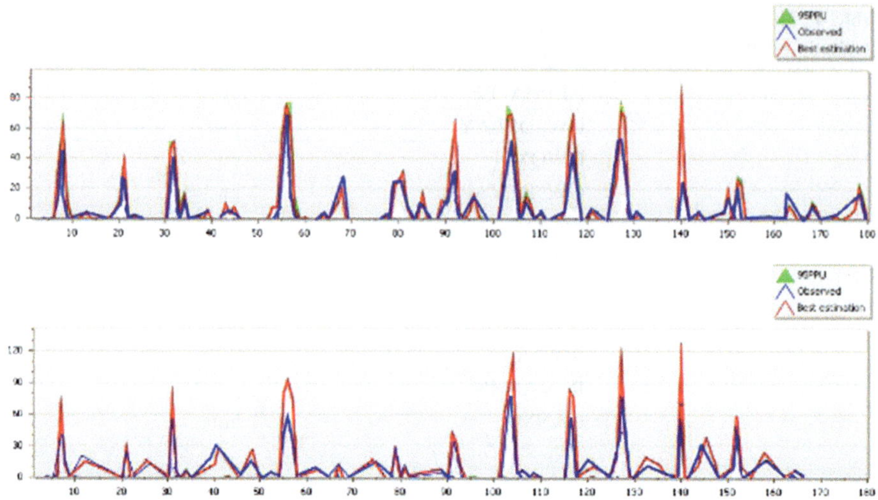

Fig. 2 Simulation results by SWATCUP-2012

Fig. 3 Observed and simulated monthly runoff during calibration

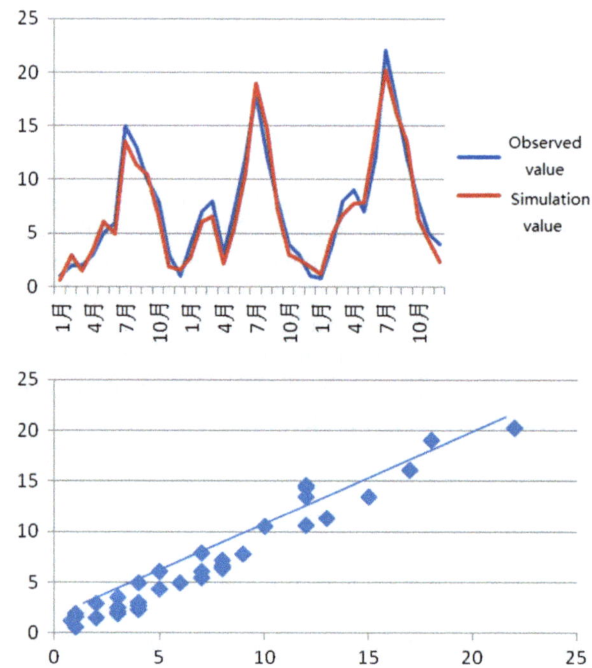

Table 5 The final parameter values

Parameter	Format	Modification method	Value
CN2	.mgt	Multiply	0.58
ALPHA_BF	.gw	Assignment	0.04
GW_DELAY	.gw	Assignment	55
GWQMN	.gw	Assignment	18
GW_REVAP	.gw	Assignment	0.16
ESCO	.hru	Assignment	0.21
CH_N2	.rte	Assignment	0.09
CH_K2	.rte	Assignment	93.1
SOL_AWC	.sol	Multiply	0.38
SOL_K	.sol	Multiply	-0.53
REVAPMN	.gw	Assignment	34
CANMX	.hru	Assignment	0.6

6 The Impact of Climate Change on the Model

Changes in climate conditions were very sensitive to the effects of rainfall and surface runoff. In the basin, the temperature and rainfall changes caused by climate change were the important topics at present. However so far, there is no method for forecasting climate change that is reliable and can be directly used. Therefore, the method in the study of climate change on runoff used an indirect method to obtain possible scenarios of future climate change. At present, the indirect research methods of climate change are generally as follows: the assumption of the climate scenario method, based on the atmospheric circulation patterns model output method; and the relevant statistical method, the relevant statistics, and the comprehensive method. In this paper, the impact of climate change on the runoff generation in Dagu River basin is simulated by the assumption of the climate scenario method.

On the basis of the initial meteorological data, the temperature changes in the basin are: reduce 2 °C, reduce 1 °C, constant, increasing 1 °C and increasing 2 °C. Basin rainfall changes are as follows: reduce 20%, reduce 10%, unchanged, increase of 10% and increase of 20%.

Land use and land cover data are still using the original data. The scenario simulation combination is as follows (Table 6).

Based on the 1986 land use data and climate data from 2000 to 2010, the runoff of the Dagu River basin was simulated with 25 different climatic factors. Then the impact of climate change scenarios on the Dagu River were determined (Table 7).

Table 6 25 kinds of scenario simulation combinations

		Rainfall variation				
		P × (1 − 20%)	P × (1 − 10%)	P	P × (1 + 10%)	P × (1 + 20%)
Temperature variation	T − 2 °C	S11	S12	S13	S14	S15
	T − 1 °C	S21	S22	S23	S24	S25
	T °C	S31	S32	S33	S34	S35
	T + 1 °C	S41	S42	S43	S44	S45
	T + 2 °C	S51	S52	S53	S54	S55

Table 7 The output of 25 scenarios

		P × (1 − 20%)	P × (1 − 10%)	P	P × (1 + 10%)	P × (1 + 20%)
Runoff (m^3/s)	T − 2 °C	10.42	11.72	13.17	16.03	19.44
	T − 1 °C	10.32	11.66	12.89	15.6	18.31
	T °C	10.08	11.52	12.6	15.16	17.76
	T + 1 °C	9.91	11.42	12.45	14.99	17.32
	T + 2 °C	9.65	10.93	12.17	14.39	17.01
Runoff variation (m^3/s)	T − 2 °C	−2.18	−0.88	0.57	3.43	6.84
	T − 1 °C	−2.28	−0.94	0.29	3	5.71
	T °C	−2.52	−1.08	0	2.56	5.16
	T + 1 °C	−2.69	−1.18	−0.15	2.39	4.72
	T + 2 °C	−2.95	−1.67	−0.43	1.79	4.41
Runoff variation rate	T − 2 °C	−17.32	−6.99	4.56	27.23	54.32
	T − 1 °C	−18.12	−7.43	2.34	23.78	45.33
	T °C	−19.98	−8.57	0	20.33	40.94
	T + 1 °C	−21.34	−9.34	−1.23	18.94	37.45
	T + 2 °C	−23.43	−13.23	−3.43	14.23	34.98

7 Conclusion

(1) The basin DEM map, land use maps, soil maps, weather stations and meteorological data were imported into the SWAT model, after the division of sub-basin hydrological response unit, land use and soil cover, import of meteorological data, and editing of the input file and run. The SWAT hydrological model was established. In the process of tuning the parameters, the use of the SWAT CUP2012 program concluded that the model of Nancun village station's index R2 is 0.85; NS coefficient is 0.79; the model of Zhangjiayuan station's index R2 is 0.80; NS coefficient is 0.73; and the simulation result was good.

(2) In this paper, authors set up 25 simulated scenarios under different climate scenarios. Through the analysis of the results obtained following are the conclusions: ① As temperatures rise, river runoff decreases; with the reduction of the temperature, river runoff increases. ② With the increase of rainfall, river runoff increases; contrary to reduced rainfall, river flow is reduced. ③ The increase or decrease of rainfall has a larger influence than the temperature. ④ For the river runoff, climate change is more influential than land use change.

Acknowledgements This research was financially supported by Key research and development project of Hebei Province (16263604D).

References

1. Freeze, R.A., Harlan, R.L.: Blueprint for a physically-based-digitally simulated hydrological response model. J. Hydrol. **9**, 237–258 (1969)
2. Hewlett, J., Troendle, C.: Non-point and diffused water sources: a variable source area problem. In: Watershed Management Proceeding Symposium of American Society of Civil Engineers. New York held in Logan, Utah (1975)
3. Beven, K.J., Kirkby, M.J.: A physically based, variable contributing area model of basin hydrology. Hydrol. Bull. **24**, 43–69 (1979)
4. Abbott, M.B., Bathurst, J.C., Cunge, J.A., et al.: An introduction to the European hydrological system-system hydrologique European. "SHE", 1: history and philosophy hydrology of a physically-based distributed modeling system. J. Hydrol. **87**(1–2), 45–59 (1986)
5. Abbort, M.B., Bathurst, J.C., Cunge, J.A., et al.: An introduction to the European hydrological system-system hydrologique European. "SHE", 2: structure of a physically based distributed modeling system. J. Hydrol. **87**(1–2), 61–77 (1986)
6. Bathurst, J.C., Wiek, J.M., O'Connell, P.E.: The SHE/SHESED basin scale water flow and sediment transport modeling system. Colorado water flow Water Resources (1995)
7. Neitsch, S.L., Arnold, J.G., Kiniry, J.R., et al.: Soil and water assessment tool's theoretical documentation, version 2000. TWRI Report TR-191, Texas Water Resources Institute, College Station, Texas (2002)
8. Srinivasan, R.: Integration of a basin-scale water quality model with GIS. Water Resour. Bull. **30**(3), 453–462 (1994)
9. Arnold, J.G., Srinivasan, R., Muttiah, R.S.: Large-area hydrologic modeling and assessment-Part 1: model development. J. Am. Water Resour. Assoc. **34**(1), 73–89 (1998)
10. Arnold, J.G., Allen, P.M.: Automated methods for estimating base flow and groundwater recharge from stream flow. J. Am. Water Resour. Assoc. **35**(2), 411–424 (1999)
11. Rosenthal, W.D., Hoffman, D.W.: Hydrologic modeling/GIS as an aid in locating monitoring sites. Trans. ASAE **42**(6), 1591–1598 (1999)
12. Harmel, R.D., Richardson, C.W., King, K.W: Hydrologic response of a small watershed model to generated precipitation. Trans. ASAE **43**(6), 1483–1488 [28] (2000)
13. Chanasyk, D.S., Map Fumo, E., Wilhns, W.: Quantification and simulation of surface runoff from fescue grassland watersheds. Agric. Water Manag. **59**(2), 137–153 (2003)
14. Bingner, R.L.: Runoff simulated from Goodwin Creek watershed using SWAT. Trans. ASAE **39**(1), 85–90 (1996)
15. Andrea, W., Fohrer, N., Moler, D.: Long-term land use change in a mesoscale watershed due to socio-economic factors-effects on and landscape structures functions. Ecol. Model. **140** (1–2), 125–140 (2001)

16. Kannan, N., White, S.M., Worrall, F.: Sensitivity analysis and identification of the best evapotranspiration and runoff options for hydrological modeling in SWAT2000. J. Hydrol. (2006)
17. Pandey, V.K., Panda, S.N., Sudhakar, S.: Modelling of an agricultural watershed using remote sensing and a geographic information system. Biosyst. Eng. **90**(3), 331–347 (2005)
18. Eckhardt, K., Haverkamp, S., Fohrer, N., Frede, H.G.: SWAT-G a version of SWAT99.2 modified for application to low mountain range catchments. Phys. Chem. Earth **27**(9/10), 641–644 (2002)
19. Fontaine, T.A., Cruickshank, T.S., Arnold, J.G., et al.: Development of a snowfall–snowmelt routine for mountainous terrain for the soil water assessment tool (SWAT). J. Hydrol. **262**, 209–223 (2002)

Study on Water Threshold Value of Protection of Lake Wetland Ecosystem Health—Using Xinghai Lake as an Example

Jian-Wei Wang, Hong-Zhen Ni, Lin Wang, Yu-Fei Zhang, Gen-Fa Chen and Si-Yu Liu

Abstract A lake wetland is the production of humans and nature. It has the functions of a natural lake wetland and an artificial reservoir. Lake wetland ecosystem health is an important symbol of humans and nature harmoniously using water. The study of water quantity threshold is important to the management of a lake wetland. Therefore, this paper explores the calculation method of water quality threshold value of protection of lake wetland ecosystem health from two aspects of non-biological and biological water which are formed by a lake, using Xinghai Lake as an example and using methods of water balance and minimum water level. Finally, the threshold value of water supply is 2462.35 million m^3 and the minimum water requirement is 2778 million m^3. Water quality requirements must be better than Class III.

Keywords Lake wetland · Ecosystem health · Water quality threshold Xinghai Lake

---

J.-W. Wang · H.-Z. Ni (✉) · L. Wang · G.-F. Chen · S.-Y. Liu
Department of Water Resources, China Institute of Water Resources and Hydropower Research, Beijing 100038, China
e-mail: nihongzhen@163.com

J.-W. Wang
e-mail: wangjianwei199063@163.com

L. Wang
e-mail: 373361982@qq.com

G.-F. Chen
e-mail: chengf02@163.com

S.-Y. Liu
e-mail: 382417889@qq.com

J.-W. Wang
Research Center for Water Ecological Civilization & Social Governance of Hebei Province, Handan 056038, China

Y.-F. Zhang
College of Environment, Hohai University, Nanjing 210098, China
e-mail: zyfhhu1996@163.com

W. Dong et al. (eds.), *Sustainable Development of Water Resources and Hydraulic Engineering in China*, Environmental Earth Sciences,
https://doi.org/10.1007/978-3-319-61630-8_11

1 Introduction

As the important carrier of water resources, lakes are the crucial part of the natural ecological system. However, due to the global climate warming, unreasonable exploitation and utilization of lake resources and a large number of pollutant emissions, problems arise, such as lake dryness and withering, severe blockage, water ecosystem degradation, etc. To maintain regional water security and ecological and environmental safety, and to function as an ecosystem service, protection and restoration of lake ecosystem has become imminent, especially in ecologically fragile areas, such as arid Northwest China [1–4].

At present, research on the theory and method of ecosystem health tends to diversity [5]. At home and abroad, the research on lake ecosystems is mainly focused on the following three aspects: (1) to construct the index system of lake ecosystem health assessment, and to evaluate the health status of the lake ecosystem. Li et al. [6] summarized and reviewed ecological system health assessment methods before pointing out the key issues in the evaluation process. Shear et al. [7] built an ecosystem evaluation index system of Great Lakes. (2) To calculate lake ecological water requirement and environmental capacity. This paper calculates assimilative capacity under the premise of maintenance of the lake ecosystem health for ecological water requirement and lake health from the perspective of water balance, water quality balance, normal ecosystem function, etc. Zhang et al. [8] studied the ecological water requirement of 10 important wetlands, including the seven Caspian Sea, Baiyangdian and Hengshui Lake in the Haihe River Basin. Zeng et al. [9] calculated the water environment capacity of CODcr, ammonia nitrogen, total nitrogen and total phosphorus at Xitiaoxi in Zhejiang Province by using a water environmental capacity model. (3) To build a lake water ecological model, from water consumption to water quality, and then to the hydrodynamic coupling simulation and perfect it from one-dimensional to two-dimensional to three-dimensional. Xi et al. [10] constructed a two-dimensional hydrodynamic and water quality coupled model of Poyang Lake, and simulated flow and mass transport process. Johnson et al. [11] summarized an evaluation model of the ecological system, and gave out the criteria for selection.

Through the analysis and calculation of the above three aspects, a basic understanding of the lake ecosystem health status comes into being. However, theories and techniques in practical operation to the management and restoration of lake ecosystem system have not yet been formed. In addition, research on the artificial lake wetland still has some deficiency, especially research on the lake wetland. As a product of the interaction of humans and nature, a lake wetland is not only a link to coordinate the relationship between humans and nature, but also an important part of maintaining ecosystem health. Therefore, this paper carries out research on water threshold values of a lake wetland ecosystem health from two angles—the non-biological water body and biological water body. Using Xinghai Lake as an example, this paper calculates water threshold value of protection of

Xinghai Lake ecosystem health so as to offer reference for scientific management and restoration of lake ecosystem.

2 Research Methods

A healthy ecosystem is the original intention of ecological environment management [12]. As an important part of the ecological system, a lake wetland plays a pivotal role in the improvement of a regional ecological environment and social economic development. Thus, exploring the threshold value of a regional ecosystem health is of great significance for the protection and restoration of ecosystem health. As far as it is concerned, the lake ecosystem is composed of two main systems, including organisms in lakes and water-based environment, which cannot be separated, linking organically and interacting with each other [13].

Therefore, this paper will analyze the representative index threshold value according to the two elements of which the water ecosystem is composed (non-biological water and biological water) so as to determine the health threshold value of Xinghai Lake wetland ecosystem as a whole.

2.1 Threshold Value of Non-biological Water Environment

Water is not only a necessary condition for the existence of lake ecosystem, but also the key factor to the health of the lake ecosystem. On the one hand, the right amount of water provides a habitat for the living things in the lake; on the other hand, the right amount of water can maintain and improve water quality. Thus, maintaining the long-term dynamic stability of lake water is the precondition of a lake ecosystem. This paper uses the principle of water balance, takes into account all the inputs and outputs under natural and artificial conditions of lakes, calculates water requirements for the maintenance of the lake water dynamic balance (i.e. The minimum water requirement that keeps the lake area from degenerating), and determines the minimum threshold value of water requirements for maintaining environmental health of a non-biological water body. Calculation is shown in formula 1:

$$W = O_n + O_h - (I_n + I_h) \quad (1)$$

In the formula, “O” represents the output term; “I” represents the input term; and “*n*” represents the natural condition; “*h*” represents the artificial condition.

2.2 Threshold Value of Biological Water Environment

Aquatic organisms are also the integral part of the lake ecosystem. However, the survival of different living creatures requires different water environments. Meanwhile, the composition of the biological community in different regions is various. The living creatures of the lake ecological system include: zoobenthos, macrophytes, phytoplankton, zooplankton, emergent water plants, fish, birds, etc. From the point of view of the food chain, the animal feeds on the plant, while the advanced animal feeds on the lower animal. Fish is the key species to connect birds with other organisms in water. That is, fish feeds on other animals and plants for food; at the same time, it provides food sources for the birds (as shown in Fig. 1). Thus, fish plays an important role in the aquatic food chain.

Therefore, considering requirements of fish for the aquatic environment as environment threshold value of the lake organism water body, this paper understands the most suitable water depth for fish through the investigation of the living habits of fish; uses the minimum water level method, combined with water storage capacity curve; and finally obtains the threshold value of a biological water body of lakes, and regards the minimum requirements of the most sensitive fish for water quality as the threshold value.

Fig. 1 Food chain of lake

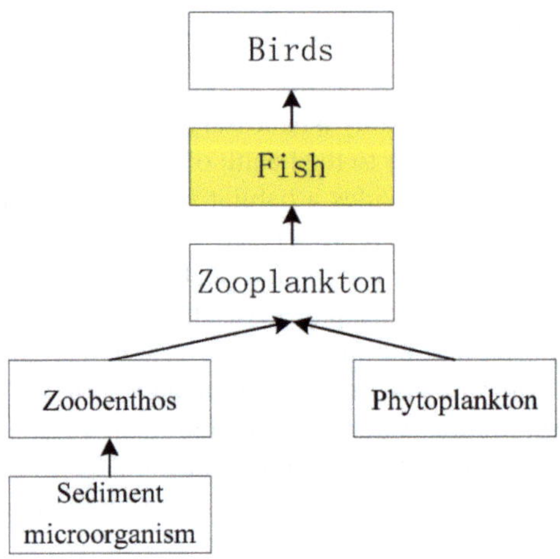

3 The Calculation of Threshold Value of Xinghai Lake Ecosystem Health

3.1 Overview of Study Area

Natural Geography. The location of Shizuishan City is illustrated in Fig. 2. Xinghai lake belongs to Shizuishan City which is showed in Fig. 3. And it is located at the eastern foot of Helan Mountain (105°58′–106°59′ E, 38°22′–39°23′ N) with high terrain in the southwest and lower in the northeast, belongs to the temperate arid climate, which is controlled by the continental wind flow, presenting the characteristics of a continental climate. Xinghai Lake is composed of the northern region, East region, domain, south region, western region, and crescent sea. The wetland protection area covers 43 km^2, of which the water area covers 24.5 km^2. The four regions, including south region, the domain, the western region, and north region, are flood retaining banks, storing flood waters from Ruji ditch, windy gully, wind ditch, guide ditch, leek ditch, Dawukou ditch, of which the drainage area covers about 1061.3 km^2. East domain is the detention area, of which water area covers 5.8 km^2. The flood control capacity of the flood reservoir is 4478.17 million m^3, whose maximum water storage capacity is 6176.53 million m^3. When a flood comes, the jointed dispatch flood retaining banks and detention area ensure the safety of flood control at urban industrial and mining enterprises, the second farm canal, Baotou Lanzhou railway, farmlands, and safety of life and

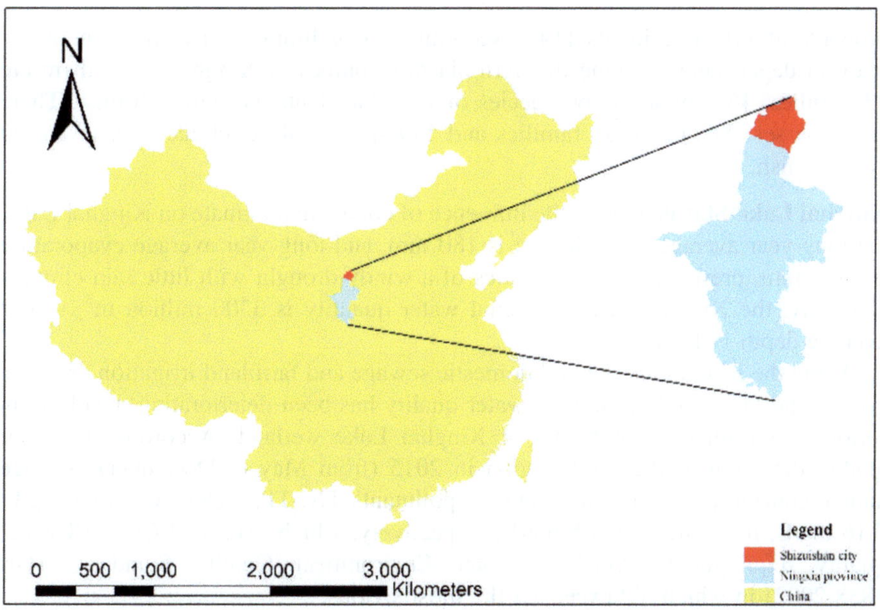

Fig. 2 Location map of Shizuishan City

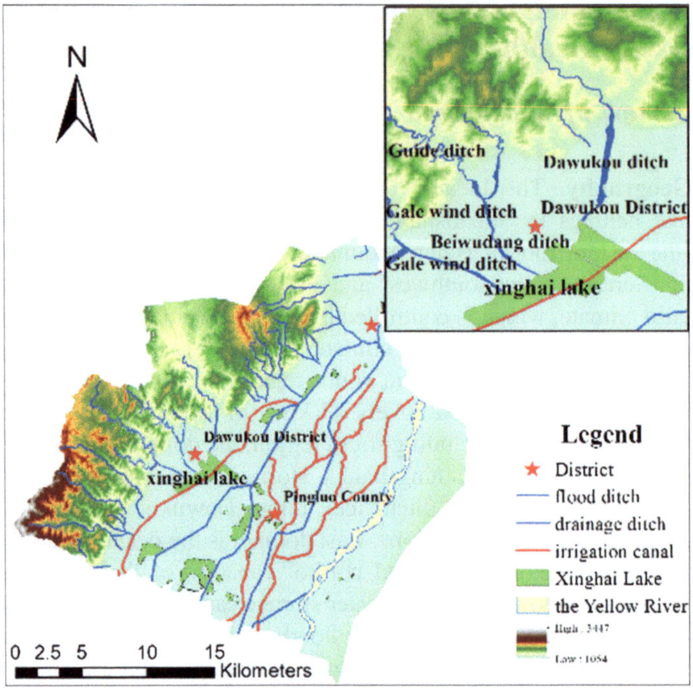

Fig. 3 Outline map of Xinghai Lake

property of urban residents [14]. According to preliminary investigation of the relevant departments, among the germplasm resources of Xinghai Lake, there are 23 families, 46 genera and 66 species of vascular plants, (varieties, forms). There are 5 classes, 51 orders, 51 families and 144 species of vertebrates. There are 29 kinds of fish.

Xinghai Lake Status. Due to the influence of continental climate on Xinghai Lake, the long-year average precipitation is 180 mm, and long-year average evaporation is 1291 mm, presenting characteristics of a windy drought with little rain climate. Currently, the Xinghai Lake perennial water quantity is 3200 million m^3, whose average depth is 1.3 m.

Along the lake, industrial and domestic sewage and farmland irrigation drainage go into the lake, so that the lake water quality has been deteriorating, resulting in serious environmental problems in Xinghai Lake wetland. According to water quality detection of the north region in 2015 (from May to December), average annual concentrations of representative pollutants TN, TP, COD were 1.64 mg/L, 0.16 mg/L, in contrast, 32.05 mg/L, respectively, which exceeded Class III water quality requirements "Surface Water Environment Quality Standard" (GB 3838-2002) in which TP exceeded the most serious.

The perennial average replenishment in Xinghai Lake from The Yellow River is about 20 million m^3. Due to different rainfall conditions, the annual water supply from the Yellow River varies. In addition, the Yellow River and the mountain flood carry large amounts of sediment, which leads to deposition at many places in lake replenishment port, resulting in capacity decreases, flat lake bottom, and insufficient hydrodynamic condition, causing poor water flow and water eutrophication aggravation.

Xinghai Lake, a national scenic area, is an integral part of the ecological civilization construction in Shizuishan City. In addition, Xinghai Lake, together with Sha Lake and the surrounding wetland, constitutes an important bird habitat in the Northwest China, which is not only an important transfer station of the Western migratory route for Chinese migratory birds, but also an important ecological corridor in the upper and middle reaches of the Yellow River as well as one of the most important symbols of the Yellow River ecosystem health. Therefore, protecting the health of Xinghai Lake is very necessary and has important significance.

3.2 Threshold Value Calculation of Xinghai Lake Ecosystem Health

Using meteorological and hydrological data, this paper calculates input and output items of the Xinghai Lake, investigates and accesses relevant information, understands water species and its characteristics of life in Xinghai Lake, and determines threshold value of the security of Xinghai Lake ecosystem health.

Threshold Value of Xinghai Lake Abiotic Water Environment. As for its ecological functions position, Xinghai Lake is not only a flood control reservoir which takes on the task of flood control in flood season, but also detention reservoir that protects farmland irrigation water in the surroundings during dry season as well as ecological aquaculture lakes that protects the water quality of the lake and brings economic income for Shizuishan City. Combining the Xinghai Lake ecological function with formula 1, this paper analyzes the input and output items of Xinghai Lake as follows: under natural conditions, water input of Xinghai lake includes lake precipitation and mountain flood caused by precipitation, while the output water includes lake evapotranspiration water and surface lake water inflow and groundwater seepage water; under artificial conditions, Xinghai Lake water input includes water supply from the Yellow River, field irrigation and drainage, and the surrounding industrial and domestic sewage, while the output water includes irrigation water during dry season.

The above analysis covers the input and output water of Xinghai Lake whenever possible. Under the artificial conditions, water input will cause the pressure of the water pollution in lakes in addition to the Yellow River water supply. Meanwhile, Xinghai Lake irrigation function has not been given full play. Therefore, this paper only calculates the required water (W1) that keeps Xinghai Lake area from shrinking under natural conditions. Formula 1 is specified into formula 2.

$$W_1 = D + E + G - (R + I) \tag{2}$$

In the formula, "D" represents the surface lake water quantity; "E" represents the lake evapotranspiration water; "G" represents the groundwater seepage water quantity; "R" represents the rainfall; "I" represents the flood water inflow caused by precipitation.

This paper collects meteorological, hydrological and reservoir data according to the above formula. The long-year average precipitation in Xinghai Lake area is 180 mm; the long-year average evaporation is 1291 mm; the long-years average runoff is 2123 million m^3, among which monthly runoff is 1831.4 million m^3 in flood season from June to September. Taking flood control into account, flood water is considered as lakes outflow water. During other time periods, there is no runoff output in Xinghai Lake; the groundwater infiltration is calculated according to the total storage capacity of 1%. Xinghai Lake perennial water volume is 32 million m^3. According to formula 1, conclusions can be drawn that threshold value of the non-biological water environment in Xinghai Lake is 2462.35 million m^3. On the other hand, flood limited water level of reservoir in Xinghai Lake is 1098 m, while the corresponding storage capacity is 1161.8 million m^3. Limited water level of flood storage capacity is 1096.5 m; the corresponding storage capacity is 536.56 million m^3. Therefore, to ensure safety in the flood season, the water quantity in Xinghai Lake should not exceed the corresponding storage capacity of limited flood level.

Threshold Value of Xinghai Lake Biological Water Environment. From the angle of the food chain, the biological communities in the Xinghai Lake area are sediment microorganisms, phytoplankton, zooplankton, fish and birds, etc., whose predator-prey relationship is shown in Fig. 1.

Therefore, by using the key species method, this paper chooses the key fish species that connects water with land as the research object, analyzes its request of water environment, and determines the threshold value of Xinghai Lake biological water environment by combining with the lowest water level method. To this end, this paper carries out literature research on the life habits of protected fish, famous fish and economic fish in autonomous region [15, 16]. The statistical results are shown in Table 1.

According to Table 1, protected fish in Xinghai Lake are *Squaliobarbus curriculus*, yellow catfish, etc.; famous fish are culter, mandarin fish, snakehead, etc.; economic fish are grass carp, silver carp and Bighead Carp, etc., among which most fish live in the water bottom, and like static or slow flow. Their breeding period concentrates in late spring and summer. There is no such fish that is particularly sensitive to water quality. Thus, Class III water quality can meet the needs of all fish for survival. Currently, the ecological fishery pattern "Fish protects water" has been taken in Xinghai Lake. Only the water with good quality can ensure healthy growth

Table 1 Statistics of living habits of main fishes in Xinghai Lake

Classification	Name	Living water layer	Water flow requirement	Breeding period	Suitable water depth (m)	Water quality
Protected fish	*Squaliobarbus curriculus*	Middle and lower layer	Static or slow	6–7 months		Class III
	The Yellow River catfish	lower layer	slow	4–6 months		Class III
Famous fish	Culter mongolicus	Middle and upper layer	slow	5–7 months		Class III
	Mandarin fish	lower layer	Static or slow	5–8 months		Class III
	Snakehead	lower layer	Static or Micro flow	4–8 months		Class III
Economic fish	Yellow carp	lower layer	Running water	3–8 months	1–1.5	Class III
	Grass carp	Middle and lower layer and nearshore		4–7 months	2–2.5	Class III
	Silver carp	Middle and upper layer		3–7 months	2.5	Class III
	Bighead	Middle and upper layer		4–6 months	2.5	Class III
	Crucian carp	lower layer	Running water or still water	3–8 months	2–2.5	Class III

of fish. At the same time, the amount of water needs to meet certain requirements. According to the experience, the suitable survival water depth for wild fish is 1 m. According to Table 1, the optimum water depth for cultured fish is 1–2.5 m. In the full view of the living conditions of wild and cultured fish, the minimum water depth for the key species of fishes under artificial condition is 2.5 m. According to the minimum water level method, water depth can be transformed into the corresponding water level before calculating water consumption by water storage capacity curve. Due to the various lake elevations in Xinghai Lake, water depth must be ensured at the areas of the minimum lake elevation of 2.5 m from the point view of ecological water saving. At this time, according to Table 2 the corresponding water level is 1098.9 m. Nearly all regions in Xinghai Lake will be connected except east regions. After that, the water threshold is 1098.9 m. East region is for a detention basin, undertaking the task of consumptive flood, so the minimum depth for the survival of fish is 1 m, whose corresponding water level is 1096.2 m.

Table 2 Threshold value of the minimum water level in Xinghai Lake

Region	Lake elevation (m)	Minimum water level threshold (m)	Water level after being connected	Notes
North region	1096.6	–	1098.9	Flood retaining bank
Inter-domain	1096.4	1098.9	1098.9	
The western regions	1096.6	–	1098.9	
Southern region	1097.3	–	1098.9	
The eastern region	1095.2	1096.2	1096.2	Detention basin

This paper checks water level-storage capacity curve and detention area in Xinghai Lake through the water level threshold value. According to the above mentioned, the corresponding water demand at the flood retaining bank in Xinghai Lake is 22.08 million m^3, and water demand of east region is 580 million m^3. Thus, a total is 2788 million m^3.

4 Conclusions and Suggestions

Through the two aspects of the lake (non-biological and biological water body), this paper systematically analyzes the water threshold value of protection of the lake wetland ecosystem health. Using the water balance method, considering all input and output items in the lake and reservoir under the natural and artificial conditions, this paper calculates water threshold value of the non-biological water body in the lake. Using the method of key species, from the point of view of the food chain, this paper selects fish as key species, and analyzes their life habits, so as to determine water threshold value of the biological water body in lakes and reservoirs. Using Xinghai Lake as a case study, conclusions are drawn that the water threshold value of non-biological water body is 2462.35 million m^3, and threshold value of biological water body is 2788 million m^3. The water quality requirements are more than Class III. As for the non-biological water body, annual average water supply from the Yellow River in Xinghai Lake is about 20 million m^3, which cannot meet the water threshold value requirements of the abiotic water environment. As for the biological water body, actual water consumption in Xinghai Lake is 32 million m^3, which can meet the water demand of the aquatic organism, but it is far from reaching Class III water quality. In addition, in order to ensure the normal and healthy reproduction of fish, it is necessary to ensure the water quality and water consumption, especially in the late spring and summer. In view of the water shortage and serious water pollution in Xinghai Lake, this paper gives advice from the following two aspects to solve the problem. First, control pollutant input around the lake, and improve the efficiency of pollutant treatment, especially purification of the irrigation return water so as to meet multiple water discharge standards and achieve multiple water supply for Xinghai Lake. Second, enhance the self-purification ability of Xinghai Lake, and appropriately use biological measures to increase lake self-repair and the pollutant carrying capacity under the premise of not destroying the existing ecological system.

In this paper, the following problems still exist in the calculation process:

(1) Computational nature of the non-biological water environment threshold value is water balance. Thus, this paper should consider input and output of all water consumption of the research object in theory. However, this paper does not consider farmland return water of which pollutant concentration is higher and municipal wastewater because Xinghai Lake water quality requirements are higher.
(2) The confirmation of the threshold value of biological water environment is determined by the requirements of key species for the water environment. Due to the data limitation, the determination of the threshold value of the water body environment needs further consideration and improvement.

Acknowledgements This research was financially supported by the researchers who would like to extend their thanks to the Chinese National Natural Science Foundation (No. 51279207, No. 51409266) and National Science Foundation of study on the total loss accounting of Taihu River flood (No. 51309244).

References

1. Chen, L.: Enhancing the management and protection of lakes and promoting the healthy and sustainable development of lakes—the keynote report on the first China lake forum, water policy and resource, A026-8 (2011)
2. Yang, Y., Cai, Y., Bai, Y., et al.: A dynamic evaluation of regional ecosystem health using a multiple index system: a case study of Maoji biosphere reserve. Acta Ecol. Sin. **36**(14), 1–9 (2016). https://doi.org/10.5846/stxb201412112455
3. Li, S., Dou, H., Shu, J., et al.: Water environment problems of lakes and water ecosystem recovery in China. China Water Resour. **13**, 14–17 (2006). DOI 1000-1123(2006)13-0014-04
4. Wilkins, D.A.: Assessing ecosystem health. Trends Ecol. Evol. **13**(10), 69 (1998)
5. Liu, Y., Peng, J., Wang, A.N., et al.: New research progress and trends in ecosystem health. Acta Ecol. Sin. **35**(18), 5920–5930 (2015). https://doi.org/10.5846/stxb201401060032
6. Li, B., Yang, G., Wan, R.: Progress on evaluation methods of lake ecosystem health. Adv. Sci. Technol. Water Resour. **06**, 98–106 (2014). https://doi.org/10.3880/j.issn.1006-7647.2014.06.019
7. Shear, H., Stadler-Salt, N., Bertram, P., et al.: The development and implementation of indicators of ecosystem health in the great lakes basin. Environ. Monit. Assess. **88**(1–3), 119–152 (2003). https://doi.org/10.1023/A:1025504704879
8. Zhang, H., Guo, L., Bai, X., et al.: Study on ecological water demand and securing measures for wetlands of great significance in Haihe river basin. China Water Resour. **7**, 33–36 (2016). DOI 1000-1123(2016)07-0033-04

9. Zeng, Y., Wang, X.: Water environmental capacity model and its parameters sensitivity analysis for Xitiaoxi River in Zhejiang Province, China. China Environ. Sci. **30**(12), 1627–1632 (2010). DOI 1000-6923(2010)12-1627-06
10. Lai, X., Jiang, J., Huang, Q., et al.: Two-dimensional numerical simulation of hydrodynamic and pollution transport for lake Poyang. J. Lake Sci. **23**(06), 893–902 (2011)
11. Johnson, J.B., Omland, K.S.: Model selection in ecology and evolution. Trends Ecol. Evol. **19**(2), 101–108 (2004)
12. Costanza, R.: Ecosystem health and ecological engineering. Ecol. Eng. **45**(8), 24–29 (2012)
13. Wang, S., Dou, H.: Lakes of China, pp. 58–160. No. 16 North East Huang Chenggen in Beijing. Science Press (1998)
14. Zhao, M., Zhang, B., Li W., et al.: Comprehensive utilization of mountain flood water based on flood control safety. Agri. Eng. **3**(1), 52–53 (2013). DOI 2095-1795(2013)01-0052-02
15. Wang, X.: Vertebrate fauna of Ningxia, pp. 68–96. No. 139 Beijing East Road of Xingqing District in Yinchuan. Ningxia People's Publishing House (1990)
16. Liao, Y.: Fauna Sinica. Science Publishing Company (1997)

Construction and Application Analysis of SWMM Model in Beijing Future Science Park

Lanshu Jing, Qinghua Luan, Haichao Wang and Xuerui Gao

Abstract Aiming at more frequent urban waterlogging problems in China, a system of rainwater drainage pipe network by SWMM (Storm Water Management Model) was constructed, using Beijing Future Science Park as the study area. The empirical runoff coefficient method is used for model parameter calibration and validation of "6.23" and "7.21" rainfall runoff processes to build the regional SWMM model and simulate regional rainfall runoff processes in different return periods. Comprehensive runoff coefficient and flood peak flow are used as indicators to analyze the above simulation results. In terms of the whole area, both elements increase with longer design return period, the same as regional flood risk. In terms of the local area, comprehensive runoff coefficient of sub-catchment area increases with a longer design return period, but different with various underlying surface conditions. Among them, the comprehensive runoff coefficient of green land with better permeability is only 0, while the corresponding value of road area with permeability of 90% is 0.89, to indicate the differences in spatial distribution of regional flood risk along with the underlying surface conditions. The above research results provide the technical support for the construction of regional low

L. Jing · Q. Luan
Hebei University of Engineering College of Water Conservancy and Hydropower, Handan 056001, Hebei, China
e-mail: 15032021864@163.com

Q. Luan
e-mail: carol97011202@163.com

L. Jing · Q. Luan
Research Center for Water Ecological Civilization & Social Governance of Hebei Province, Handan 056001, Hebei, China

H. Wang (✉)
China International Engineering Consulting Corporation, Beijing 100044, China
e-mail: wanghaichao@ciecc.com.cn

X. Gao
Northwest A&F University, Yangling 712100, Shanxi, China
e-mail: gaoxuerui666@163.com

W. Dong et al. (eds.), *Sustainable Development of Water Resources and Hydraulic Engineering in China*, Environmental Earth Sciences,
https://doi.org/10.1007/978-3-319-61630-8_12

impact development measures (Low Impact Development, LID) and urban flood warning.

Keywords SWMM · Rainstorm flood simulation · Integrated flow coefficient Designing precipitation · Beijing Future Science Park

1 Introduction

In recent years, the extreme climate change has led to significant urban waterlogging problems. In the summer of 2016, "to see the sea in the city" occurred in Nanjing, Wuhan, Handan, Beijing and other regions, not only bringing great inconvenience to people's lives and traffic, but also increasing the huge burden for urban flood control work. Establishment of the urban rain flood management model, effective storm water management and prediction of urban rainfall runoff has become one of the urban flood management works. The powerful function and the characteristics of the free and easy-to-use model, SWMM has been widely used both at home and abroad by Song et al. [1]. The urban drainage area is selected to construct the SWMM model by Palla and Gnecco [2] to increase the Low Impact Development (LID) measures and simulate runoff processes in different return periods (2, 5 and 10 years). The results show that the LID measure scheme is effective for heavy rain events with the return period of 10 years, which can decrease the peak runoff and provide technical support for the evaluation of LID control performance. Seine in France is selected to build the SWMM model by Versini et al. [3] and long-series precipitation data are used to analyze the green roof runoff in different coverage. The results indicate that green roofs can effectively relieve the precipitation at ordinary times, but there is no significant effect on the heavy rain. If combined with infrastructure, green roofs can make important contributions to urban water management. In China, frequent flood areas in Dahongmen, Beijing are selected to build SWMM model by Zhao et al. [4] to analyze the flood characteristics of different precipitation events and provide technical support for flood control decision-making for this region. The typical communities in Beijing are selected to calculate the community drainage effects, waterlogging and slope flow conditions in different frequencies of design rainstorm by Cong et al. [5] to simulate storm floods under different conditions and evaluate the impact, providing important auxiliary measures for urban flood control.

In this paper, SWMM model of Beijing Future Science Park is established to analyze the rainfall runoff processes in different return periods (10, 20 and 50 years) and runoff coefficient variation, providing technical support for urban flood control and the set of LID measures.

2 Model and Methods

2.1 SWMM Model

At present, urban rain flood model has became the important tool and means of rain flood management, bringing great convenience to the layout of the urban drainage pipe network, urban flood early warning and the sponge city construction. For example, Mike urban, Inforworks-CS and SWMM model all can be used for rainfall runoff simulation. The advantages and disadvantages of SWMM, Inforworks CS and MOUSE model are analyzed to point out the common existing problems, introduce the prospects for the development of the urban rain flood model, and provide the choice basis for current research on the urban rain flood model [6]. SWMM model is developed by the United States Environmental Protection Agency (USEPA) as the first comprehensive urban runoff analysis model, suitable for the construction of pipeline hydraulic model, characterized by more selective runoff generation and concentration model and more intuitive display of dynamic results. The complete network data of Beijing Future Science Park is used with SWMM model to build the rainfall runoff model of Beijing Future Science Park, which can be more intuitive to reflect the runoff processes corresponding to different rainfall processes. Thus, the model is selected in this paper.

2.2 Runoff Coefficient Method

The Beijing Future Science Park lacks measured runoff data, so the runoff coefficient method is selected to calibrate model parameters in this paper. The rainfall runoff model calibration idea of runoff coefficient method is to use runoff coefficient as the objective function of model parameter calibration, especially to compare the in priori comprehensive view path coefficient used for rainwater pipe network design with the runoff coefficient simulated by computer model for model parameter calibration [7]. In this paper, the empirical value of urban comprehensive runoff coefficient in the concise drainage design manual [8] is selected as the standard for model parameter calibration, see Table 1.

Table 1 The experience value of comprehensive runoff coefficient

Unit type	Impervious percentage (%)	The integrated runoff coefficient
Central dense habitations	>70	0.6–0.8
Dense habitations	50–70	0.5–0.7
Sparse habitations	30–50	0.4–0.6
Very sparse habitations	<30	0.3–0.5

3 Modeling in the Study Area

3.1 Study Area

In this paper, Beijing Future Science Park is selected for modeling. This region is the new urban area with a flat terrain, prone to waterlogging, belonging to Wenyu River Basin with planned total area of about 10 km^2. The planning area is bounded by Wenyu River and Dingsi Road to divide the Beijing Future Science Park into the south and north district, which is located in the southeast of Xiaotangshan Town with an area of 2.19 km^2. The south district is located in the eastern part of Beiqijia Town with an area of 4.46 km^2. Green space of 3.54 km^2 exists between the two districts. The planned two districts include R&D industry land, the land for the public service facilities, supplementary residential land, municipal land for traffic infrastructure and green space. On this basis, according to the rainwater control and utilization planning, this district can be divided into the north rain zero-discharge area, the north rain discharge control area, the riverside rain zero-discharge area, Beijing-Chengde rain zero-drainage area, the southeast rain discharge control area and the southwest rain discharge control area (Fig. 1). Due to the zero-discharge area with no runoff, the north rain discharge control area, the southeast rain discharge control area and the southwest rain discharge control area are only generalized in the model.

3.2 Drainage Pipe Network Generalization

Drainage pipe network is generalized based on the south and north pipe network data, rain well data and river channel section data in Rain and Sewage Discharge Planning in Beijing Future Science Park (South District), Rain and Sewage Planning Outside Beijing Future Science Park and Relevant River Regulation Project Planning in Beijing Future Science Park. After considering pipeline direction, the rainwater pipe networks in the south and north districts are generalized in a reasonable manner. There are generalized 196 rain nodes and 194 rain pipelines, including unclosed drainage channels. In addition, according to the flow direction data in the planning, the water in the southwest rainwater drainage control area, the southeast rainwater drainage control area and the north rainwater drainage control area is discharged to Wenyu River Bay, Lutong West Gully and Wenyu River along the planning river drains, respectively. Three outlets are set, including one in the downstream outlet of Wenyu River downstream and two in Lutong West Gully.

The sub-catchment area is divided according to the regional division data in Controlling Planning Scheme Details of Beijing Future Science Park. The south and north districts of Beijing Future Science Park are divided into 226 sub-catchment areas, including the permeable zone with an area of 2.06 km^2, accounting for 31%

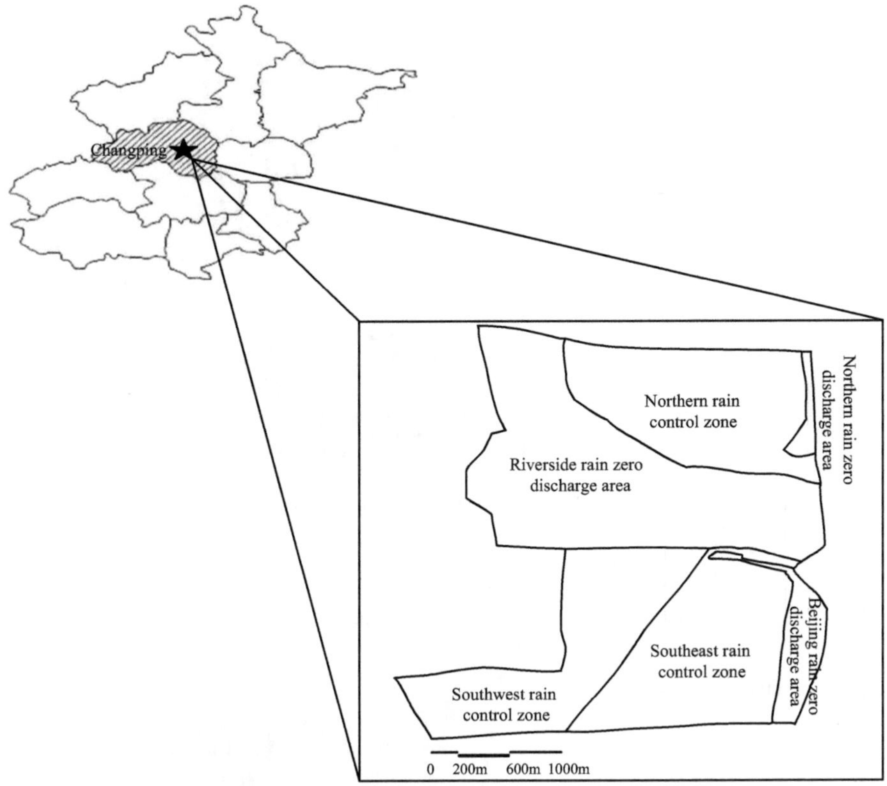

Fig. 1 The distribution of Beijing Future Science Park drainage area

of the total area, and impervious area with an area of 4.59 km^2, accounting for 69% of total area, as shown in Fig. 2a, b.

3.3 Parameter Determination

Model parameters mainly include sub-catchment area parameters, infiltration parameters and pipe parameters. Except for the characteristic width of sub-catchment area, the rest of the parameters refer to the user manual of SWMM model [9].

The sub-catchment area parameters include the characteristic width of sub-catchment area, depression storage quantity of permeable and impermeable area, and manning roughness of permeable and impermeable area. Among them, Eq. 1 is calculated to characteristic width of each sub-catchment area according to

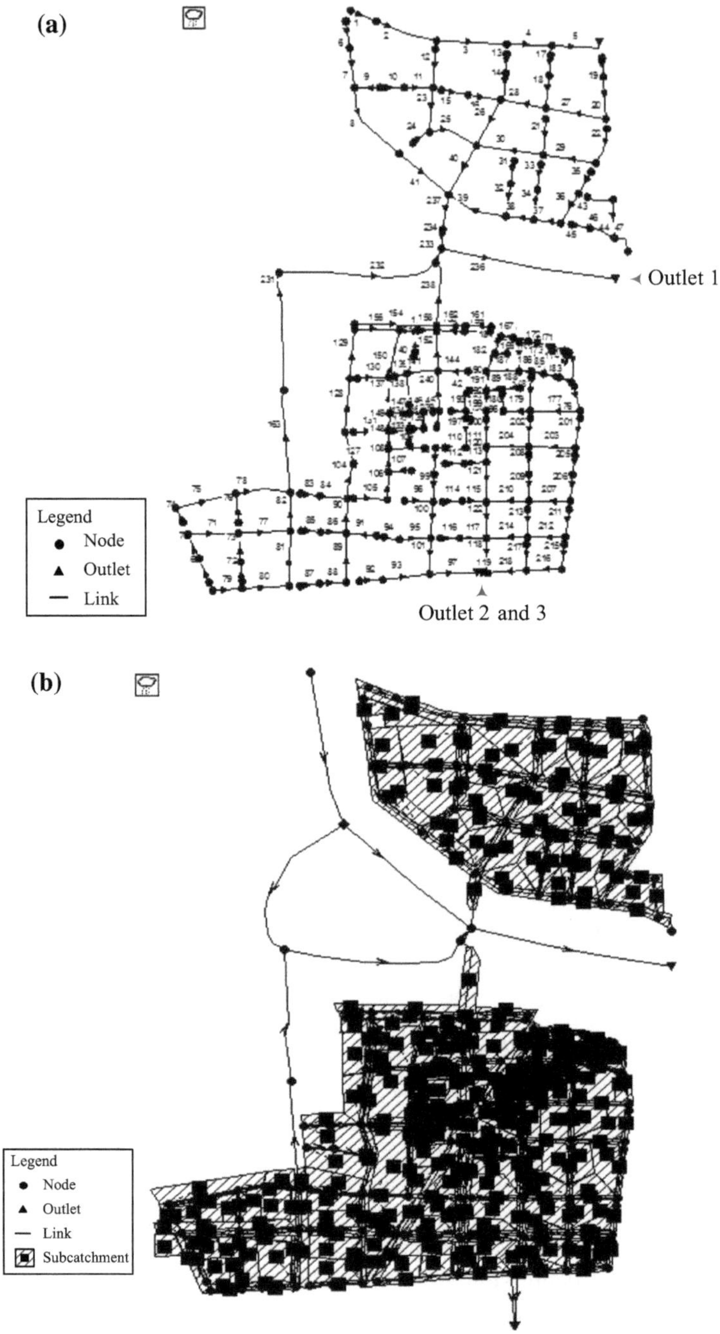

Fig. 2 **a** Drainage pipe and outlet distribution; **b** Generalized drainage system

Table 2 The final calibration results

Parameter	Value
Manning coefficient of the flooded area	0.15
Manning coefficient impermeable area	0.012
Permeable zone depression storage volume	25.4 mm/h
Impervious area depression storage volume	1.27 mm/h
The biggest infiltration rate	25.4 mm/h
Minimum infiltration rate	0.5 mm/h
Closed catheter manning coefficient	0.013
Attenuation coefficient	4
Sunny days period	7d

the calculation method in the literature [10]. After field survey, the regional permeable area is a lawn or short grass land, while the impervious area is the cement surface. Therefore, the depression storage quantity of permeable and impervious area is selected as 3.81 mm/h and 1.27 mm/h, respectively; and the manning coefficient of permeable and impervious area is 0.15 and 0.012, respectively.

$$\mathrm{W} = \mathrm{K} * \mathrm{Sqrt}\,(\mathrm{area})\,(0.02 < \mathrm{K} < 5),\ \mathrm{K} = 1 \tag{1}$$

The infiltration parameters include maximum infiltration rate, minimum infiltration rate, attenuation constant and sunshine time. According to the survey results of Rainwater Utilization Planning Scheme of Beijing Future Science Park, the area belongs to silty clay. Therefore, maximum infiltration rate, minimum infiltration rate, attenuation constant and sunshine time are selected as 25.4 mm/h, 0.5 mm/h, 4 and 7 d, respectively.

Pipe parameters refer to the manning coefficient of pipe. According to field investigation and External Rain and Sewage Planning, pipelines belong to asbestos cement pipes, and the manning coefficient of closed conduit is selected as 0.013.

Due to lack of the measured data for validation, the runoff coefficient method is used to input the "6.23" rainfall process of the model parameter calibration in this paper. After adjusting parameters, the comprehensive runoff coefficient corresponding to this regional rainfall event is 0.43, consistent with the empirical value of comprehensive runoff coefficient in the specifications. Therefore, the final calibration results are shown in Table 2.

According to the results of parameter calibration, the "7.21" measured rainfall process is input to the model to simulate and obtain the comprehensive runoff coefficient of this rainfall event as 0.52, consistent with the empirical value of comprehensive runoff coefficient in the specifications, showing that this model can be used in the simulation of rainfall runoff process in the region.

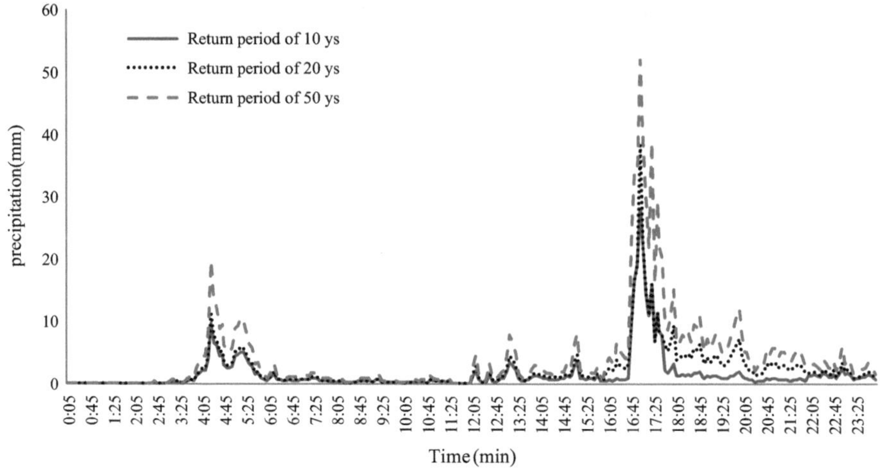

Fig. 3 Different design rainstorm processes in different return period

3.4 *Simulation Result Analysis*

Due to the basin area of less than 300 km^2, the rainfall spatial difference is smaller; and the rainfall at the design point can be used to replace that on the design surface. Based on "Beijing Hydrologic Handbook (First Edition of Rainstorm Atlas)" [11] and "Design Rainstorm Runoff Calculation Standard of Urban Rain System Planning" [12], the rain-type duration distribution is carried out. According to the minimum period of 5 min and the total time of 1440 min, the design rainstorm processes with the hydrological return period of 10, 20 and 50 years are derived (Fig. 3).

The same rain-type distribution is used for three design rainstorm processes. The rains lasted for 24 h, and all were the bimodal type. The first and second peak time occurred at 4 h 20 min and 17 h after precipitation. The double peak and total cumulative rainfall values in each return period are shown in Table 3.

Table 3 The rainfall statistics under different return period

Return period (yrs)	The depth of first peak (mm)	The depth of second peak (mm)	The total rainfall (mm)
10	9.63	27.06	391.23
20	11.06	30.06	641.47
50	19.16	51.86	1064.73

4 Result Analysis

According to the set parameters, the design rainstorm processes in different return periods are input to the model for simulation, and the obtained rainfall runoff processes are shown in Fig. 4a–c. From the angle of flood peak flow, the rainfall processes in three different return periods are analyzed with the same rain-type for duration distribution, belonging to the bimodal type with short peak time and big rainfall intensity. The simulated runoff process also shows shorter peak time, and the runoff process presents a steep rise and decline trend. Particularly, the runoff process corresponding to the rainfall process with a return period of 50 years has the flood peak flow of 6.07 m^3/s as 2–3 times over the other two kinds of precipitation situation. However, the three runoff processes have the same convergence time and underlying surface conditions, proving that the flood with a return period of 50 years will increase the load of urban drainage pipe network, intensifying the regional flood risk. Analyzing from regional comprehensive runoff coefficient, the increase of urban construction density will lead to the rising of impervious rate so that the regional comprehensive runoff coefficient increases. According to Rainwater Control and Utilization Special Planning of Beijing Future Science Park, it can be known that the runoff coefficient is about 0.15 before regional development, while the simulation results show that the comprehensive runoff coefficient of the rainfall with the return period of 10 years, 20 years and 50 years is 0.52, 0.53 and 0.58, respectively, as shown in Table 4. It is indicated that with the increase of return period, regional comprehensive runoff coefficient increases, the same as regional flood risk.

Based on the above analysis, in case of greater design return period, the flood peak flow and comprehensive runoff coefficient will increase, adding the discharge load to urban drainage pipe network, resulting in more urban flood risks.

As viewed from local sub-catchment area, different return periods correspond to different comprehensive runoff coefficients of each sub-catchment area. In this paper, nine typical sub-catchment areas are selected, i.e., according to the underlying surface conditions, three areas with poor permeability, three weak permeable areas and three green space areas are selected to simulate the contrast in comprehensive runoff coefficient (Table 5).

According to Table 5, the impervious rate of No. 13, No. 91 and No. 108 sub-catchment area can reach 90%, corresponding to larger comprehensive runoff coefficient, where the comprehensive runoff coefficient is greater than that in the whole area, prone to local waterlogging problems. The impervious rate of No. 41, No. 124 and No. 152 sub-catchment area is 45%, where comprehensive runoff coefficient is larger than green space area, close to regional comprehensive runoff coefficient. In case of a heavy rain, a waterlogging risk exists. By contrast, No. 204, No. 224 and No. 223 sub-catchment areas belong to green space areas, so comprehensive runoff coefficient is close to zero, which is not prone to waterlogging. Through comparing comprehensive runoff coefficients under three kinds of underlying surface conditions, comprehensive runoff coefficient is larger in the area with smaller impervious rate, and

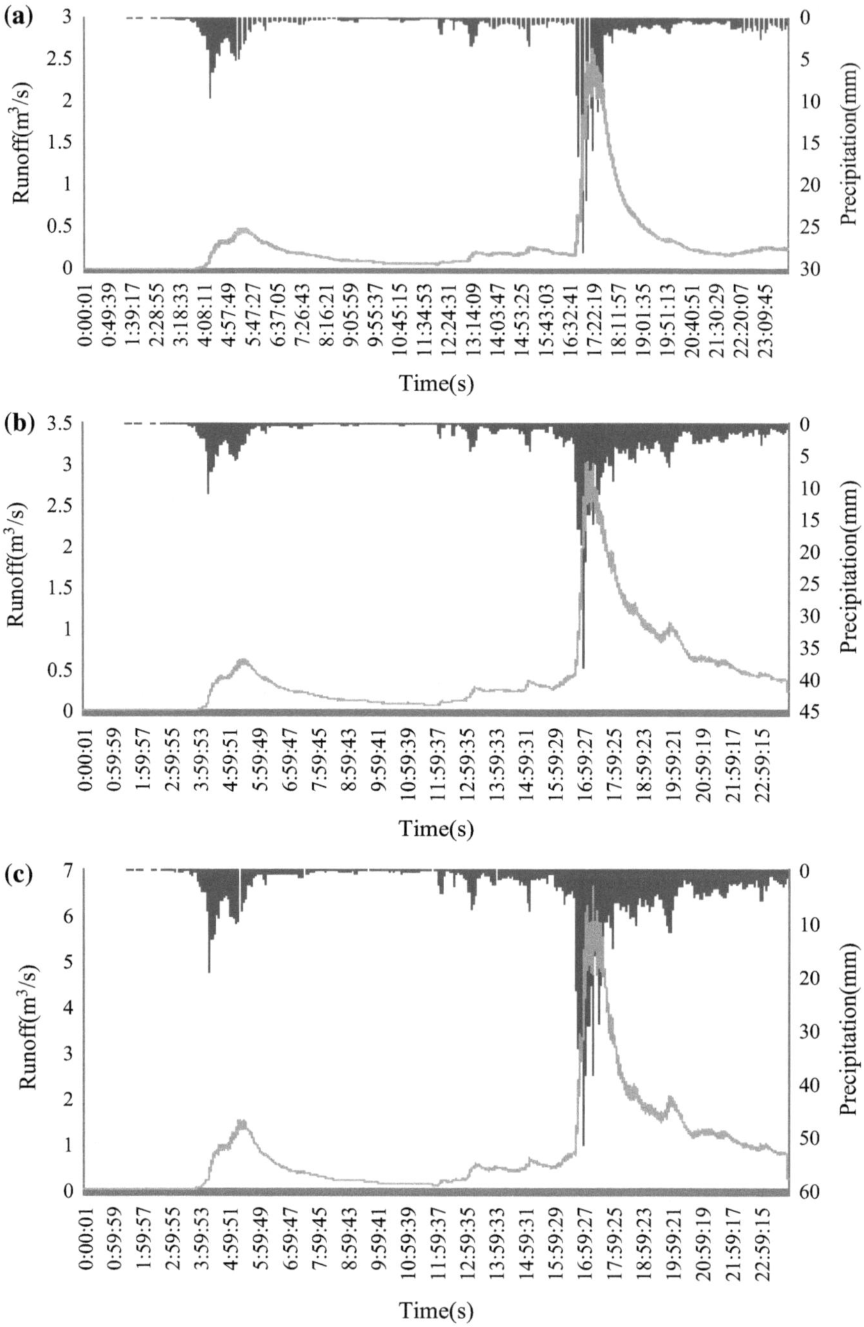

Fig. 4 The runoff processes in different periods **a** 10 yrs **b** 20 yrs **c** 50 yrs

Table 4 The comparing of different simulation result of return period runoff

Return period (yrs)	The comprehensive runoff coefficient	Peak discharge(m^3/s)
10	0.52	2.43
20	0.53	3.01
50	0.58	6.07

Table 5 The comprehensive runoff coefficient of different return period

Sub-catchment node	10 yrs	20 yrs	50 yrs	Impervious rate
13	0.85	0.862	0.89	90
91	0.836	0.849	0.88	90
108	0.822	0.835	0.868	90
41	0.381	0.399	0.458	45
124	0.38	0.398	0.455	45
152	0.385	0.404	0.47	45
204	0	0	0.065	0
223	0	0	0.048	0
224	0	0	0.085	0

the greater the return period is, the larger the comprehensive runoff coefficient will be. On the contrary, comprehensive runoff coefficient is close to zero in the area with low impervious rate, due to uneven distribution of urban underlying surface conditions and impervious rates. Thus, runoff confluence conditions are different in the sub-catchment areas, leading to different runoff generation, reflecting the spatial difference in regional flood risk.

5 Summary

SWMM model was used to build rainfall runoff model of Beijing Future Science Park of Changping in Beijing. The runoff coefficient method is selected to validate the model and simulate rainfall scenarios in different return periods (10, 20 and 50 years) after validation. In terms of the whole area, two factors of flood peak flow and regional comprehensive runoff coefficient were analyzed to show that the increase of comprehensive runoff coefficient will intensify the urban waterlogging risk. In terms of the local area, the variations in the comprehensive runoff coefficient of sub-catchment area over the increase of return period under different underlying surface conditions were analyzed to reflect the spatial difference in urban regional flood risk along with the underlying surface conditions.

Under the condition of no measured validation data, the runoff coefficient method is used for parameter calibration to create the model in a certain precision, reduce the requirement of the model for the observed data of rainwater pipe network, and expand the applicability of this method to the areas in the absence of observed runoff data.

Due to lack of regional land use type and the application analysis of rain flood utilization measures, future research should be based on this model to combine different land use types and rain flood utilization measures for studying the setting of regional low impact development measures.

Acknowledgements The researchers would like to extend their thanks to the Chinese National Natural Science Foundation (No. 51409275, No. 51522907, No. 51739011).

References

1. Song, C.P., Wang, H.C., Tang, D.S.: Research progress and development trend of storm water management model. China Water Wastewater **16**, 16–20 (2015)
2. Palla, A., Gnecco, I.: Hydrologic modeling of low impact development systems at the urban catchment scale. J. Hydrol. **528**, 361–368 (2015). https://doi.org/10.1016/j.jhydrol.2015.06.050
3. Versini, P.A., Ramier, D., Berthier, E., et al.: Assessment of the hydrological impacts of green roof: from building scale to basin scale. J. Hydrol. **524**, 562–575 (2015). https://doi.org/10.1016/j.jhydrol.2015.03.020
4. Zhao, G., Pang, B., Zongxue, X.U., et al.: Simulation of urban storm at Dahongmen drainage area by SWMM. J. Beijing Norm. Univ. **05**, 452–455 (2014)
5. Cong, X.Y., Guang-Heng, N.I., Hui, S.B., et al.: Simulative analysis on storm flood in typical urban region of Beijing based on SWMM. Water Resour. Hydropower Eng. **2006**(04), 64–67 (2006). https://doi.org/10.3969/j.issn.1000-0860.2006.04.021
6. Wang, H., Chen, J., Zhang, S., et al.: Application status and comparative analysis of urban storm water models. Water Resour. Hydropower Eng. **42**(11), 10–13 (2011). https://doi.org/10.3969/j.issn.1000-0860.2011.11.003
7. Liu, X.: Parameter calibration method for urban rainfall-runoff model based on runoff coefficient. Water Wastewater Eng. **35**(11), 213–217 (2009). https://doi.org/10.3969/j.issn.1002-8471.2009.11.056
8. Beijing General Municipal Engineering Design & Research Institute: Concise Drainage System Design Manual. Beijing Building Materials Press, Beijing (1990)
9. Rossman, L.A.: Storm Water Management Model User's Manual Storm Water Management Model User's Manual Version 5.0. United States Environmental Protection Agency (2010)
10. Zhou, Y., Ming-Hui, Y.U., Chen, Y.X.: Estimation of sub-catchment width in SWMM. China Water Wastewater **22**, 61–64 (2014)
11. Beijing Water Bureau: Hydrologic Handbook of Beijing City—Rainstorm Atlas (1999)
12. Beijing Academy of Urban Planning & Design: Standard of rainstorm runoff calculation for urban stormwater drainage system planning and designing. DB11/T 969-2013. (2013)

Source Apportionment of $PM_{2.5}$ in Handan City, China Using a Combined Method of Receptor Model and Chemical Transport Model

Zhe Wei, Litao Wang, Liquan Hou, Hongmei Zhang, Liang Yue, Wei Wei, Simeng Ma, Chengyu Zhang and Xiao Ma

Abstract Handan is one of the top polluted cities in China, characterized by high concentration of fine particulate matter ($PM_{2.5}$). In this paper, a receptor model, i.e., the Positive Matrix Factorization (PMF) model, and a chemical transport model, i.e., the Mesoscale Modeling System Generation 5 (MM5) and Models-3/Community Multiscale Air Quality (CMAQ) model, are both applied to apportion the sources of

Z. Wei · L. Wang (✉) · L. Hou · S. Ma · C. Zhang · X. Ma
Department of Environmental Engineering, School of City Construction, Hebei University of Engineering, Handan 056038, Hebei, China
e-mail: wanglitao@hebeu.edu.cn

Z. Wei
e-mail: kyzyweizhe002@163.com

L. Hou
e-mail: 1357908565@qq.com

S. Ma
e-mail: masimeng1004@163.com

C. Zhang
e-mail: 425573920@qq.com

X. Ma
e-mail: maxiao1992_air@163.com

H. Zhang
School of Economics and Management, Hebei University of Engineering, Handan 056038, Hebei, China
e-mail: Zhanghongmeili2008@126.com

L. Yue
Environmental Monitoring Center of Handan, Handan Environmental Protection Bureau, Handan 056001, Hebei, China
e-mail: hjjc2009yue@163.com

W. Wei
Department of Environmental Science, Beijing University of Technology, Beijing 100124, China
e-mail: weiwei@bjut.edu.cn

W. Dong et al. (eds.), *Sustainable Development of Water Resources and Hydraulic Engineering in China*, Environmental Earth Sciences,
https://doi.org/10.1007/978-3-319-61630-8_13

$PM_{2.5}$ in Handan. It is concluded that regional sources contribute 36.0% of $PM_{2.5}$, and within local sources, the contributions of major emission sectors are: 22.3% from coal combustion, 10.7% from metal smelting, 7.3% from Zn-OC-Ba, 18.5% from industry, 11.3% from transportation, 10.6% from biomass burning, and 19.2% from dust emissions. It indicates that regional joint air pollution controls should be emphasized in the future control strategy, and local source controls on coal combustion and industries are the key points to mitigate the severe $PM_{2.5}$ pollution in Handan.

Keywords $PM_{2.5}$ · Source apportionment · PMF · MM5-CMAQ
Handan

1 Introduction

Urban air pollution, especially the pollution of fine particulate matter ($PM_{2.5}$) is one of the most urgent environmental problems to be resolved in China and has raised wide public, governmental, and academic concerns. Hebei province, located in northeastern China, is the top polluted province according to the reports on air quality status publicized by the Ministry of Environmental Protection (MEP) [1, 2]. In 2013 and 2014, seven out of the ten top polluted cities in China are within Hebei province [1, 2]. Handan city, located at the southern edge of Hebei, has an area of 1.2×10^4 km^2 and a population of 9.0 million, with about 1.5 million living in the urban area of 419 km^2. It is a heavily industrialized city with a large amount of productions of iron and steel, cement, and coke [3]. Air pollution in Handan is very severe and it ranked as the third top polluted city in China in 2013 and the fifth in 2014 [1, 2].

Handan is also a city with complex regional sources surrounding it. It is in the intersection of four provinces, Hebei, Henan, Shandong, and Shanxi (see Fig. 1a). All of them are heavily populated, industrialized, and urbanized and their major air pollutants emissions, e.g., $PM_{2.5}$, PM_{10} (atmospheric aerosols with aerodynamic diameter less than or equal to 10 μm), sulfur dioxide (SO_2), and nitrogen oxides (NO_X) are about as high as one fourth of the total emissions in China [4]. It results in a complex problem of source apportionment of $PM_{2.5}$ in Handan, which is urgently required by the local government to design an effective control strategy.

Receptor models and chemical transport models are both effective methods for $PM_{2.5}$ source apportionment [5, 6]. The Positive Matrix Factorization (PMF) model, as one of the receptor models recommended by US Environmental Protection Agency (US EPA), has been successfully applied in many cities in Asia [7–11]. Some studies used multiple source apportionment approaches to improve the robustness of the results and understand the spatial characteristics [12, 13]. Contini et al. [13, 14] compared the source contributions of PM_{10} or $PM_{2.5}$ at different sites in Venice Lagoon. However, multi-site samplings are not available in Handan before 2015. The Mesoscale Modeling System Generation 5 (MM5) and the

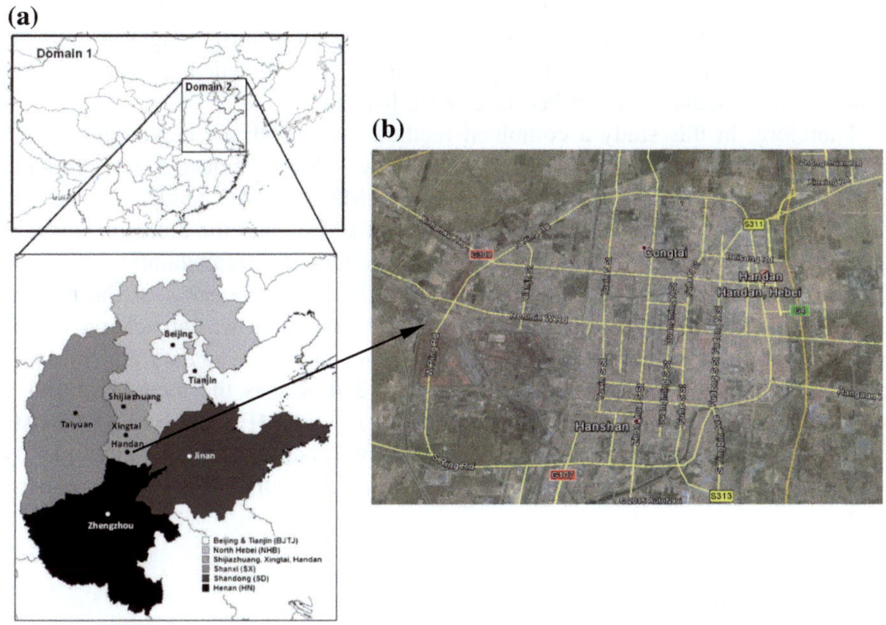

Fig. 1 **a** CMAQ modeling domains at a horizontal grid resolution of 36 km over East Asia (Domain 1 with 164 × 97 cells) and 12 km over an area in northern China (domain 2 with 93 × 111 cells); **b** location of the $PM_{2.5}$ sampling site in Handan (the red star, referred to as HEBEU site)

Models-3/Community Multiscale Air Quality (CMAQ) model, as one of the source oriented, chemical transport model, has been proved to be an effective tool to identify the sources of air pollution in Chinese cities [15–20]. $PM_{2.5}$ source apportionment of $PM_{2.5}$ in Handan use of those tools was reported in several recent studies [4, 21, 22]. However, neither of them can produce enough detailed and reliable results to fully support the policy making of local government in Handan, because the receptor models cannot easily distinguish the spatial origins of each source sector [5, 6], unless multiple-site sampling data and more advanced analysis methods (e.g., microscopic methods) are employed or trajectory models using detailed meteorological data are employed, which were not available in Handan for this study. As mentioned above, and the reliability of the results from the source-oriented models are highly dependent on the precision of the emission inventory, which is far from developed for the sources in Hebei at this time. Bove et al. [23], applied a hybrid method using the PMF model and the chemical transport model of Comprehensive Air Quality Model with Extensions (CAMx) in the city of Genoa in Italy and found the two methods gave consistent results even the source categories were different. In their study, PAST (Particulate matter Source Apportionment Technology) was used to calculate source contributions [23]. In this paper, the Brute-force method, a source sensitivity method, is applied to calculate source contributions of $PM_{2.5}$. This method, although computationally expensive, is straightforward and can simulate the non-linear relationship among PM and PM precursors under large emission perturbations. The detailed explanation on choosing this method is described elsewhere [4].

Therefore, in this study a combined method using both a receptor model, the PMF, and a chemical transport model, the MM5-CMAQ, is applied to $PM_{2.5}$ source apportionment in Handan, e.g., the results by MM5-CMAQ model are used to distinguish the regional or local contributions of $PM_{2.5}$, and the receptor model is used to apportion the sectoral contributions to limit the uncertainties from the MM5-CMAQ introduced by the uncertainties and incompleteness of the present emission inventory. This study might be the best estimation we can achieve on the source apportionment of $PM_{2.5}$ in Handan, given the present data and research basis. This paper is organized as below: Sect. 2 describes the method of $PM_{2.5}$ sampling and analysis, configurations and inputs of the PMF and the MM5-CMAQ model. Section 3 introduces the results of the two methods and the combination analysis. Section 4 summarizes the major findings and conclusions of this paper.

2 Methodology

2.1 $PM_{2.5}$ Sampling and Analysis

The $PM_{2.5}$ sampling site is located on the top of a 12 m high building at Hebei University of Engineering (HEBEU) in Handan, Hebei, China (36°34′ N, 114°29′ E) (See Fig. 1b). This location is at the south edge of the urban area of Handan, representing a cultural area, and it is also one of the four official online monitoring sites in Handan. There is no obvious emission source around the site. Four months in 2013, i.e., January, April, July, and October, representing the four seasons, were selected to collect $PM_{2.5}$ samples continuously from 8:00 to 7:30 a.m. on the next day, using a high volume $PM_{2.5}$ air sampler (Thermo Scientific Co., model number: HIVOL-CVBLD) with 20.3 × 25.4 cm quartz filters. The flow rate was 1.13 m^3 min^{-1}. First, the sampling filters were heated at 350 °C to remove the organic material in the filters. Then they were reserved in constant-temperature and humidity chamber for 24 h before the pre-weighting and sampling. The sampled filters were kept in the chamber for another 24 h before the post-weighting. The differences in the two weights are used to calculate sampled $PM_{2.5}$ concentrations (the sampling concentrations of 23.5 h were converted to daily concentration). Those sampled filters were stored in a refrigerator under −20 °C until chemical analysis. In total, 120 samples were collected in this study.

Water-Soluble Ions (WSI). Water-soluble cations and anions were measured by ion chromatography (Dionex, DX–600, Dionex, ICS–2100, USA). First, a small section of the quartz filter (1 cm^2) was put into 10 mL ultrapure water (18.2 MΩ cm) for extracting water-soluble inorganic ions. The extracted solution was filtered through a microporous membrane filter (0.45 μm of pore size), and stored in a refrigerator at −18 °C until sending into ion chromatography. The measured ions include Cl^-, NO_3^-, SO_4^{2-}, Na^+, K^+, Mg^{2+}, Ca^{2+}, and NH_4^+. Strictly quality control during all procedures avoid contamination of the samples and ensure the accuracy of chemical analysis. The method detection limit (MDL) of NH_4^+, Na^+, K^+, Mg^{2+}, Ca^{2+}, SO_4^{2-}, Cl^-, and NO_3^- were 0.01, 0.001, 0.001, 0.004, 0.005, 0.01, 0.005, 0.01, and 0.01 μg ml^{-1}, respectively.

Organic Carbon (OC) and Elemental Carbon (EC). OC and EC were analyzed by the method of Thermo-optical Reflectance (TOR) which used a Thermal/Optical Carbon Analyzer (Model 2001A) produced by the Desert Research Institute (DRI). First, OC1, OC2, OC3, and OC4 were measured at 140 °C, 280 °C, 480 °C, and 580 °C in an environment with pure He gas without O_2, respectively. Second, the sampled filter was further heated to 580 °C, 740 °C, and 840 °C to measure EC1, EC2, and EC3 in an environment with 2% O_2 and 98% He, respectively. The precision of OC and EC were <5% and <10% along with the MDL of 0.82 μg cm^{2-} for TOC and 0.20 μg cm^{2-} for TEC, respectively.

Elements. Inductively coupled plasma mass spectrometry (ICP–MS) was used to analyze the concentrations of trace elements, i.e., As, Ba, Cr, Cu, Mn, Pb, Sr, Ti, V and Zn, in $PM_{2.5}$. A piece of quartz filter (1 cm^2) was digested for 25 min at 190 °C with 8 ml concentrated nitric acid (BV–III) and 0.5 ml H_2O_2 (30%) in a Teflon microwave digestion tank (CEM, MARS). Then, the bulk solution was diluted to 100 ml, of which 10 ml were put into ICP-MS to measure the concentrations of the elements. ICP-MS was calibrated by standard injection each several samples analysis. Relative standard deviation (SD) was controlled below 5%, and internal standard recovery was controlled at the range of 80–120%. All of them aim at ensuring the accuracy of detection.

2.2 The Receptor Model—PMF

The PMF model, as a receptor model, is a data analysis technique of factor analysis based on multivariate statistical methods, of which the fundamental ability is to resolve identities and source contributions of components in an unknown mixture. PMF uses the chemical components of $PM_{2.5}$ to calculate the source contributions without knowledge of emissions inventory [24]. PMF decomposes a matrix of speciated sample data into two matrices: factor contributions (G) and factor profiles (F). These factor profiles need to be interpreted by the user to identify the source types that may be contributing to the sample using measured source profile information, and emissions or discharge inventories. The key equations of this method are listed as below as Eqs. (1) and (2). A speciated data set can be viewed as a data matrix X of i by j dimensions as in the Eq. (1), where i and j represents the number of samples and chemical species, respectively; g_{ik} is the contribution of the pth source to the ith sample; and f_{kj} is the concentration of the jth species in the pth source.

$$X_{ij} = \sum_{k=1}^{p} g_{ik} f_{kj} + e_{ij} \quad (1)$$

$$Q = \sum_{i=1}^{n} \sum_{j=1}^{m} \left[\frac{e_{ij}}{u_{ij}}\right] \quad (2)$$

where e_{ij} is the residual for each sample/species, and u_{ij} is an uncertainty term introduced to facilitate a statistical solution of the mass balance as opposed to an analytical mathematical solution. The PMF model is constrained to non-negative species concentrations and source contributions, which makes the results meaningful. Also, PMF applies uncertain matrix (u_{ij}) to minimize the Q (2) to optimize the results. The PMF model is described in detail in [25, 26]. A brief description can be found in Wei et al. [22].

PMF has been widely used for both regulatory and research applications [8, 22, 27–31], and is recommended by US EPA [32]. In this paper, the PMF version 5.0 was selected to calculate the source contributions of $PM_{2.5}$ in Handan in 2013. Ten percent of the concentrations of each species was used as the uncertainty matrix in the dataset, following the suggestion by Zhang et al. [33], considering that the sampling technique (e.g., high volume sampling with quartz filters may lead to larger uncertainties [34]), analytical method applied, and the number of samples analyzed. If the concentration is less than or equal to the method detection limit (MDL), 1/2 MDL is used to replace the concentration and 5/6 MDL is used as uncertainty [35]. Missing data were substituted with the median value and their uncertainties were replaced by four times the median value. Signal to noise (S/N) was used to review the dataset (Eq. 3). S_{ij} indicates standard deviation of ith sample and jth chemical species. The species were categorized as 'strong' when S/N $\geq$ 2, 'weak' (0.2 $\leq$ signal-to-noise ratio $\leq$ 2), or 'bad' (signal-to-noise ratio < 0.2) species. In this study, all species were categorized as "strong". A fundamental problem using this method is to identify a reliable number of factors. The number of factors is determined as following: (1) most of the residual matrix is within −3.0 to 3.0; (2) the results tend to be stable when adjusting the number of factors. In this paper, the concentration data of Cl^-, NO_3^-, $SO^{4-}{}_2$, Na^+, K^+, Mg^{2+}, Ca^{2+}, NH_4^+, OC, EC, V, Cr, Mn, Cu, Zn, As, Sr, Ba, and Pb were used to run the PMF model because of all these species were available for the four months that were studied. As for estimating uncertainty of PMF solutions, this paper selected classical bootstrap (BS) error analysis to estimate the PMF results [36, 37]. Default value was applied to run BS error estimation in Base run and Block size option in PMF model. One hundred BS runs were performed to ensure the robustness of the statistics. R-value was set as 0.6. The first, the second, the third and the fourth factor were mapped with BS in 99, 77, 87, and 97% of the runs, while other factors were mapped 100% of runs. Therefore, the authors were convinced that 8 factors were selected to explain the source contribution of $PM_{2.5}$.

$$\frac{S}{N} = \frac{1}{2}\sqrt{\sum\nolimits_{i=1}^{n} x_{ij}^2 / \sum\nolimits_{i=1}^{n} s_{ij}^2} \tag{3}$$

2.3 *The Chemical Transport Model—MM5-CMAQ*

The CMAQ model, a 3-D chemical transport model developed by US EPA [24], has been increasingly used to simulate regional air quality in China in the past decade [15–20]. The MM5 model, a regional mesoscale meteorological model developed by the National Center for Atmospheric Research (NCAR) and then Penn State University [38], is applied to generate meteorological field to drive the CMAQ model. The source apportionment method used in this study is the so-called Brute-Force method [39], that is, the source contributions of each source region or sector are estimated by calculating the differences between the baseline simulation and the sensitivity simulation in which the emissions from that source region or sector are set as zero. This method has been proven to be an effective way to quantify the source contributions to support the policy making in air pollution controls in China in recent years [15–20].

The detailed descriptions on the model configurations, performance evaluation, simulation scenarios and results can be found in [40, 41]. In brief, the modeling domains are seen in Fig. 1a. Two nested domains are applied over East Asia and North China at 36 and 12 km horizontal grid resolutions, respectively. The model configurations and inputs are summarized in Table 1. Besides the base case, twenty-four emission reduction scenarios are simulated, including the emission zero-out simulations of the nine source regions and the five emission sectors in the three cities, Handan, Shijiazhuang and Xingtai (in this study only the results for Handan are used). The nine source regions (see Fig. 1a) are Handan (the local sources), Shijiazhuang, Xingtai, the northern Hebei (other Hebei area besides the above three cities), Beijing and Tianjin (BJTJ), Shanxi (SX), Shandong (SD), Henan (HN), and the other regions (OTH, other regions in Domain 1 besides the above eight). The five emission sectors locally include power plant (PO), industrial (IN), domestic (DO), transportation (TR), and agriculture (AG). Therefore, the local (anthropogenic emissions within Handan) and regional (emissions outside Handan) source contributions to $PM_{2.5}$ in Handan, and the contributions of each emission sector within Handan are qualified by using those simulations, which is absolutely necessary for policy making in air pollution controls of the local government.

Table 1 Configurations and inputs of MM5-CMAQ modeling in this study

Item	Configuration and input	References and notes
Simulation period	Jan. and Jul., 2013	5-spin up days
Domain	East Asia (36 km), northern China (12 km)	–
Horizontal resolution	36 and 12 km	–
Vertical resolution	23 layers (MM5) and 14 layers (CMAQ) from 1000 to 100 mb	–
Nesting	Two way for MM5; one way for CMAQ	–
MM5	Version 3.7	–
CMAQ	Version 4.7.1	–
Anthropogenic emissions	MEIC	[http://www.meicmodel.org/]
Cumulus scheme	Kain–Fritsch	Kain and Fritsch [42]
PBL scheme	Blackadar	Zhang and Anthes [43]
Moisture scheme	Mixed phase (Reisner 1) explicit	Reisner et al. [44]
Longwave and shortwave radiation	Cloud atmospheric radiation scheme	Dudhia [45]
Surface scheme	Force/restore (Blackadar)	Blackadar [46], Deardorff [47]
FDDA	Wind, temperature, and water vapor mixing ratio in and above PBL	–
Gas phase chemistry	SAPRC-99 with aqueous and aerosol extensions	Carter [48, 49]
Aerosol phase model	AERO5	Binkowski and Shankar [50]
Aqueous-phase chemistry	Updated regional acid deposition model (RADM)	Chang et al. [51], Walcek and Taylor [52]
Land use data	US geological survey terrain and land use data	–
Meteorological initial and boundary conditions	National center for environmental prediction (NCEP) final (FNL) operational global analysis data	–
FDDA input	NCEP automated data processing (ADP) surface and upper air data	–
Chemical initial and boundary conditions	Clean profile in CMAQ for 36 km domain; nested from 36 to 12 km domain	–
Photolysis rates processor (JPROC) input	Ozone column data from the Ozone Measurement Instrument (OMI) on the Aura	[http://toms.gsfc.nasa.gov/ozone/ozone_v8.html]
Emission scenarios	Zero-out anthropogenic emissions in Handan, Shijiazhuang, Xingtai, northern Hebei (NHB), Beijing and Tianjin (BJTJ), Shanxi (SX), Shandong (SD), Henan (HN), and other area (OTH) Zero-out emissions from power plant (PO), industrial (IN), domestic (DO), transportation (TR), and agriculture (AG), one by one, in Handan, Shijiazhuang, and Xingtai	Total 24 scenarios, only results of sectoral zero-out scenarios in Handan are used in this study

3 Results and Discussions

3.1 Chemical Composition of $PM_{2.5}$ in Handan

Table 2 presents the concentrations of $PM_{2.5}$ and WSI, OC, and EC in $PM_{2.5}$ in the four representative months in 2013. It can be seen that the highest monthly-average concentration of $PM_{2.5}$ was 309.2 μg m^{-3} in January, followed by 213.3 μg m^{-3} in October, 103.2 μg m^{-3} in April, and 85.6 μg m^{-3} in July. In general, the WSI are the most abundant species in $PM_{2.5}$, accounted for 43.2% of $PM_{2.5}$ in January, 48.9% in April, 55.9% in July, and 36.2% in October. The highest fraction of WSI appears in July due to the rapid reaction speed in secondary aerosol formation, e.g., NO_3^-, SO_4^{2-} and NH_4^+, led by higher temperature. OC accounted for 15.1% of $PM_{2.5}$ in January, followed by 13.5% in October, 10.6% in April, and 7.9% in July.

Table 3 presents the concentrations of trace elements in $PM_{2.5}$ in Handan. They only account for less than 1.0% of mass concentrations in $PM_{2.5}$. However, the heavy metal elements should be paid more attention to because of their negative effects on human health and the capability of serving as catalyst in the formation of secondary aerosols in the atmosphere. Among those elements, the concentration of Zn is the highest of 314.0 ng m^{-3}, followed by Pb (221.9 ng m^{-3}), Mn (64.4 ng m^{-3}), Ti (49.7 ng m^{-3}), Ba (34.0 ng m^{-3}), As (30.9 ng m^{-3}), Cu (21.4 ng m^{-3}), Sr (8.8 ng m^{-3}), and Cr (6.4 ng m^{-3}). The concentrations of Ca has the highest

Table 2 The concentrations of $PM_{2.5}$ and WSI, OC, and EC in $PM_{2.5}$ in Handan in 2013

		$PM_{2.5}$ (μg m^{-3})	Cl^-	NO_3^-	SO_4^{2-}	Na^+	K^+	Mg^{2+}	Ca^{2+}	NH_4^+	OC	EC
Jan.	Ave.	309.2	7.3	23.2	36.3	1.3	3.2	0.2	0.5	23.8	33.3	14.8
	Min	62.4	1.0	5.0	7.0	0.5	1.1	0.0	0.0	5.6	8.5	4.2
	Max	643.0	17.3	43.4	78.2	2.4	6.4	0.4	2.5	43.1	63.7	33.8
	S.D.	133.3	3.9	8.4	14.7	0.5	1.2	0.1	0.7	7.8	13.8	7.6
Apr.	Ave.	103.2	3.2	17.2	16.2	0.5	1.1	0.1	2.0	7.3	10.2	3.9
	Min	34.8	0.8	3.2	4.9	0.0	0.5	0.0	0.3	2.5	3.5	1.6
	Max	224.2	9.2	38.7	29.4	1.3	2.3	0.3	4.4	13.2	18.0	7.9
	S.D.	42.3	2.1	9.2	7.0	0.3	0.5	0.1	1.0	3.1	3.2	1.3
Jul.	Ave.	85.6	1.1	12.5	24.8	0.4	0.8	0.1	0.4	9.0	6.1	2.6
	Min	36.7	−0.2	2.1	7.7	−0.1	0.2	0.0	0.0	2.1	2.1	1.1
	Max	190.5	4.0	37.3	59.7	1.4	1.5	0.2	0.9	19.1	8.8	3.8
	S.D.	37.6	1.2	8.0	12.4	0.4	0.3	0.0	0.2	3.7	1.9	0.8
Oct.	Ave.	213.3	5.8	31.6	23.6	0.6	2.2	0.2	1.1	11.5	28.5	4.4
	Min	38.6	1.4	4.1	5.3	0.2	0.3	0.0	0.1	3.0	2.3	1.0
	Max	402.6	9.9	76.4	51.1	1.4	3.6	0.3	2.4	21.6	67.2	6.4
	S.D.	95.0	2.3	14.8	10.3	0.3	0.8	0.1	0.6	4.2	14.1	1.0

S.D. indicates standard deviation

Table 3 The concentration (ng m^{-3}) of elements in $PM_{2.5}$ in Handan in 2013

	January	April	July	October	Ave.
As	30.5	29.3	28.2	35.6	30.9
Ba	11.1	47.7	22.9	54.1	34.0
Cr	10.1	5.5	4.5	5.3	6.4
Cu	28.1	24.0	11.2	22.6	21.4
Mn	80.2	70.9	33.3	73.1	64.4
Pb	350.2	155.3	151.0	231.2	221.9
Sr	5.4	18.6	3.7	7.5	8.8
Ti	13.2	124.5	16.1	45.1	49.7
V	5.9	5.0	0.4	1.7	3.2
Zn	339.0	259.5	246.4	411.2	314.0

seasonal variation, with the highest value appearing in April which resulted from dust blowing weather that occurred over northern China.

3.2 *The Results of the PMF Model*

Eight factors are identified by PMF model, as shown in Fig. 2. The first factor was recognized as dust/road dust source, because Ca^{2+} has high loadings in this factor. Ca^{2+} may also be associated with steel production, cement/concrete, construction activities, and earth crustal material. EC and V emitted from the combustion of diesel oil and deposited into soil. The contaminated soils can be resuspended by mobile transportation or wind blowing. The source of Mn and Ba is related to brake parts on the vehicles. Therefore, this factor was regarded as dust/road dust source, which contributed 12.3% of $PM_{2.5}$.

The second factor has high loadings of Cr and Na^{+}. Cr may be from the production process of iron and steel. Because Cr was usually used to produce stainless steel. Iron and steel is a pillar industry in Handan, which produced 51 million tons steel and 45 million tons crude steel in 2013 (http://www.handannews.com.cn/new_epaper/hdrb/html/2014/03/25/content_10718.htm), as well as a large amount of stainless steel (http://www.hgjt.com.cn/index.aspx). Na usually exists in iron ore. As for SO_4^{2-} and As of the high loadings, they may be from coal combustion because coal was the major energy source for iron and steel industry. Therefore, this factor was identified as metal smelting source, which contributed 11.0% of $PM_{2.5}$.

K^{+} is a very important marker of biomass burning. The third factor has high loading of K^{+} and medium loadings of OC and Cl^{-}, so that it is attributed to biomass burning source, with the contributions of 10.7% in $PM_{2.5}$. However, this paper found that Mn, Zn, Pb, and Cu have relative high loadings in this factor, which were usually from the industrial processing that may be influenced by multi-colinearity of PMF model. Therefore, further analyses should be encouraged.

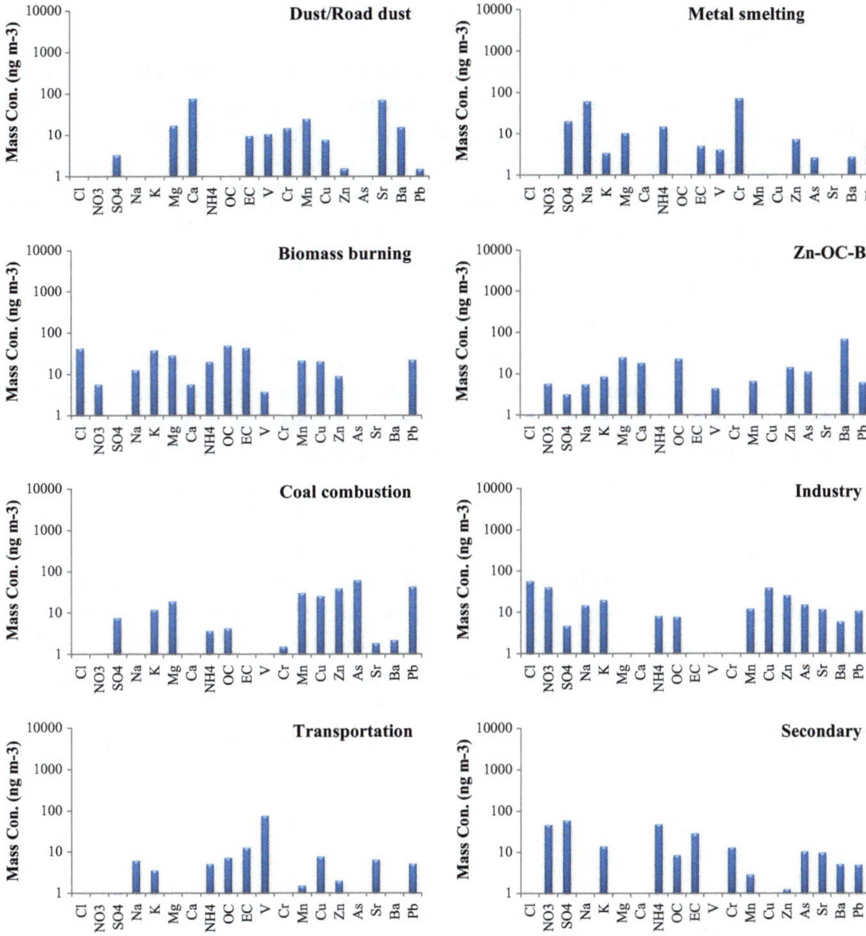

Fig. 2 Source profiles in mass concentrations of the factors found by the PMF model

The fourth factor was seen as a mixed source. The reason is that the factor has high loadings of Ba which may be from the brake parts on vehicles or construction activity. At the same time, this factor has close correlations with Zn, Ca^{2+}, Mg^{2+}, OC and As. In general, As is seen as the marker element of coal combustion. OC was also from the coal combustion. Zn was usually made into various product, including brake of vehicles. Ca^{2+} and Mg^{2+} may be from steel production, cement/concrete, construction activities, and earth crustal material as mentioned above. This factor was influenced by dust/road dust. Therefore, this factor could be seen as a mixed source with Zn-OC-Ba, which contributes 7.6% of $PM_{2.5}$.

The fifth factor is coal combustion sources, which has obvious relationship with As, an important maker of coal combustion. Additionally, coal combustion would emit amounts of organic matter and SO_2, which also show high loadings in this factor.

Zn and Pb may be from the coal combustion. The source contribution of coal combustion to $PM_{2.5}$ is 13.8%.

Mn, Cu, Zn, As, Sr, Pb and Ba have high loadings in the sixth factor. Zn and Pb is usually from rolling mills and industry processing [9, 35]. In Handan, the iron and steel industries are very important components of its industries. Mn and Cu are in close correlation with metal smelting industry. Sr and Ba were used for the ceramic industry because of their dielectric properties. Therefore, this factor was identified as an industrial source, which contributes 17.5% of $PM_{2.5}$.

The seventh factor was characterized by high loading of V, EC, and moderate loadings of Mn, Zn, and Pb. V is an auxiliary material used by oil products. EC could be emitted from the vehicle engines due to incomplete combustion of diesel oil. Mn, Zn and Pb are typically related to the brakes of motors. Additionally, Zn is a component of a common fuel detergent and anti-wear additive which also exists in the diesel emissions, as well as in tires, brake linings and pads. This factor is regarded as a transportation source, which contributes 11.2% to $PM_{2.5}$.

The eighth factor, with high loadings of NH_4^+, SO_4^{2-} and NO_3^-, is seen as the secondary sources, because those species are produced in the atmosphere from the primary pollutants of SO_2, NO_X and NH_3. It contributes 16.0% in $PM_{2.5}$. A secondary source should be identified easily because secondary components are produced in the atmosphere. However, OC, EC, and Cr have high loadings in this factor. There are some limitations for source apportionments of $PM_{2.5}$ with the PMF model, such as the colinearity of receptors. Further study and analysis should be needed.

The above results on $PM_{2.5}$ source apportionment in Handan are summarized in Fig. 3.

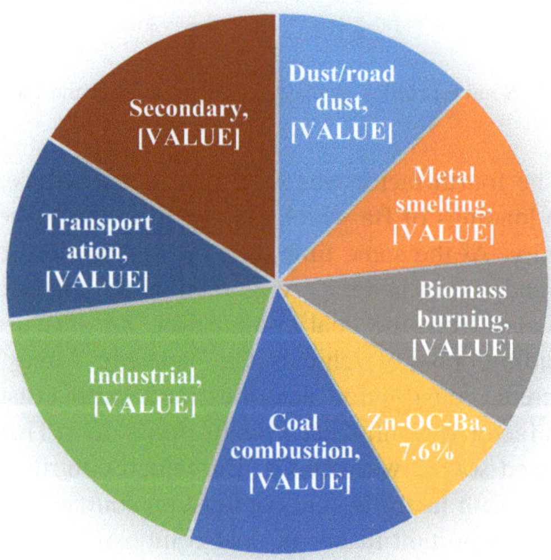

Fig. 3 The source contributions of $PM_{2.5}$ in Handan by the PMF model

Table 4 The source contributions (%) of regional and local sector emissions to $PM_{2.5}$ in Handan by the MM5-CMAQ model[a]

	% in total $PM_{2.5}$	% in local contributions	% in SO_4^{2-}	% in NO_3^-	% in NH_4^+	% in total SIA
Regional	40.4		51.1	65.1	55.1	59.1
PO	0.9	1.5	1.3	−0.6	0.6	0.4
IN	35.9	59.5	34.1	1.5	23.5	20.6
DO	13.5	22.3	6.7	3.1	5.4	4.8
TR	3.7	6.2	1.3	2.9	1.3	1.5
AG	6.2	10.3	4.8	38.5	20.4	17.8
Total local	60.2	99.8	48.2	45.4	51.2	45.2

[a]The sum of local and regional contributions is not 100% due to the non-linear response of $PM_{2.5}$ to its precursors emissions

3.3 The Results of MM5-CMAQ Model

The contributions of each source region and sector to the secondary $PM_{2.5}$ are discussed in detail as well in [40], because the receptor model can recognize one source as "secondary aerosol", i.e., NO_3^-, SO_4^{2-}, and NH_4^+, as discussed in Sect. 3.2, but cannot distinguish their spatial and sectoral source of origins [5, 6]. The modeling results show that in the total secondary inorganic aerosols (SIA) the contribution of regional sources is 59.1% on annual average, and the local contributions of PO, IN, DO, TR, and AG are 0.4%, 20.6%, 4.8%, 1.5%, and 17.8% [40], respectively. Those results are summarized in Table 4. Those treatments are not pursued for secondary organic aerosols (SOA) because of the insignificance of OC in the "secondary aerosol" recognized by the PMF model (see Fig. 2), and the weakness of the CMAQ model in predicting SOA concentrations at present.

3.4 Combined Analysis and Results

3.4.1 Regional Contributions

The receptor model apportions the $PM_{2.5}$ concentrations by the representative trace species of the recognized sources; so, it is not easy to distinguish the spatial origins of those sources unless data from multiple sampling locations can be employed, which was not available in Handan for this study. Therefore, the results of the MM5-CMAQ model are used to quantify the regional contributions in Handan. It indicates 40.4% of $PM_{2.5}$ in Handan are from regional sources. However, it should be noted that this number may overestimate the regional contributions because of the lack of dust emissions in the MM5-CMAQ simulations in this study (the online dust module has not been involved in CMAQ until version 5.0), as the dust emissions contribute not a small amount in $PM_{2.5}$ (12.3%), indicated by the PMF model. Here the assumed dust emissions are all from the local area, given the reasons that in 2013 only one dust event occurs on March 6–9th in Handan, which is not within our sampling period; and the urban fugitive dust emissions are relatively large particles and not so active in long-range transport as other aerosols. Therefore, the regional contributions in $PM_{2.5}$ are revised as 36.0% (40.4%/(1 + 12.3%)). It is still a considerable amount so that the regional-joint air pollution controls need to be fully considered when designing control strategies for Handan.

Although the results from the PMF model are given priority in the estimation of the local sectoral contributions, the results from the CMAQ model are still necessary in two purposes: (1) distinguishing the contributions from the regional sources and the local sources, as discussed in Sect. 3.4.1; and (2) quantifying the contributions from each local sector to the sector of "secondary aerosols" identified by the PMF model. For the purpose (2), the sectors in the CMAQ model need to be transformed to the sectors of the PMF model. Therefore, Eq. (4) is used to map the contributions from the sectors in the CMAQ simulations, i.e., PO, IN, DO, TR, and AG, to the sectors identified by the receptor model, i.e., coal combustion, industry, transportation and biomass burning:

$$S_j = \sum_{i=1}^{n} \sum_{k=1}^{m} \left(S_{i,k} \times \frac{E_{j,i,k}}{E_{i,k}} \right) \tag{4}$$

where S_j is the source contributions of the local sector j to $PM_{2.5}$ (the j sectors are those identified by the PMF model). $S_{i,k}$ is the contributions from the local sector i, which are the sectors in the simulations (PO, IN, DO, TR and AG, n = 5), to the chemical species k in $PM_{2.5}$. Those species include SO_4^{2-}, NO_3^-, NH_4^+, OC, EC, and other $PM_{2.5}$ (m = 6). $E_{j,i,k}$ are the emissions of species k or the precursors of k (SO_2 to SO_4^{2-}, NO_X to NO_3^-, NH_3 to NH_4^+) from sector j in the sector i, and $E_{i,k}$ are the total emissions of the k from the sector i. For example, the contributions of coal combustion within the IN sector to sulfate in $PM_{2.5}$ are calculated by multiplying the sulfate contributions from the IN and the fractions of SO_2 emissions by coal combustion within the IN sector to the total IN SO_2 emissions (note the IN activity includes coal combustion, other fuel combustion and industrial processes). Then the coal combustion contributions from each sector, e.g., PO, IN, DO, to each species in $PM_{2.5}$ are added together to obtain the total contributions of coal combustion. The $E_{i,k}$ and $E_{j,i,k}$ are calculated according to the MEIC inventory and other references on Hebei emissions [40, 54, 55]. Table 5 summarizes the $E_{j,i,k}/E_{i,k}$ for the sectors of PO, IN, and DO to be split into the sectors by the PMF model in Handan.

Table 5 The value of $E_{j,i,k}/E_{i,k}$ in the Eq. (4) for primary $PM_{2.5}$ and its major precursors for the sector mapping

Sectors in CMAQ	Sectors in PMF	Primary $PM_{2.5}$	SO_2	NO_X	NH_3	VOC
PO[a]	Coal combustion	100	100	100	100	100
IN	Coal combustion	12	76	51	100	0
	Other fuel and processes	88	24	49	0	100
DO	Coal combustion	49	93	52	95	23
	Biomass burning	51	7	48	5	77

[a]Power plants in Handan are all coal-fueled

3.4.2 Unity the Local Source Sectors

As for estimating the source contributions from the local sectors using the combined results from the receptor model and the chemical transport model, it firstly should be noted that there are consistencies in sector split in the two modeling system, i.e., the sectors recognized by the PMF model are coal combustion, industrial, transportation, dust/road dust, metal-smelting, biomass burning, Zn-OC-Ba, and secondary; however, those in the CMAQ model are PO (power plant), IN (industrial, including industrial coal combustion, other fuel combustion, and industrial process emissions), DO (domestic, including domestic coal combustion and other fuel combustion), TR (transportation), and AG (agriculture). In this study, the results from the PMF model are given priority on the estimation of the local sectoral contributions in Handan, for the three reasons: (1) the sectors identified by the PMF are more detailed than from the CMAQ model; (2) the local sectoral contributions estimated by the CMAQ model are highly dependent on the precision of the emission inventory, which is still far from developed for the sources in Handan; and (3) according to the guidance on the $PM_{2.5}$ source apportionment released by the MEP [53], the sectors recognized by the PMF model are acceptable.

The model performance evaluation is presented in [4, 40] in detail. In brief, the meteorological predictions are evaluated against the National Climate Data Center (NCDC) integrated surface database. The statistics include the mean bias (MB), the root mean square error (RMSE), the normalized mean bias (NMB), and the normalized mean error (NME), for the parameters of temperature at 2 m (T2), relative humidity at 2 m (RH2), wind speed at 10 m (WS10), wind direction at 10 m (WD10), and daily precipitation. The air quality observations used in model evaluation are the online observations published by the China National Environmental Monitoring Center (CNEMC). The statistics for $PM_{2.5}$ and PM_{10} are MB, NMB, NME, the mean fractional bias (MFB), and the mean fractional error (MFE). Those evaluation database, protocols, and results are explained in detailed in [4, 40]. Here only the source apportionment results were presented from the MM5-CMAQ modeling. It is concluded that the regional source contributions to

Table 6 The source contributions (%) of local sectors to each species in $PM_{2.5}$ by MM5-CMAQ after the sector mapping[a]

	SO_4^{2-}	NO_3^-	NH_4^+	OC	EC	Others	$PM_{2.5}$	In local SIA
Coal combustion	33.5	1.7	29.3	6.9	13.5	14.6	16.9	79.8
Industry	8.1	0.7	0.0	19.4	21.4	48.2	26.0	13.5
Transportation	0.7	2.9	1.3	5.9	20.1	2.1	3.7	2.4
Domestic biomass burning	0.5	1.5	0.2	23.7	10.8	6.6	7.3	4.4
Sum	42.7	6.8	30.8	55.9	65.9	71.5	54.0	100

[a]This table presents the results of Eq. (4) for $PM_{2.5}$ as well as each species in $PM_{2.5}$. The $E_{j,i,k}/E_{i,k}$ of SO_2, NO_X, and NH_3 are used for the sector mapping for the species of SO_4^{2-}, NO_3^-, and NH_4^+, respectively. The $E_{j,i,k}/E_{i,k}$ of primary $PM_{2.5}$ is used for OC, EC and Others in $PM_{2.5}$

$PM_{2.5}$ in Handan are 40.6% [40], which is a non-negligible fraction indicating that policy makers should pay more attention to the regional joint controls of air pollution in Handan. The contributions of local emissions of PO, IN, DO, TR, and AG are 0.9%, 35.9%, 13.5%, 3.7%, and 6.2% in total $PM_{2.5}$ concentrations, respectively, as 1.5%, 59.5%, 22.3%, 6.2%, and 10.3% in local contributions [40], respectively.

The local sectoral contributions from the CMAQ simulations after sector mapping are summarized in Table 6, for each species in $PM_{2.5}$, as well as in the total local SIA. Those species, i.e., sulfate, nitrate, and ammonia, are formed through the oxidation, condensation, and nucleation of the precursor gaseous pollutants including SO_2, NO_x, and NH_3.

3.4.3 Calculation of Local Sectoral Contributions

Table 7 presents the calculation process using the combined results of the PMF model and the MM5-CMAQ model. First, it is started from the results of the PMF model (see line 1 in Table 7). Then, it is assumed that the dust emissions are all from local sources, for the reasons mentioned in the above Sect. 3.4.1, and the regional contribution estimations were applied from the MM5-CMAQ model. After that, the contributions of the regional and local dust emissions are determined as 36.0 and 12.3%, as other local source contribution is 51.7% in total (step 1 in Table 7).

To estimate each local sectoral contribution (except for the "secondary aerosols"), it is assumed that their relative importance in the regional contributions and in the local contributions is similar, e.g., the total industrial contribution (17.5%) is higher than those from coal combustion (13.8%); then the local industrial contribution is higher than the local coal combustion for the same ratio, as well as those from regional.

The secondary aerosols are treated differently because they are formed in the atmosphere from gaseous precursors and need to be further identified as their spatial

Table 7 Calculation of source contributions (%) from regional and local sector emissions in Handan using the combined method of PMF and CMAQ

	Coal combustion	Industry	Metal smelting	Zn-OC-Ba	Motor	Biomass burning	Secondary	Dust	Regional
PMF model	13.8	17.5	11.0	7.6	11.2	10.7	16.0	12.3	–
Step 1[a]	51.7							12.3	36.0
Step 2[b]	8.5	10.9	6.8	4.7	6.9	6.6	7.2	12.3	36.0
% in total $PM_{2.5}^{c}$	14.3	11.8	6.8	4.7	7.3	6.8	–	12.3	36.0
% in local contributions	22.3	18.5	10.7	7.3	11.3	10.6	–	19.2	–
% in local contributions by CMAQ	25.3	38.9	–	–	5.5	10.9	–	19.2[d]	–

[a]The results after using the regional contribution calculated by CMAQ and assuming all the dusts are from local emissions
[b]The results after using the apportionment factor for secondary inorganic $PM_{2.5}$ calculated by CMAQ and supposing the relative importance of the local sectors is the same as the original PMF model results
[c]The results after attributing the local secondary $PM_{2.5}$ to each original local sector
[d]This number is following the above analysis in this table, not from the CMAQ model due to the lack of dust module

and sectoral origins to support the future controls. In Table 4 the factions to apportion the secondary aerosols to each sector are provided; therefore, firstly Table 4 can be used to estimate the local source contributions in the total secondary aerosols, that about 45.2% can be attribute to local sources, as shown in step 2 in Table 5 (45.2% × 16.0% = 7.2%). Then, the contributions are further apportioned from each sector to secondary aerosols using the data in the last column of Table 6, and obtain the results of final contributions of each sector, e.g., the contribution from coal combustion is 8.5% + 7.2% × 79.8% = 14.3%, as the line of "% in total $PM_{2.5}$" in Table 7. In addition, the contribution percentages of each sector in the total local contributions are calculated, as 22.3% from coal combustion, 10.7% from metal smelting, 7.3% from a mix source of Zn-OC-Ba, 18.5% from other industry, 11.3% from transportation, 10.6% from biomass burning, and 19.2% from dust emissions.

Table 7 also shows the estimations from the CMAQ model calculated according to Table 6 (considering the dust contribution as 19.2%), in the last line. It can be seen that its estimation on coal combustion is 25.3%, which is very close to a combined result of 22.3%, considering that the Zn-OC-Ba is a mixed source of coal combustion, construction activity, etc. The total contribution from coal combustion and industry is approximate 64.1% (25.3 + 38.9) from the CMAQ model, comparable to the combined results of 58.8% (22.3 + 18.5 + 10.7 + 7.3). The biomass contributions are also quite close from the two estimations, but the transportation contribution is estimated higher by the combined method than that from the CMAQ, which may attribute to the tendency of the CMAQ model of underestimating the transportation contributions.

It should be mentioned that the Primary Component Analysis (PCA) model has also been applied in Handan for the four months of 2013 in the study of [41]. It identified four source factors and concluded that in the local source of Handan, the source contributions from coal and biomass combustion, industrial, dust, and transportation are 29.9%, 26.8%, 18.9%, and 8.8%, respectively [41]. This estimation on dust is quite close to the above conclusion (18.9% vs. 19.2%). The contributions from motor vehicles (8.8%) are between the estimations from the combination method (11.3%) and the CMAQ model (5.5%). It did not distinguish the contributions from coal combustion and biomass burning. Therefore, if the total contributions are compared from coal combustion, industrial, and biomass burning, the PCA result is 56.7% (29.9 + 26.8), lower than the 69.4% (22.3 + 18.5 + 10.7 + 7.3 + 10.6) by the combined method, and the 75.1% (25.3 + 38.9 + 10.9) by the CMAQ model. This may be attributed to the large fraction (around 15.6%) of unidentified sources by the PCA model. Therefore, the PCA results are only presented here as a comparison in this study.

After the above comparisons, more confidence could be placed on the contribution estimations of some sectors, e.g., although the CMAQ model cannot calculate the dust contribution, the PMF model and the PCA model provide quite similar estimations (19.2% vs. 18.9%); the contributions of coal combustion and biomass burning from the combined method (22.3% and 10.6%, respectively) are quite close to those from the CMAQ model (25.3% and 10.9%, respectively); the industrial contribution, if the sum of industrial, metal smelting, and Zn-OC-Ba is taken as a rough estimation of the combined method, it is also consistent to the estimation of the CMAQ model (36.5% vs. 38.5%). At the same time, a rough estimation on the uncertainties could be achieved on the other sectors, e.g., the transportation contribution might be overestimated by the combination method (11.3%) compared to the PCA model (8.8%) and the CMAQ model (5.5%).

4 Conclusions

In this study, a combined method using a receptor model, PMF, and a chemical transport model, MM5-CMAQ, is applied on source apportionment of $PM_{2.5}$ in Handan, Hebei, China. First, $PM_{2.5}$ samples were collected daily in Handan in the four months in 2013, i.e., January, April, July, and October, and their chemical compositions were analyzed including water soluble ions, OC and EC, and trace elements. Those composition data were then used in the PMF model to identify the $PM_{2.5}$ sources and their contributions in Handan. At the same time, regional-scale air quality simulations using the MM5-CMAQ modeling system were pursued over North China to quantify the spatial and sectoral contributions to $PM_{2.5}$ in Handan using the Brute-Force method. Finally, the two sets of results were combined to better answer the questions on $PM_{2.5}$ sources in Handan to support the policy making of the local government, given the reasons that the receptor model cannot easily distinguish the spatial origins of each source sector under the present status, and the chemical transport model may not be able to give as detailed sectoral apportionment as the receptor model at present, due to the incompleteness and uncertainties in the emission inventory for Hebei.

It is found that the annual average concentration of $PM_{2.5}$ in Handan in 2013 is 177.8 $\mu g\ m^{-3}$, as SO_4^{2-}, NO_3^-, NH_4^+, OC, and EC are major chemical compositions with the percentages of 18.0%, 14.0%, 8.7%, 11.8%, and 4.1%, respectively. The source apportionment of $PM_{2.5}$ by the PMF model are: 13.8% from coal combustion, 17.5% from industry, 11.0% from metal smelting, 7.6% from a mixed source of Zn-OC-Ba, 11.2% from motor vehicles, 10.7% from biomass burning, 12.3% from dust, and 16.0% of secondary sources. The MM5-CMAQ model indicates that the regional source contributions to $PM_{2.5}$ in Handan is 40.6%, and within the local contributions, PO, IN, DO, TR, and AG contribute 1.5%, 59.5%, 22.3%, 6.2%, and 10.3%, respectively (note dust emissions are not included that will result in an overestimation of the regional contributions).

The combination of the two results shows that the regional source contributions are 36.0% in total $PM_{2.5}$ concentrations, and in the local sources; the contributions from each sector are: 22.3% from coal combustion, 10.7% from metal smelting, 7.3% from a mix source of Zn-OC-Ba, 18.5% from other industry, 11.3% from transportation, 10.6% from biomass burning, and 19.2% from dust emissions. The regional joint air pollution controls should be emphasized in future control strategies, and the local source controls on coal combustion and industries, especially the iron and steel industry, are the key points to mitigate the $PM_{2.5}$ pollution in Handan. Although some limitations exist in this paper, e.g., the dust emissions are not included in MM5-CMAQ modeling, it still provides a method to do $PM_{2.5}$ source apportionment in the cities like Handan, where it still lacks of both a bottom-up emission inventory with high precision and long-term $PM_{2.5}$ chemical observation data so that neither receptor model nor chemical transport model can singly produce source apportionment results reliable and detailed enough to support the policy making of the local government.

Acknowledgements This study was sponsored by the National Natural Science Foundation of China (No. 41475131), Hebei Science Fund of Distinguished Young Scholars (No. D2017402086), the Program for the Outstanding Young Scholars of Hebei Province, the Hebei Support Program of Hundred Outstanding Innovative Talents from Universities (SLRC2017025), the Hebei Support Program of Hundred Outstanding Innovative Talents from Universities (SLRC2017025), Hebei Cultivating Project of Talent Development (A2016002022), the Innovation Team Leader Talent Cultivation Fund of Hebei University of Engineering.

Conflicts of Interest The authors declare no conflict of interest.

References

1. MEP: 2013 report on the state of the environment in China. MEP, Beijing, China (2014). http://jcs.mep.gov.cn/hjzl/zkgb/2013zkgb/201406/t20140605_276521.htm. Accessed 20 Aug 2015. (in Chinese)
2. MEP: 2014 report on the state of the environment in China. MEP, Beijing, China (2015). http://jcs.mep.gov.cn/hjzl/zkgb/2014zkgb/201506/t20150608_303142.htm. Accessed 20 Aug 2015. (in Chinese)
3. Handan, B.S.: (Handan Bureau of Statistics): Handan Statistical Yearbook 2013. China Statistical Press, Beijing, China (2014)
4. Wang, L.T., Wei, Z., Yang, J., et al.: The 2013 severe haze over southern Hebei, China: model evaluation, source apportionment, and policy implications. Atmos. Chem. Phys. **14**, 3151–3173 (2014). https://doi.org/10.5194/acp-14-3151-2014
5. Tang, X.Y., Zhang, Y.H., Shao, M.: Atmospheric Chemistry. Higher Education Press, Beijing, China (2006)
6. Zhang, Y.J., Zheng, M., Cai, J., et al.: Comparison and overview of $PM_{2.5}$ source apportionment methods. China Sci. Bull. **60**, 109–121 (2015). https://doi.org/10.1360/N972014-00975. (in Chinese)
7. Chueinta, W., Hopke, P.K., Paatero, P.: Investigation of sources of atmospheric aerosol at urban and suburban residential areas in Thailand by positive matrix factorization. Atmos. Environ. **34**, 3319–3329 (2000). https://doi.org/10.1016/S1352-2310(99)00433-1
8. Lee, E., Chan, C.K., Paatero, P.: Application of positive matrix factorization in source apportionment of particulate pollutants in Hong Kong. Atmos. Environ. **33**, 3201–3212 (1999). https://doi.org/10.1016/S1352-2310(99)00113-2
9. Yang, L.X., Cheng, S.H., Wang, X.F., et al.: Source identification and health impact of $PM_{2.5}$ in a heavily polluted urban atmosphere in China. Atmos. Environ. **75**, 265–269 (2013). https://doi.org/10.1016/j.atmosenv.2013.04.058
10. Belis, C.A., Karagulian, F., Larsen, B.R., et al.: Critical review and meta-analysis of ambient particulate matter source apportionment using receptor models in Europe. Atmos. Environ. **69**, 94–108 (2013). https://doi.org/10.1016/j.atmosenv.2012.11.009

11. Viana, M., Pandolfi, M., Minguillón, M.C., et al.: Inter-comparison of receptor models for PM source apportionment: case study in an industrial area. Atmos. Environ. **42**, 3820–3832 (2008). https://doi.org/10.1016/j.atmosenv.2007.12.056
12. Hopke, P.K., Ito, K., Mar, T., et al.: PM source apportionment and health effects: 1. intercomparison of source apportionment results. J. Expo. Sci. Environ. Epidemiol. **16**, 275–286 (2006). https://doi.org/10.1038/sj.jea.7500458
13. Contini, D., Belosi, F., Gambaro, A., et al.: Comparison of PM10 concentrations and metal content in three different sites of the Venice Lagoon: an analysis of possible aerosol sources. J. Environ. Sci. **24**(11), 1954–1965 (2012). https://doi.org/10.1016/S1001-0742(11)61027-9
14. Contini, D., Cesari, D., Donateo, A., et al.: Characterization of PM10 and PM2.5 and their metals content in different typologies of sites in South-Eastern Italy. Atmosphere **5**, 435–453 (2014). https://doi.org/10.3390/atmos5020435
15. Chen, D.S., Cheng, S.Y., Liu, L., et al.: An integrated MM5-CMAQ modeling approach for assessing trans-boundary PM10 contribution to the host city of 2008 Olympic summer games-Beijing. China Atmos. Environ. **41**, 1237–1250 (2007). https://doi.org/10.1016/j.atmosenv.2006.09.057
16. Streets, D.G., Fu, J.S., Jang, C.J., et al.: Air quality during the 2008 Beijing Olympic games. Atmos. Environ. **41**, 480–492 (2007). https://doi.org/10.1016/j.atmosenv.2006.08.046
17. Fu, J.S., Streets, D.G., Jang, C.J., et al.: Modeling regional/urban ozone and particulate matter in Beijing, China. J. Air Waste Manag. Assoc. **59**, 37–44 (2009). https://doi.org/10.3155/1047-3289.59.1.37
18. Wang, L.T., Hao, J.M., He, K.B., et al.: A modeling study of coarse particulate matter pollution in Beijing: regional source contributions and control implications for the 2008 Summer Olympics. J. Air Waste Manag. Assoc. **58**, 1057–1069 (2008). https://doi.org/10.3155/1047-3289.58.8.1057
19. Xing, J., Zhang, Y., Wang, S.X., et al.: Modeling study on the air quality impacts from emission reductions and a typical meteorological conditions during the 2008 Beijing Olympics. Atmos. Environ. **45**, 1786–1798 (2011). https://doi.org/10.1016/j.atmosenv.2011.01.025
20. Zhou, Y., Fu, J.S., Zhuang, G.S., et al.: Risk-based prioritization among air pollution control strategies in the Yangtze River delta. China. Environ. Health Perspec. **118**(9), 1204–1210 (2010). https://doi.org/10.1289/ehp.1001991
21. Meng, C.C., Wang, L.T., Su, J., et al.: Chemical compositions and source apportionment of PM2.5 in Handan City, Hebei Province. Environ. Sci. Technol. **39**(2), 57–64 (2016). https://doi.org/10.3969/j.issn.1003-6504.2016.02.010. (in Chinese)
22. Wei, Z., Wang, L.T., Chen, M.Z., et al.: The 2013 severe haze over the Southern Hebei, China: PM2.5 composition and source apportionment. Atmos. Pollut. Res. **5**, 759–768 (2014). https://doi.org/10.5094/APR.2014.085
23. Bove, M.C., Brotto, P., Cassola, F., et al.: An integrated PM2.5 source apportionment study: positive matrix factorisation vs. the chemical transport model CAMx. Atmos. Environ. **94**, 274–286 (2014). https://doi.org/10.1016/j.atmosenv.2014.05.039
24. Pekney, N.J., Davidson, C.I., Robinson, A., et al.: Major source categories for PM2.5 in Pittsburgh using PMF and UNMIX. Aerosol Sci. Technol. **40**, 910–924 (2006). https://doi.org/10.1080/02786820500380271
25. Paatero, P.: Least squares formulation of robust non-negative factor analysis. Chemometr. Intell. Lab. Syst. **37**, 23–35 (1997). https://doi.org/10.1016/S0169-7439(96)00044-5
26. Paatero, P., Tapper, U.: Positive matrix factorization: a non-negative factor model with optimal utilization of error estimates of data values. Environments **5**, 111–126 (1994). https://doi.org/10.1002/env.3170050203
27. Saba, T., Su, S.: Tracking polychlorinated biphenyls (PCBs) congener patterns in Newark Bay surface sediment using principal component analysis (PCA) and positive matrix factorization (PMF). J. Hazard. Mater. **260**, 634–643 (2013). https://doi.org/10.1016/j.jhazmat.2013.05.046

28. Baumann, K., Jayanty, R.K.M., Flanagan, J.B.: Fine particulate matter source apportionment for the chemical speciation trends network site at Birmingham, Alabama, using positive matrix factorization. J. Air Waste Manag. Assoc. **58**, 27–44 (2008). https://doi.org/10.3155/1047-3289.58.1.27
29. Brown, S.G., Frankel, A., Raffuse, S.M.: Source apportionment of fine particulate matter in phoenix AZ, using positive matrix factorization. J. Air Waste Manag. Assoc. **57**, 741–752 (2007). https://doi.org/10.3155/1047-3289.57.6.741
30. Huang, S., Rahn, K.A., Arimoto, R.: Testing and optimizing two factor-analysis techniques on aerosol at Narragansett. Rhode Island. Atmos. Environ. **33**, 2169–2185 (1999). https://doi.org/10.1016/S1352-2310(98)00324-0
31. Pekey, H., Pekey, B., Arslanbas, D.: Source apportionment of personal exposure to fine particulate matter and volatile organic compounds using positive matrix factorization. Water Air Soil Pollut. **224**, 1403 (2013). https://doi.org/10.1007/s11270-012-1403-2
32. U.S. EPA (U.S. Environmental Protection Agency): EPA positive matrix factorization (PMF) 3.0 fundamentals & user Guide. U.S. EPA, Research Triangle Park, NC, USA (2008)
33. Zhang, J.G., Wang, Y.S., Wang, S., et al.: Source apportionment of non-methane hydrocarbon in Beijing using positive matrix factorization. Environ. Sci. Technol. **32**(5), 35–39 (2009). (in Chinese)
34. Viana, M., Chi, X., Maenhaut, W., et al.: Influence of sampling artefacts on measured PM, OC, and EC Levels in carbonaceous aerosols in an urban area. Aerosol Sci. Technol. **40**(2), 107–117 (2007). https://doi.org/10.1080/02786820500484388
35. Yu, L.D., Wang, G.F., Zhang, R.J. et al.: Characterization and source apportionment of PM2.5 in an urban environment in Beijing. Aerosol Air Qual. Res. **13**, 574–583 (2013). https://doi.org/10.4209/aaqr.2012.07.0192
36. Brown, S.G., Eberly, S., Paatero, P., et al.: Methods for estimating uncertainty in PMF solutions: examples with ambient air and water quality data and guidance on reporting PMF results. Sci. Total Environ. **518–519**, 626–635 (2015). https://doi.org/10.1016/j.scitotenv.2015.01.022
37. U.S. EPA: EPA positive matrix factorization (PMF) 5.0 fundamentals and user guide (2014)
38. NCAR: PSU/NCAR mesoscale modeling system tutorial class notes and user's guide: MM5 modeling system version 3. National Center for Atmospheric Research, USA (2003)
39. Dunker, A.M., Morris, R.E., Pollack, A.K., et al.: Photochemical modeling of the impact of fuels and vehicles on urban ozone using auto oil program data. Environ. Sci. Technol. **30**, 787–801 (1996). https://doi.org/10.1021/es950175m
40. Wang, L.T., Wei, Z., Wei, W., et al.: Source apportionment of PM2.5 in top polluted cities in Hebei, China using the CMAQ model. Atmos. Environ. **122**, 723–736 (2015). https://doi.org/10.1016/j.atmosenv.2015.10.041
41. HEBEU: Report on PM2.5 source apportionment in Handan. HEBEU, Handan, Hebei, China (2015)
42. Kain, J.S., Fritsch, J.M.: Convective parameterization for mesoscale models: the Kain-Fritsch scheme. In: Emanuel, K.A., Raymond, D.J. (eds.) The Representation of Cumulus Convection in Numerical Models, p. 246. American Meteorological Society (1993). https://doi.org/10.1007/978-1-935704-13-3_16
43. Zhang, D.L., Anthes, R.A.: A high-resolution model of the planetary boundary layer-sensitive tests and comparisons with SESAME-79 data. J. Appl. Meteorology **21**, 1594–1609 (1982). https://doi.org/10.1175/1520-0450(1982)021<1594:ahrmot>2.0.co;2
44. Reisner, J., Rasmussen, R.J., Bruintjes, R.T.: Explicit forecasting of supercooled liquid water in winter storms using the MM5 mesoscale model. Q. J. Royal Meteorological Soc. **124B**, 1071–1107 (1998). https://doi.org/10.1002/qj.49712454804
45. Dudhia, J.A.: A non-hydrostatic version of the Penn State/NCAR mesoscale model: validation tests and simulation of an Atlantic cyclone and cold front. Monthly Weather Rev. **121**, 1493–1513 (1993). https://doi.org/10.1175/1520-0493(1993)121<1493:anvotp>2.0.co;2

46. Blackadar, A.K.: Modeling the Nocturnal Boundary Layer. In: Proceedings of the Third Symposium on Atmospheric Turbulence, Diffusion and Air Quality, American Meteorological Society, Boston, MA, pp. 46–49 (1976)
47. Deardorff, J.W.: Efficient prediction of ground surface temperature and moisture, with inclusion of a layer of vegetation. J. Geophys. Res. **83**, 1889–1903 (1978). https://doi.org/10.1029/JC083iC04p01889
48. Carter, W.P.L.: A detailed mechanism for the gas-phase atmospheric reactions of organic compounds. Atmos. Environ. **24**, 481–518 (1990). https://doi.org/10.1016/0960-1686(90)90005-8
49. Carter, W.P.L.: Implementation of the SAPRC-99 chemical mechanism into the models-3 framework. Report to the U.S. EPA, prepared by Cater, W.P.L., Statewide Air Pollution Research Center, University of California, Riverside, CA (2000)
50. Binkowski, F.S., Shankar, U.: The regional particulate model. 1. Model description and preliminary results. J. Geophys. Res. **100**, 26191–26209 (1995). https://doi.org/10.1029/95JD02093
51. Chang, J.S., Brost, R.A., Isaksen, I.S.A., et. al.: A three-dimensional acid deposition model: physical concepts and formulation. J. Geophys. Res. **92**, 14681–14700 (1987). https://doi.org/10.1029/jd092id12p14681
52. Walcek, C.J., Taylor, G.R.: A theoretical method for computing vertical distributions of acidity and sulfate production within cumulus clouds. J. Atmos. Sci. **43**, 339–355 (1986). https://doi.org/10.1175/1520-0469(1986)043<0339:atmfcv>2.0.co;2
53. MEP: Guidance on the report of source apportionment of atmospheric aerosols. MEP, Beijing, China (2015)
54. Cao, G.L., Zhang, X.Y., Gong, S.L., et al.: Emission inventories of primary particles and pollutant gases for China. Chin. Sci. Bull. **56** (2011). https://doi.org/10.1007/s11434-011-4373-7. (in Chinese)
55. Guan, D.B., Liu, Z.: The truth of haze: PM2.5 pollution source analysis and reduction strategy research. China Environmental Press, Beijing, China (2014)

The Review of River Health Assessment in China

Cun Liu, Shi-Lu Zhang, Wen-Yan Cui, Chao Lin and Jun Zhang

Abstract This paper makes a systematic overview of the development process of evaluation on China river health growing out of nothing and summarizes the explanations of the river health concept made by famous experts and scholars in China to obtain the essence of river health evaluation. This paper compares the domestic typical evaluation methods and summarizes the common advantages and disadvantages of these methods. Based on the above-mentioned summary, this paper concludes that the research direction of health evaluation is the urban river health evaluation, biological habitat evaluation, research on the river ecological base flow and evaluation on the influence of inter-basin water transfer. Finally, this paper gives a brief prospect of China's river health evaluation work in the future.

Keywords River health · Evaluation method · Evaluation index
Research direction

1 Introduction

China is a country with extremely scarce water resources per capita and the total length of rivers in China is more than 400,000 km. These rivers play an important role in developing the national economy and guaranteeing people's livelihood. However, due to the rapid growth in the urban population in China and quick

C. Liu
College of Water Conservancy and Hydropower, Hebei University of Engineering,
Handan 056000, China

C. Liu · S.-L. Zhang · W.-Y. Cui · J. Zhang (✉)
Ministry of Water Resources, Haihe River Water Conservancy Commission,
Tianjin 300170, China
e-mail: zhangjun506@163.com

C. Lin
Zhanghe Uptream Management Bureau, Haihe River Water Conservancy Commission,
Handan 056006, China

W. Dong et al. (eds.), *Sustainable Development of Water Resources and Hydraulic Engineering in China*, Environmental Earth Sciences,
https://doi.org/10.1007/978-3-319-61630-8_14

development of industrial and agricultural production, the demands on water resources have increased sharply, leading to a common blind excessive development phenomenon. Due to the relatively scarce river protection consciousness during the preliminary development period, the ecological river water is occupied, and many rivers show the functional degradation trend. Frequent occurrence of “water problems” such as the reduction of river water, water pollution, river blockage, sharp decline in biodiversity and water and soil loss, etc. have caused serious influence to the people’s normal life and the sustainable development of the social economy. River health issues are gradually brought to the attention of the Chinese river management, while river health assessment has become a main content of river ecosystem restoration and river management and also has gradually become a hot issue in domestic related research.

2 River Health Evaluation

2.1 Domestic Research Process

River health evaluation in China started relatively late, Yang et al. [1], using the Ephemeroptera, Trichoptera, Plecoptera (EPT) and the Family Biotic Index (FBI) to assess Jiuhua River of Anhui province, is the beginning of the domestic research in 1992; however, the assessment did not get much attention. Until Tang et al. [2] wrote an article entitled “The river ecosystem health and its evaluation” published by the Chinese Journal of Applied Ecology in 2002, which advanced domestic research of river health assessment. After that, many experts and scholars such as Liu and Liu [3], Dong [4], Geng et al. [5], Wu et al. [6], etc. had made great efforts to produce targeted research on the concept and connotation of river health, evaluation method and the construction of an index system. Zhao and Yang [7] and Wang et al. [8] have evaluated and researched the urban river health status according to the acceleration of China’s urbanization process and the present situation of the rapid expansion of urban scale. China’s three major river basin administrative agencies also successively put forward the concepts of “Healthy Yellow River” [9], “Healthy Yangtze River” [10] and “Green Pearl River” [11] according to the actual situation of the water basins. With the expansion of the research scope, Zhang et al. [12] and Wang et al. [13] also successively evaluated and researched some small rivers, such as the rivers in Shenzhen city, East River in the Three Gorges Reservoir Region, etc. In recent years, domestic researchers, combined with river health evaluation with the management science, statistics, etc., led to the occurrence of river health evaluation based on the projection pursuit—extension set theory [14], prediction—comprehensive index evaluation model [15] and river health evaluation based on the entropy weight comprehensive health index method [16]; therefore, the thought and scope of research was gradually expanded.

On the whole, this concept was only regarded as a kind of river management and development tool when it was just proposed, and then gradually turned into an indicator to measure the river ecosystem conditions and social service function to implement the river sustainable management policy in a targeted manner.

2.2 *Concept of River Health*

Chinese experts and scholars pay more attention to its vitality and functional health and emphasize the harmony between humans and nature, compared with the understanding of foreign researchers [17, 18]. Dong [19] suggested to regard the "sustainable utilization of ecological good river" as the interpretation of river health. The academicians Liu and Liu [3] believe that "healthy rivers are the rivers with balanced social function and nature function in a certain period". Wu [20] believes that "river health refers to that the river shall not only have a complete ecological structure, but also provides reasonable social service functions". The academicians Wen et al. [21] give full consideration to the structural stability and social service of rivers and emphasize the balance between development and protection, which embodies the important idea of sustainable development and human-water harmony. At present, the domestically universally recognized definition is: river health is the unity of ecological value and human service value, and health not only means the integrity in the sense of ecology, but also emphasizes the normal exertion of river ecosystem service [7]. This has similarities with Meyer's [22] explanation that: in addition to maintaining the ecosystem structure and function, river health shall also include the social value of the ecosystem and cover the ecological integrity and service value to "mankind in the health concept".

2.3 *Essence of River Health Evaluation*

In accordance with the research process at home, the essence of river health evaluation is to establish a complete evaluation system by using the scientific tools and methods and to research the ecosystem status of rivers and the change trend of rivers' service value to the human society under the dual drive function of natural factors and disturbance of human activities. The evaluation results can not only intuitively reflect the river ecological status and the potential service value to human society to provide the basis of river development and protection for related departments, but also can realize the transparency of information, so as to cause the public to understand the current situation of rivers and reflect the fusion of science and life.

3 Domestic Evaluation Methods

Because of the limitation of actual conditions, the emphasis of domestic work lies in the water quality improvement and water environment recovery. River management from the angle of river health was started in recent years, and the research of river health evaluation is still in the beginning stage. The current evaluation methods are to establish the complete evaluation system based on related standards and to determine the weight and score of various indicators of rivers such as the hydrology and water resources, biological and physical structure, chemical characteristics, water quality, social service function, etc. The river health index (RHI) would be calculated through weighting and then the health degree of rivers would be determined according to the value of RHI.

On the basis of drawing lessons from foreign advanced technology, China's three major river basin administrative agencies [9–11] and some researchers [3, 5, 7, 23] have initially established the health evaluation system for some rivers by fully combining the current domestic river situation, insisting on the "human-water harmony" concept and giving overall consideration to the ecosystem and social service function of rivers. See Table 1.

Speaking from the point of view of evaluation principle, the evaluation methods included in Table 1 belong to the multiple-indicators comprehensive evaluation method which is the domestic common research focus. This method has many reference factors compared with the model prediction method; and it not only considers the ecological status of rivers, but also pays attention to the social service function of rivers, which can comprehensively, accurately and objectively reflect the river health status and has the characteristics of multiple indicators, multiple spaces and multiple time scales; therefore, it is the development direction of the river health evaluation. Through comparison of the evaluation indicators in Table 1, it can be seen that different methods basically include the evaluations in aspects of organisms, river form and water quality, etc., while different evaluations may have different focus points of the research area and also have different ideas about the classification of river health class. However, from the point of view of current researches, there are certain defects in the multiple-indicators comprehensive evaluation method, such as the weight of indicators and score excessively relying on the experts' opinions, non-uniform evaluation criteria, deficient evaluation precision, evaluation result unable to reflect the single index status and covering of poor index, etc. All of these problems need to be solved.

4 Research Direction

China is very vast, the difference in geographical conditions, ecosystem, economy, and society is very obvious between different river basins, different rivers, even upstream and downstream of the same river. Meanwhile, the strength of human

Table 1 Comparison of domestic river health evaluation methods

Main proposer	Evaluation region	Evaluation index and method
Yangtze River water resources commission	Yangtze River	Establish the 4-level evaluation system consisting of general target layer, systematic layer, status layer and element layer, including 18 evaluation indicators such as the water and soil resources and water environment condition, river integrity and stability, aquatic biodiversity, flood storage capacity, service ability, etc. [9]
Yellow River water resources commission	Yellow River	Including 9 indicators, such as low-limit flow, bank full discharge, flood carrying capacity of the channel, water quality classification of the water quality category, wetland size, aquatic ecosystems, the supply amount of the Yellow River to human beings, etc. [10]
Pearl River water resources commission	Pearl River	The evaluation system consists of the general layer, systematic layer, status layer and index layer, including 14 indicators such as the river bank, water quantity, water quality, flood prevention, power generation, hydrology, etc. [11]
Chang-ming Liu	Yellow River	Use the water sand channel, water quality, river bed form, river ecology and river base flow as the indicative factors [3]
Lei-hua Geng	Lancang River basin	Cover the functions in aspects of social service of rivers, environment, flood prevention, development, utilization and ecology, and include 25 indicators such as the urban water supply probability, irrigation probability, water and soil loss ratio, change rate of flood carrying capacity, water resources utilization rate, natural vegetation rate, etc. [5]
Nan Zhang	Liao River	Use the 10 indexes (such as the BOD_5, DO, total nitrogen, integrity of benthonic animal) reflecting the habitat quality, aquatic organism and physical - chemical characteristics of the water body as the evaluation indicators [7]
Yan-wei Zhao	Trunk stream of Yong River, Yan River, Xi River and Fuxi river in Ningbo	Include 5 basic elements (water quantity, water quality, aquatic organism, physical structure and riparian zone) and 17 standards such as the development and utilization rate, the index of water quality pollution, fish biological integrity, etc. [23]

disturbance to rivers is also not identical. When an evaluation of a specific actual river is made, the target of management and assessment could be a big difference, so the research of river health assessment presents different trends. However, the basic directions include three areas: urban river health evaluation, evaluation on the habitat of living things, evaluation on the influence of inter-basin water transfer.

4.1 Urban River Health Evaluation

As the key carrier of resources and the environment in the process of urban formation and development, a river is an important factor affecting urban style and beautifying the urban environment; it relates to the urban survival and restricts the urban development [24]. Along with the constant advancement of China's urbanization process, more and more populations are concentrated in cities and the urban scale expands unceasingly, leading to the continuous increase of industrial water demands and the sewage produced. Generally speaking, an urban river is the important local water supply source, and the water quantity and quality of rivers directly influence the living of local residents and social economic development. Therefore, it is of practical significance to strengthen the urban river health evaluation and provide the theoretical basis for the future development of cities.

4.2 Evaluation on the Habitat of Living Things

River habitat is an important part of the river ecological system and is also the combination of physical, chemical and biotic conditions required to provide living environment for the river organisms. It relates to the biological food chain and energy flow process, and directly reflects the river health status [25, 26]. With the expansion of the scope of human activities and strengthening of human activities in recent years, the area of habitat has shrunk and fragmented gradually; and the biological diversity of habitat continues to decline and even be extinct. Enhancement of evaluation on the quality of the biological habitat will be favorable to reasonably controlling the utilization of water resources and also an important step to realize the ecological restoration of rivers and implement the ecological civilization construction of rivers.

4.3 Evaluation on the Influence of Inter-basin Water Transfer

At present, problems such as uneven spatial and temporal distribution of water resources, regional water resources shortage, etc. exist in China, especially in

Beijing, Tianjin and Hebei in north China that are facing the severe water shortage problem. The inter-basin water transfer project is one of the important means to solve the regional water problem and enhance the optimum configuration capacity of water resources. With the successive construction of the water diversion project from Yellow River to Shanxi province, the water diversion project from the Yellow River into Qingdao, Water Diversion Project from Luanhe River to Tianjin City, east route and central route of the South-to-North Water Diversion Project and Hanjiang-to-Weihe River Water Transfers Project, etc., the practice has proved that inter-basin water transfer projects will influence the ecological environment of the water transfer region, water storage region, area along the water transfer line and water-receiving area and will also change the original ecological environmental system [27]. Strengthening the evaluation on the water quality, water quantity and biological diversity, etc. in the water source area, water-receiving area and areas along the line of the water transfer projects are also an important part of the river health evaluation. Evaluation and timely adjustment of the water transfer scale and time can form a virtuous cycle of water resources in related areas and better promote the development of the local economy and society.

5 Prospect of River Health Evaluation

China is a vast area, so the natural geographical conditions and economic and social development situation between different river basins will be obviously different and show a typical regional characteristic. Different river basins also suffer different strengths of disturbance of human activities and show different development trends. In future work, the connotation of river health should be constantly enriched, the appropriate evaluation indexes selected based on the river characteristics, and the appropriate weights determined to establish a comprehensive evaluation system, so as to make objective and accurate evaluation of the river health status. Related river management departments should establish a more comprehensive monitoring system and strengthen the collection of basic data to provide more information for river health evaluation. Meanwhile, they should also develop the correct solutions, implement the development, protection and ecological recovery properly and ensure that the needs of human social development are met under the premise of sustainable utilization of rivers.

Acknowledgements The researchers would like to extend their thanks to The National Key Research and Development Program of China (No. 2016YFC0401401-6) & Hebei Provincial Research Center of Water Ecological Civilization &Social Governance.

References

1. Yang, L.-F., Li, Y.-W., Qi, D.-G., et al.: The assessment of aquatic insect community structure and biological water quality in Jiuhua River. Acta Ecol. Sinica **12**(1), 8–15 (1992). https://doi.org/10.3321/j.issn:1000-0933.1992.01.010
2. Tang, T., Cai, Q.-H., Liu J.-K.: River ecosystem health and its assessment. Chin. J. Appl. Ecol. **13**(9), 1191–1194 (2002). https://doi.org/10.3321/j.issn:1001-9332.2002.09.031
3. Liu, C.-M., Liu, X.-Y.: Healthy river: essence and indicators. Acta Geograph. Sinica **63**(7), 683–692 (2008). https://doi.org/10.3321/j.issn:0375-5444.2008.07.002
4. Dong, Z.-R.: River health connotation. China Water Resour. **04**, 15–18 (2005). https://doi.org/10.3969/j.issn.1006-0081.2007.11.004
5. Geng, L.-H., Liu, H., Zhong, H.-P., et al.: Indicators and criteria for evaluation of healthy rivers. J. Hydraul. Eng. **37**, 253–258 (2006). https://doi.org/10.3321/j.issn:0559-9350.2006.03.001
6. Wu, A.-N., Che, Y., Xu, Q.-F., et al.: Approach for health assessment of rivers in Shanghai. J. Ecol. Rural Environ. **23**, 90–94 (2007). https://doi.org/10.3969/j.issn.1673-4831.2007.04.021
7. Zhao, Y.-W., Yang, Z.-F.: Preliminary study on assessment of urban river ecosystem health. Adv. Water Sci. **03**, 349–355 (2005). https://doi.org/10.3321/j.issn:1001-6791.2005.03.007
8. Wang, L., Gong, Z.-G., Zhang, J., et al.: Comprehensive index system for evaluation of river ecosystem health. China Water Wastewater **23**, 97–100 (2007). https://doi.org/10.3321/j.issn:1000-4602.2007.10.028
9. Li, G.-Y.: Keeping healthy life of the Yellow River-an ultimate aim of taming the Yellow River. Yellow River **26**, 1–3 (2004). https://doi.org/10.3969/j.issn.1000-1379.2004.01.001
10. Cai, Q.-H.: Maintain health of the Yangtze River, and promote the human-water harmony. Yangtze River **36**, 1–3 (2005). https://doi.org/10.3969/j.issn.1001-4179.2005.03.001
11. Yue, Z.-M.: Maintenance of the Pearl River health life, building a green Pearl River, and strive to build a harmonious society. Pearl River **2**, 1–5 (2005). https://doi.org/10.3969/j.issn.1001-9235.2005.03.001
12. Zhang, Y., Zheng, B.-H., Liu, H.-L., et al.: Indicators and evaluation of ecosystem health of typical rivers in Shenzhen City. Water Resour. Prot. **05**, 13–17, 52 (2006). https://doi.org/10.3969/j.issn.1004-6933.2006.05.004
13. Wang, Q., Yuan, X.-Z., Liu, H., et al.: Stream habitat assessment of Dong River, China, using river habitat survey method. Acta Ecol. Sinica **06**, 1548–1558 (2014). https://doi.org/10.5846/stxb201210201458
14. Wang, W., Xu, X., Dong, Z., et al.: Assessment of river health based on projection pursuit-extension set theory. J. Water Resour. Water Eng. **02**, 122–127 (2016). https://doi.org/10.11705/j.issn.1672-643X.2016.02.23
15. Xu, X., Chen, Q.-S., Dong, Z., et al.: Prediction-comprehensive index assessment model in the evaluation of river health. China Rural Water Hydropower, **05**, 23–26, 32 (2016). https://doi.org/10.3969/j.issn.1007-2284.2016.05.006
16. Zhao, J.-H., Shen, G.-H., Qin, W.: Research on Yihe River evaluation based on entropy weight comprehensive health index method. Water Resour. Dev. Manag. **01**, 49–52 (2016). https://doi.org/10.16616/j.cnki.10-1326/tv.2016.01.014
17. Simpson, J., Norris, R., Barmuta, L., et al.: Aus Riv AS National River Health Program [R] (1999)
18. Karr, J.R.: Ecological integrity and ecological health are not the same. In: Schulze, P. (ed.) Engineering Within Ecological Constraints. National Academy of Engineering, National Academy Press, Washington, DC (1995)
19. Dong, Z.-R.: Interpretation of river health. Express Water Resour. Hydropower Inf. **11**, 17–19 (2007). https://doi.org/10.3969/j.issn.1006-0081.2007.11.004
20. Wu, A.-N.: River health assessment: theory, methods and practice. Shanghai: Doctoral Dissertation of Shanghai Normal University (2008). https://doi.org/10.7666/d.y1373063

21. Wen, F.-B., Han, Q.-W., Xu, J.-X., et al.: The definition and connotation of river health. Adv. Water Sci. **01**, 140–150 (2007). https://doi.org/10.3321/j.issn:1001-6791.2007.01.023
22. Meyer, J.L.: Stream health: incorporating the human dimension to advance stream ecology. J. N. Am. Benthol. Soc. **16**, 439–447 (1997). https://doi.org/10.2307/1468029
23. Zhang, N., Meng, W., Zhang, Y., et al.: Multi-variable assessment of river ecosystem health in Liao River basin. Res. Environ. Sci. **02**, 162–170 (2009). https://doi.org/10.13198/j.res.2009.02.40.zhangn.010
24. Liu, X.-T.: Discussion on problems of city river harnessing. Planners **17**, 66–69 (2001). https://doi.org/10.3969/j.issn.1006-0022.2001.06.015
25. Yang, Y., Yan, Z.-M., Qiao, Y.: Description and review of hydraulic conditions of fish habitats. J. HOHAI Univ. (Nat. Sci.), **35**, 125–130 (2007). https://doi.org/10.3321/j.issn:1000-1980.2007.02.002
26. Shi, R.-H., Xu, S.-G.: Methods for river habitat survey and evaluation. Chin. J. Appl. Ecol. **19**, 2081–2086 (2008)
27. Li, R., Zhao, M., Chang, Y.-M.: Analysis on the influence definition and factor of trans-valley moving water on region ecological environment. Ecol. Econ. **2**, 155–157 (2009)

Dissolution Experiment of the Halite of Palaeogene Shahejie Formation, Shulu Sag, Eastern China

Langtao Liu, Chao Jin, Shuantong Shangguan, Yang Li, Chenyang Xu and Lei Li

Abstract Dissolution mining is the main production mode for well salt. Understanding the dissolution property of halite is very important controlling mining efficiency and to predict shapes of the salt cavern. The dissolution experiment of the halite of Paleogene Shahejie Formation, Shulu Sag, and Jizhong Depression is conducted in this study. The results show that, along with the rise of baume degrees, the lateral dissolution rates decline gradually from 7–8 to 1–2 mm/h; and the lateral dissolution angles increase gradually from 9° to 34°. The contents of NaCl of the brine increases linearly, from 50 to 320 g/L, while the contents of the $CaSO_4$ stay around baseline. More than 77% of the dissolution residues are smaller than 2 mm and the sinking rates of these fine residues are very little, from 2.8 to 2.0 cm/s. These results indicate in the early stage of the mining that fresh water should be injected to accelerate the forming of the caverns; and in the late stage of the mining, brine of low concentration should be injected, which will help to maintain the high dissolution rates and discharge the residues in a timely manner. This will guarantee the long-term continuous production of the salt mine.

Keywords Shulu Sag · Paleogene · Shahejie Formation · Halite
Dissolution experiment

L. Liu · C. Jin (✉)
Collaborative Innovation Center of Coal Exploitation, Hebei University of Engineering, Handan, Hebei, China
e-mail: baobei181920@126.com

L. Liu · C. Jin · Y. Li · C. Xu · L. Li
Key Laboratory of Resource Survey and Research of Hebei Province; School of Earth Science and Engineering, Hebei University of Engineering, Handan 056038, China

S. Shangguan
The Second Geology Team, Hebei Coal Geology Bureau, Xingtai 054001, China

W. Dong et al. (eds.), *Sustainable Development of Water Resources and Hydraulic Engineering in China*, Environmental Earth Sciences,
https://doi.org/10.1007/978-3-319-61630-8_15

1 Introduction

Dissolution mining is the main production mode for well salt. The principle of dissolution mining is that salt dissolves easily in water. The mining method is pumping water into the halite deposit, dissolving the halite underground and pumping brine onto the surface [1, 2]. The feature of this method is transforming halite deposit from solid to liquid, which simplifies the mining process, as well as increases mining efficiency and cuts the mining cost. The dissolution rate of halite is an important technique index of dissolution mining. The rate is concerned with dissolution property, the grade of halite deposit and the nature of solvent [2, 3]. Dissolution rate has an effect on mining efficiency, the shape of caverns and the sustainability of mining. Xinji-Ningjin Halite Mine was discovered in Xingtai, Hebei in 2010. The halite deposit hosts in the lower part of Paleogene Shahejie Formation, Shulu Sag. The prognostic resource is about 110 billion tons [4]. A dissolution experiment was performed on the halite samples of this salt mine and the detailed parameters of the dissolution property were obtained. These results offer a significant guidance to set the concentration of solvent and predict the developing shape of caverns. It will also realize the long-term continuous production of this salt mine.

2 Regional Geological Settings

Shulu Sag is in the south end of the Jizhong Depression, which belongs to the Bohai Bay Basin, eastern China. Shulu Sag is a Cenozoic basin developing on Paleozoic basement. This NNE spreading sag is a half-garben basin with a normal fault in the east and lapout in the west. The boundary between Shulu Sag and Xinhe Uplift to the east is the Xinhe Normal Fault. The boundary between Shulu Sag and Ningjin Uplift to the west is the paleo-denudation line. The boundary between Shulu Sag and northern Shenxian Sag is the Hengshui Normal Fault. The boundary between Shulu Sag and southern Xiaoliucun Uplift is a normal fault [5] (Fig. 1). In the interior of Shulu Sag, the west part is the West Slope Area and the east part is the lake basins area, which is divided into three incompletely connected parts by Taijiazhuang Uplift and Jingqiu Uplift. The three parts are North Sag, Central Sag and South Sag. Combining with the two uplifts, it presents the structure framework of “three sags alternated with two uplifts” [6, 7]. The water in the three sags gradually salinized from north to south and it developed sandstones-mudstones, marlstones and gypsum-salt rocks, respectively [8]. The Xinji-Ningjin Halite Mine is located in the South Sag (Fig. 1) and Caochang Halite Mine of this research is located in the west of the Xinji-Ningjin Halite Mine.

In Cangchang Halite Mine area, Quaternary developed alluvial-diluvial and lacustrine clay, silt-fine sands and gravels; and it is 400–550 m thick. Neogene Minghuazhen Formation developed brownish-red, brownish-yellow and purplish-red

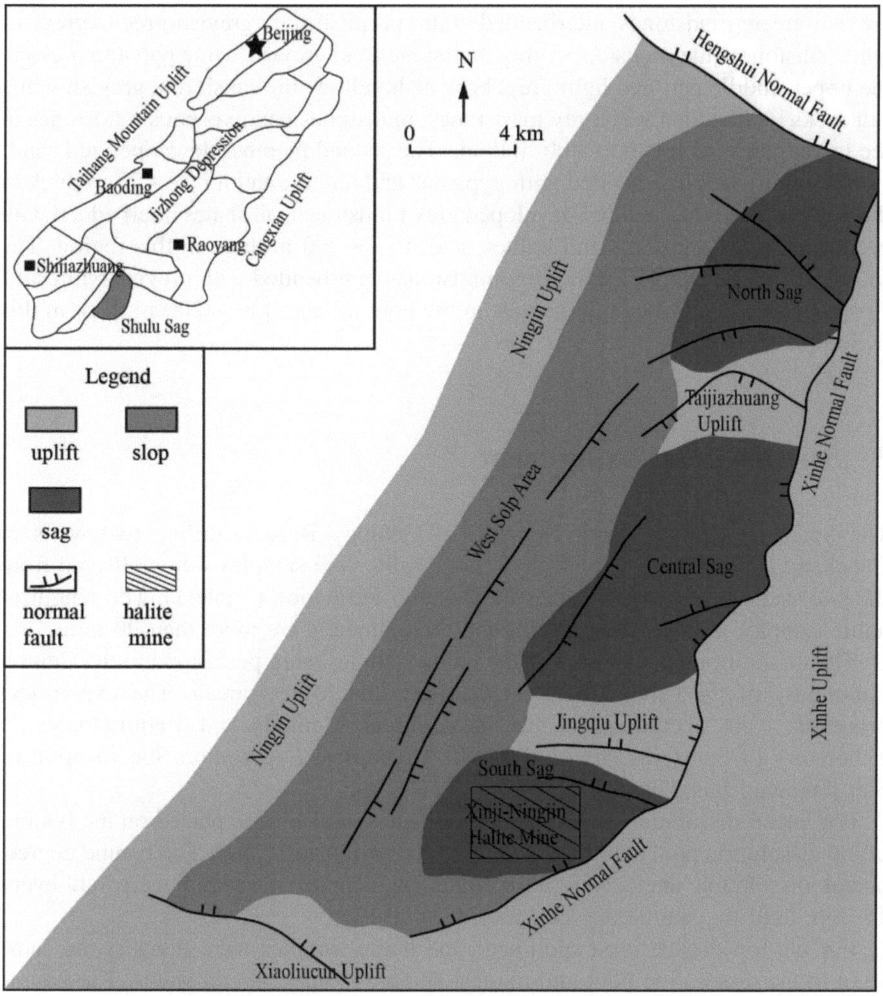

Fig. 1 Regional structures of the Shulu Sag (modified from [6, 7])

mudstones, siltstones interbedded with siltstones and fine sandstones; and it is 450–500 m thick. Neogene Guantao Formation developed purplish-red, greyish-green, brownish-yellow mudstones interbedded with grey-white, light yellow fine-medium sandstones with a layer of gravelly coarse sandstones in the bottom; and it is 450–550 m thick. Paleogene Dongying Formation developed purplish-red, greyish-green mudstones, silty mudstones interbedded with greyish-white sandstones; and it is 700–900 m thick. Paleogene Shahejie Formation can be divided into four sections, the first to the fourth member of Shahejie Formation from the top to the bottom. The first member is the main salt-bearing layer, which developed dark purple, brownish-red, greyish-green mudstones interbedded with purplish-red, greyish-green, greyish-white siltstones-fine sandstones, grey marlstones and greyish-white gypsum rocks in the upper-middle part and light grey, brownish yellow, brownish red, greyish white salt rocks interbedded with grey marlstones, mudstones and gypseous mudstones in the lower part; and it is 600–800 m thick. The second member developed red sandstones, mudstones interbedded with gypsums and halites; and it is ~450 m thick in this region. The third member developed grey mudstones, oil shales interbedded with grey sandstones, gypsums and halites; and it is ~200 m thick in this region. The fourth member developed dark grey mudstones interbedded with greyish white siltstones, with a layer of conglomerates in the bottom; and it is ~200 m thick in this region [4, 9].

3 Sampling and Experiment

The Second Geology Team, Hebei Coal Geology Bureau, drilled two wells in Caochang Halite Mine, Y-1 and Y-2. Four halite core samples were collected from the two wells in the first member of Shahejie Formation (Table 1). The length of halite samples is more than 20 cm, and the diameters are more than 90 mm.

The dissolution experiments of the halite samples were performed in the Central Laboratory of the China Chemical Geology and Mine Bureau. The experiment processes were according to the "Experiment Methods and Requirements in Laboratory of Salt Ores Dissolution" of "Geological Exploration Specification of Salt Lake and Salt Minerals" [10].

For lateral dissolution experiment, the halite samples were placed on the bottom of the dissolution pool and the test surface steeped in still water. The baume degree, lateral dissolution angle and the weights of samples measured were left every 15 min, until the baume degree reached 24 °Bé.

For the top dissolution experiment, the halite samples were hung at the same level in the dissolution pool and the test surface steeped in still water. The baume degree was measured, top dissolution depth and the weights of samples left every 15 min, until the baume degree reached 24 °Bé. The top dissolution depth is for

Table 1 Sampling information

No.	Well	Depth (m)	Strata	Lithology
CC01	Y-1	2730	The first member of Shahejie Formation	Smoky gray fine-crystalline halite
CC02	Y-1	2874	The first member of Shahejie Formation	Light smoky gray medium-crystalline halite
CC03	Y-2	2713	The first member of Shahejie Formation	Smoky gray fine-crystalline halite
CC04	Y-2	2827	The first member of Shahejie Formation	Light smoky gray medium-crystalline halite

calculating the top dissolution rate and the weight of samples left is for calculating the whole sample dissolution rate.

4 Results

Lateral dissolution rates of the four samples are basically consistent. The rates decline gradually along with the rise of baume degree of the brine, from 7–8 mm/h of the first stage to 3–4 mm/h of the fourth stage; and they match linear trend (Fig. 2, left). The rates get rid of the linear trend and decline sharply in the last stage, down to 1–2 mm/h (Fig. 2, left). The lateral dissolution angles increase gradually along with the rise of the baume degree of the brine, from ~9° in the first stage to ~34° in the fourth stage (Fig. 2, right). The angles have an indistinctive increase, about 4° in the last stage (Fig. 2, right).

The top dissolution rates are not consistent in the first stage, ranging from 9 to 13 mm/h. The rates decline gradually along with the rise of baume degree of the brine and match linear trend. The rate difference among these samples becomes small, down to consistent 1 mm/h in the last stage (Fig. 3, left). The whole sample dissolution rates are very similar to the top dissolution rates (Fig. 3, right).

By analyzing grain size of the residues of the dissolution experiments, the residues are very fine. The content of residues finer than 0.1 mm is 21–40% and the content of 0.1–2 mm is 44–67%. The total content of these two grain grades reaches 77–100% (Fig. 4).

Sinking rates of the residues decline gradually along with the rise of the baume degree of the brine, while the decline scope of 2 mm residues is very small, from 2.8 to 2.0 cm/s and the change tendency of the four samples are very consistent (Fig. 5, left). The decline scope of 5 mm residues is relatively large, from 19 to 14 cm/s, from 13 to 10 cm/s. The samples from the same well have completely the same change tendency and the samples from different wells have different change tendency (Fig. 5, right).

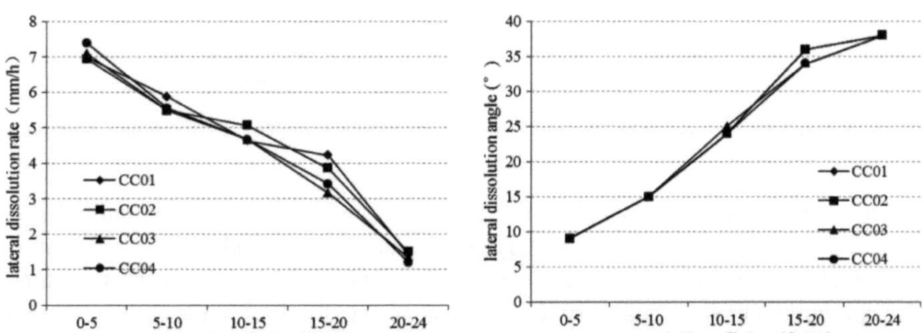

Fig. 2 Change tendency of the lateral dissolution rates (left) and lateral dissolution angles (right)

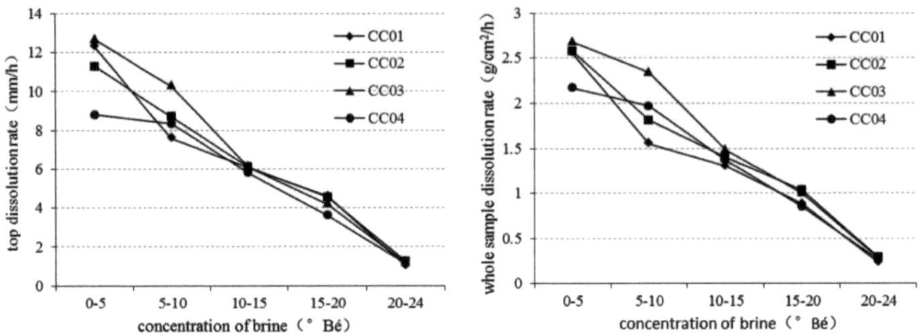

Fig. 3 Change tendency of the top dissolution rates (left) and the whole sample dissolution rates (right)

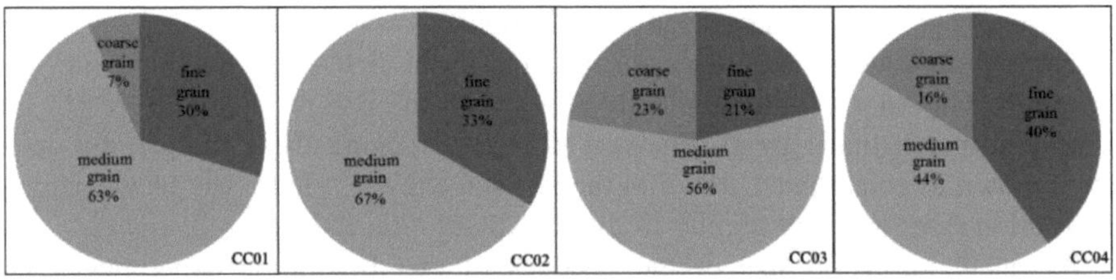

Fig. 4 Grain size constitutes of the residues (fine grain, <0.1 mm; medium grain, 0.1–2 mm; coarse grain, 2–10 mm)

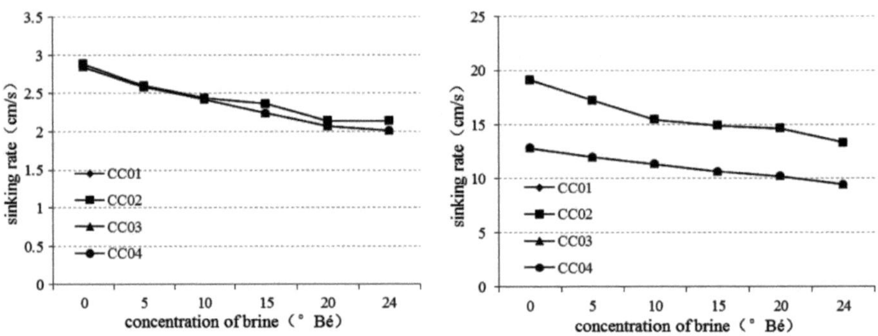

Fig. 5 Sinking rates of residues (the left is for 2 mm residues; the right is for 5 mm residues)

The component analysis of the brine with different baume degree indicates all the samples show almost a completely consistent change trend. The component of NaCl increases linearly from 50 g/L in the first stage to 320 g/L in the last stage. The component of $CaSO_4$ is very low in all stages, and it does not leave the baseline (Fig. 6).

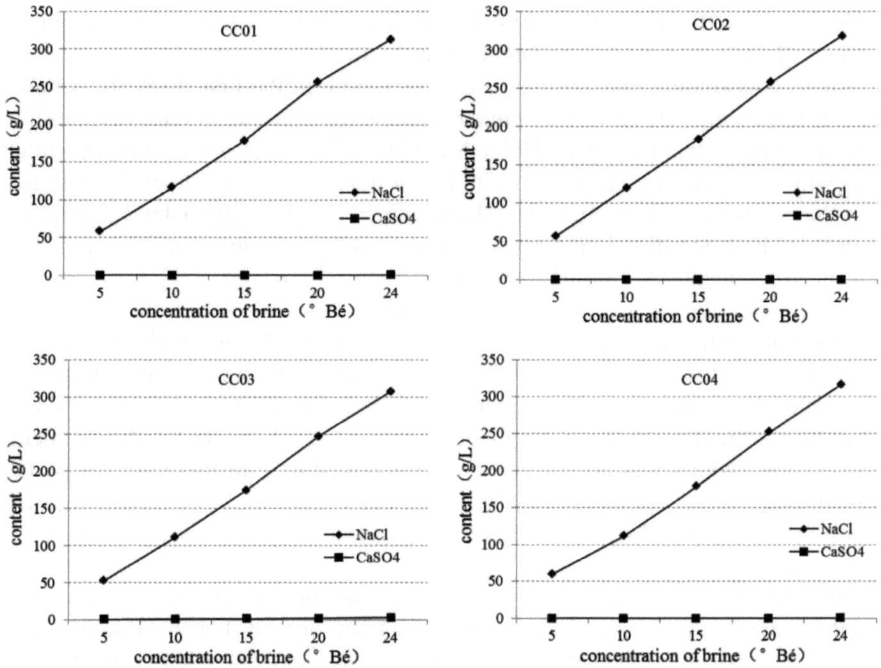

Fig. 6 Component analysis of the brine with different baume degree

5 Discussion

Along with the rise of baume degree, the lateral dissolution rates, the top dissolution rates and the whole sample dissolution rates decline remarkably, which indicate the concentration difference is the motive power of the dissolution of salt minerals [1, 11, 12]. This result is consistent with the neighboring Jichangzhuang Halite Mine [13]. The lateral dissolution rates of the high-grade halite of Yingcheng, Hubei stabilize in high level in the first-second stage; and they occur in a remarkably linearly decline in the third-fifth stage [14]. The change tendency shows a difference in this study. The lateral dissolution rates decline linearly in the first-fourth stage, while they decline non-linearly in the last stage (Fig. 2, left). The grade of the samples of Caochang Halite Mine reaches 93% [4], which means the grade of halites in the two mines is high. The remarkable differences of the lateral dissolution rates between the two mines may be caused by the high content of mirabilite in the Yingcheng Halite Mine [14].

The lateral dissolution angles in the last stage are stable, between 35 and 38° (Fig. 2, right). This is in the range of lateral dissolution angles of the Yingcheng Halite Mine, 36–52° [14], while it is very different from the range of the last stage of neighboring Jichangzhuang Halite Mine, 10–16° [13]. This indicates during the mining of this halite mine, the lateral dissolution will play an important role in the forming of caverns, which is different from the neighboring mine.

The top dissolution rates of this research are high in the first stage, 9–13 mm/h (Fig. 3, left), which are obviously higher than the top dissolution rates 6.0 mm/h of the same stage of the Yingcheng Halite Mine [14]. This indicates the top dissolution rates will also play an important part in the forming of caverns, which is consistent with the neighboring Jichangzhuang Halite Mine [13]. The large lateral dissolution angles and high top dissolution rates indicate that fresh water should be used as a solvent to accelerate the forming and enlargement of caverns.

The grain sizes of dissolution residues have close relations with the mining sustainability of halite mines [3]. Most residues of this study are smaller than 2 mm, which accounts for more than 77% of all the residues (Fig. 4). The sinking rates of this part of the residues in still water are less than 2 cm/s (Fig. 5, left). The residues larger than 2 mm have relatively high sinking rates (Fig. 5, right), but their contents are very low, which will not affect the continuous mining of this mine. This suggests during the normal mining or late mining that brine with low concentration can be used as a solvent, which will keep the mining efficiency and sustainability of this halite mine.

The content of NaCl rises remarkably along with the baume degree of the brine, and the content of $CaSO_4$ stays in the level of a trace amount (Fig. 6). This indicates the main component of the halite of this mine is NaCl and its grade is very high [4]. The linear tendency of the content of NaCl shows its content acceleration does not decline along with the slowing down of the lateral dissolution rates.

The dissolution rates of halite are influenced by lots of factors, such as the dissolution property of salt mineral, the grade of ore, the property of solvent, the flow rate of solvent, the concentration of solution, the space location of dissolution surface, the temperature and pressure of ore, etc. [2]. In this study, only the dissolution rates of halite are considered in still water in normal temperature and pressure. In the real mining process, the halite ores locate at the depth of $\sim$2800 m underground, where the temperature and pressure are very different from that on the surface, and the real dissolution rates must vary from the results of this experiment [2, 3, 12]. In addition, during real mining, the solvent flows all the time; therefore, the speed of flow will also influence the dissolution rates [1, 3]. In the future experiments, a new process will be designed to take into consideration the temperature, pressure and flow rate.

6 Conclusions

(1) The lateral dissolution rates, top dissolution rates and the whole sample dissolution rates of halite samples of the Caochang Halite Mine decline gradually along with the rise of baume degrees of the brine and the lateral dissolution angles of the samples rise gradually along with the increase of baume degrees. Both the lateral and top dissolutions play an important part in forming and developing caverns.

(2) Most dissolution residues are very small, and they sink very slowly in still water, which will not influence the continual mining. (3) Along with the rise of baume degrees of the brine, the content of NaCl increases remarkably and the content of $CaSO_4$ stays in trace amount. This indicates the main content of the halite is NaCl and its grade is very high.

Acknowledgements The authors thank Guobin Zhang, the Chief Engineer of the Second Geology Team, Hebei Coal Geology Bureau, for his strong support to this study. This research was financially supported by the Natural Science Found of Hebei Province (D2015402125).

References

1. Liang, W.G., Li, Z.P., Zhao, Y.S.: Experimental study on solution mining of salt deposit. J. Liaoning Tech. Univ. **22**, 54–57 (2003). (in Chinese)
2. Dong, Z.L., Liu, J.J., Zhang, P., Fan, C.L.: Analysis on the factors affecting the water mining recovery rate. China Well Rock Salt **45**, 16–27 (2014). (in Chinese)
3. Wang, C.R.: Experimental study of the influencing factors on dissolution rate of rock salt. Master's thesis of Chongqing University, Chongqing (2009) (in Chinese)
4. Zhang, G.B., Song, J.B.: Occurrence pattern, development and utilization evaluate of halite resources in the Xinji to Ningjin of Hebei Province. Hebei Coal **4**, 7–10 (2010). (in Chinese)
5. Kuang, H.W., Qin, H.S., Liu, J.Q., Wang, J.M.: The sequence characteristic of Member 3 of Shahejie Formation of Paleogene in the central Shulu Sag. Petro. Geol. Eng. **21**, 26–32 (2007). (in Chinese)
6. Zhang, R.F., Tian, R., Li, M., Zhu, J.Q., Wang, S.C., Tian, J.Z., Li, X.D. Zhang, W., Xiao, B.Y.: Stratigraphic sequence and types of oil and gas reservoirs in the lower part of Member 3 of Shahejie Formation in Shulu sag. Acta Petro. Sin. **36**, 10–20 (2015) (in Chinese). https://doi.org/10.7623/syxb2015S1002
7. Zou, X.H., Chen, J.Y., Wen, D.: The accumulation laws and distribution of buried hill reservoirs of Shulu Sag. J. Yang. Univ. (Nat. Sci. Ed.). **12**, 19–22 (2015) (in Chinese). https://doi.org/10.16772/j.cnki.1673-1409.2015.05.002
8. Liang, H.B., Kuang, H.W., Liu, J.Q., Guo, Y.J., Su, J.: Discussion on origin for marls of the member 3 of Shahejie Formation of Paleogene in Shulu Sag of Central Hebei depression. J. Palaeog. **9**, 167–174 (2007). (in Chinese)
9. The Second Geology Team, Hebei Coal Geology Bureau: Geological report of preliminary exploration halite resource of Caochang prospecting area of Ningjin Halite Field. Geology report. Xingtai, Hebei Province (2008) (in Chinese)
10. Ministry of land and resources of the People's Republic of China, the geological exploration specification of salt lake and salt minerals (DZ/T02122002). Industry standards (2003) (in Chinese)
11. Wang, Q.M.: The Solution Mining of Salt Minerals. Chemical Industry Press, Beijing (2003). (in Chinese)
12. Liang, W.G., Zhao, Y.S., Xu, S.G., Yu, Y.M.: Theoretical study of in-situ solution mining. J. Taiyuan Univ. Tech. **43**, 382–387 (2012). (in Chinese)
13. Gao, L., Song, J.B.: Brief analysis on the water solubility test of salt mine in Jichangzhuang halite field. Ningjin. China Well Rock Salt **46**, 17–31 (2015). (in Chinese)
14. Zhao, G.F., Yang, J.: Experimental study of the dissolution of rock salt in Yingcheng. Hubei Province. Resou. Env. Eng. **28**, 209–213 (2014). (in Chinese)

Water Resource Development and Ecological Reclaim in Chinese Fen River Basin

Xiao-Hong Zhang and He-Li Wang

Abstract In this paper, the author put forward some views and suggestions on water resources utilization and management and the ecological environment protection process in the future, through a new historical perspective of Fen River basin water resources development and the governmental efforts to solve the ecological troubles it caused. Fen River is the largest river in Shanxi Province and number two of Yellow River's first-grade tributaries. Along with the development of the social economy and changes in the natural climate environment, Fen River basin appeared to have serious ecological environment problems, an increasingly serious shortage of water resources and pollution trouble which brought about the bottleneck of ecological and economic development in Shanxi Province, and affected the water quality and flow of Yellow River. Since the founding of new China, there have been four massive, systemic governances of Fen River focusing on surface water development, utilization of groundwater, flood control for security, and recovery of clean water flow, respectively. All have made some significant achievements; however, the governances have not been able to fundamentally solve the problems of the ecological environment under the limitation of temporal governing ideas and capital. During the twelfth five-year projects Shanxi Province launched, the "two longitudinal and cross ten" large water program, which aims at facing and solving the trouble of uneven spatial and temporal distribution and guaranteeing the entire province's economic development demand of water resources by connecting the six major rivers and their large and medium-sized reservoirs in the province, taking the north trunk stream of Yellow River and Fen River as the two longitudinal trunks. The ecological restoration planning outline (2015–2030) that Shanxi Province issued in July 2015 sounded the horn of large-scale administering in Fen River basin for the fifth time. This new round of river basin ecological management must include: ecology—resources—environment—society—economy, the five elements

X.-H. Zhang (✉)
College of Geographical Science, Shanxi Normal University, Linfen 041004, China
e-mail: 13663473442@163.com

H.-L. Wang
Office of Teaching Affairs, Shanxi Normal University, Linfen 041004, China
e-mail: wangheli_2003@yahoo.com.cn

W. Dong et al. (eds.), *Sustainable Development of Water Resources and Hydraulic Engineering in China*, Environmental Earth Sciences,
https://doi.org/10.1007/978-3-319-61630-8_16

which should be balanced considerations, none of which are dispensable; from divided ruling to shared governance within basin, from passive decontamination to active anti-fouling, from focusing on engineering construction at master stream to balance controlling of whole basin, all relevant departments and personnel need to create new ideas and actions.

Keywords Fen River basin · Ecological environment · Water resource
Ecological restoration

1 Introduction

Fen River is the largest river in Shanxi Province, and the second one of the first order tributaries of Yellow River (Fig. 1) [1]. It originates from GuanQin Mountain, which is in the east village Town of Ningwu County in Shanxi Province. The total length is 695 km, of which the upstream section (217.6 km) is from the source to Lan Village stone mouth, going through the Lvliang mountains; the middle section (266.9 km) is from Lan Village to the beach of Hongtong county, which basically belongs to the river basin plain; the downstream segment (210.5 km) is from Hongtong beach to the estuary into Yellow River, as one of the main agricultural bases in Shanxi Province. The whole river basin area is 39,741 km^2, involving 45

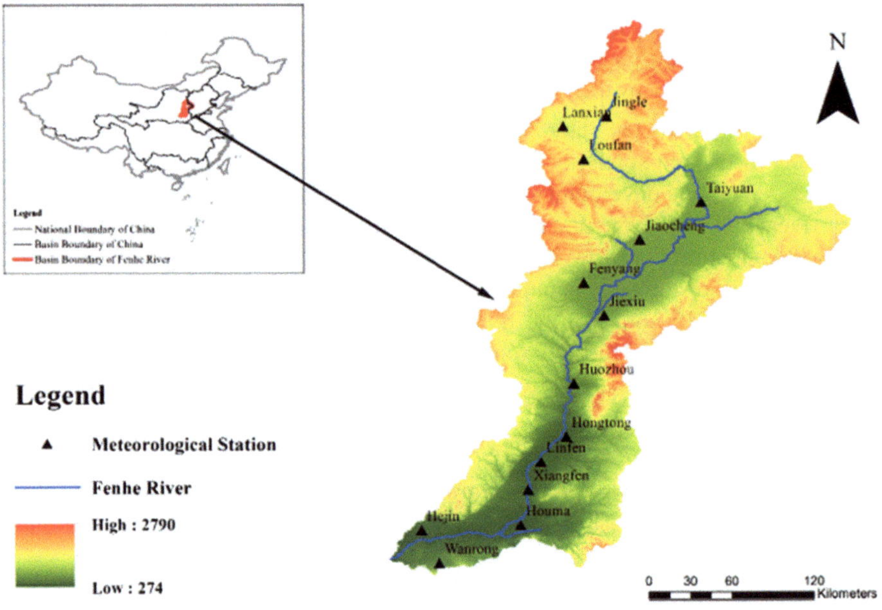

Fig. 1 Plane graph of the Fenhe River basin

counties (or city areas) of the 7 cities: Xinzhou, Taiyuan, Jinzhong, Lvliang, Linfen, Yuncheng, Changzhi. With less than a third of the entire water resources, it provides 41% of the water required for the population in the province and contributes half of the gross domestic product. Both coal reserves and coal production in Shanxi account for a quarter of China, and the province diversion volume accounts for 70% of the nation. Coal production, as the main mode of economic development for half a century, has brought about the cost of ecological environment deterioration within the scope of the whole province. As the focus of the economic development of Shanxi Province, the most developed area, Fenhe River basin has been facing a serious shortage of water resources, severe soil erosion and water pollution problems, which have become the bottleneck of the local development. To face and solve these problems correctly is crucial for the ecological economic development in Shanxi Province and even the comprehensive control of the Yellow River basin. The purpose of this article is to introduce constructive thinking and suggestions for the future management and utilization of basin water resources and ecological economic development through the observation and review of its changes in the quantity and quality of water resources and ecological environment problems co-existent with economic development, and the governmental responses to it.

This figure was quoted from the article of "The Impact of Climate Change on the Duration and Division of Flood Season in the Fenhe River Basin, China", which was written by Wang et al. [1].

2 Review of the Water Resources of Fen River Basin

The Fen River basin is often divided into two parts in the Shanxi Province water resources bulletin: the middle-upper stream and the downstream. The data that was found from 2007 and 2008 can represent the water resources status in the new century somehow (and the only shared data that could be found by a non-staff person of the water department), contrasted with an average of 45 years series (1956–2000).

2.1 Quantity of Water Resources

As shown in Fig. 2, the average total water resources quantity in Fen River basin for the 45 years series from 1956 to 2000 is 3.359 billion m^3, with a runoff of 2.067 billion m^3 and a groundwater of 2.409 billion m^3; the duplicated amount between surface water and groundwater account for 33.3% [2]. It is an important feature of surface water and groundwater to repeat for a larger proportion in the total water resources. While some large karst springs were used by wells, they are rapidly lowering, even cutoff in the clear water flow in the surface river. Compared with

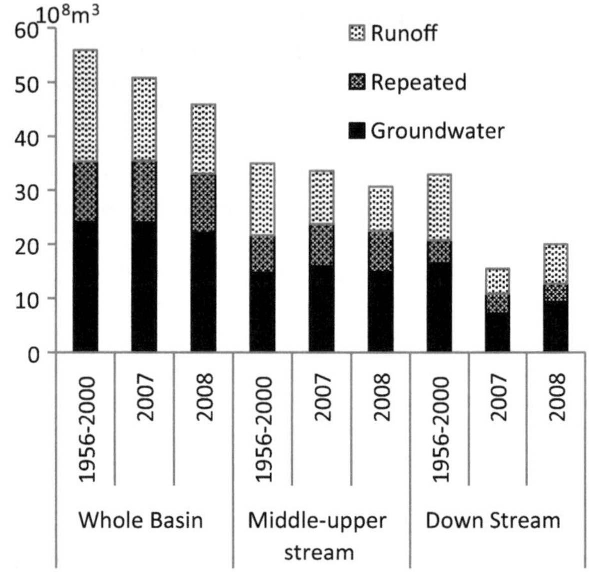

Fig. 2 The runoff and groundwater in Fen River basin

that of 1956–2000, the water resources quantity of Fen River basin in 2007 and 2008 had reductions, and the two annual amounts decreased by 17% and 28%, respectively, included in which the mid-upper stream damping were 15 and 27%, the downstream were 66 and 45% lower. The surface runoff decreased basin-wide, while the groundwater reduction only occurred severely downstream. However, in the mid-upper stream, it slightly increased.

All data were directly or indirectly from the Shanxi Province Water Resources Bulletin, which involves the quantity and quality of water resources sorted by basin and was found on the internet [2].

Basin water resources quantity and local climate, vegetation, topography and other natural conditions and social economic development are closely related. With the trend of global warming, for half a century Fenhe basin has had an obvious drying tendency; annual precipitation was in a declining trend, especially in the 1970–1990s of last century [3, 4]. This drying trend eased after entering the new century; in 2000–2013 the average rainfall was 510 mm, which rallied 10% of that in the 1990s (462.9 mm average) [5]. The selected 2007 and 2008 belong to the exceptional drought years, of which annual rainfall was only 45 and 34% of the average value in the new century, and far less compared with the average during 1956–2000 (Fig. 3).

How much of the precipitation transformed into basin water resources reflects the local underlying surface properties and human intervention on the process of water resources. From the calculation results in Table 1, compared with that of 1956–2000, the basin rainfall resource conversion rate of 2007 and 2008 were continuously improving. It was more than double the total water resources conversion, and the downstream area is 7–8 times higher due to the effect of

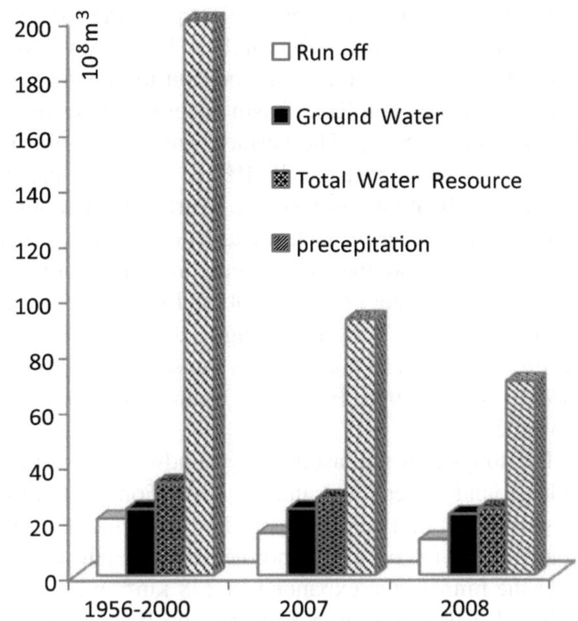

Fig. 3 Precipitation and water resources in Fen River basin [2]

Table 1 The conversion ratio of precipitation in Fen River basin (%)

		Runoff	Groundwater	Repeated amount	Total water resource
1956–2000	Whole basin	10.35	12.06	5.60	16.81
2007	Whole basin	16.59	26.03	12.39	30.23
	Middle-upper stream	20.65	33.70	16.60	37.75
	Down stream	74.50	108.59	48.21	134.88
2008	Whole basin	18.39	31.55	15.60	34.33
	Middle-upper stream	22.79	41.63	21.35	43.07
	Down stream	83.18	126.72	57.88	152.02

superposition of the upstream water. That should attribute to the continuous work of soil and water conservation and water storage management in Fen River basin after the foundation of the new nation.

2.2 *Water Resources Utilization*

Overload of water resources is the issue that has always been for Fen River basin, which is one of the causes of ecological environment deterioration. In the Chinese

Ministry of Water Resources' project planning for energy based in north China and Shanxi Water Resources Research in 1990, Fen River basin runoff utilization rate was 80% [6]. According to a study on the bearing capacity of water resources in Shanxi Province, Fen River basin water resources development has already reached its limits in 1998 [5]. The surface water utilization rates in 2007 and 2008 were between 63 and 85%; the 38.4% of 2000 was calculated by the temporal water resources utilization based on the average total water resources from 1956 to 2000 (Table 2). However, in fact, the amount of surface runoff was only 8.11 × 108 m^3 as a dry year and the actual temporal utilization rate was as high as 97.78%. Internationally recognized the rational development and utilization rate of surface water is 30–40% [7, 8]. Referring to the calculation results of the ecological water requirement [6, 9] to take 30% as the safe utilization rate for mid-upper stream and 40% for downstream, Fen River basin surface water resources was overloaded by 1–3 times.

The long-term unreasonable groundwater overdraft led to the declining of the underground water level and the formation of multiple urban funnel areas within Taiyuan basin and Linfen basin. For example, in 1965 the funnel area with closed area of 11.2 km^2 occurred and the buried depth was 16 m in Taiyuan city, while in 1987 the funnel area expanded to 298 km^2 and the buried depth dropped to 55 m. Although Taiyuan, Linfen, Lvliang, and Jincheng began to gradually implement the shut-in well and suppressing action to reduce the area of groundwater overdraft since 2003, and the groundwater level in some regions has stopped decreasing and even began increasing [10, 11], this recovery process was very slow, and had inversion because of the huge upfront overexploitation. According to Shanxi Province water resources evaluation, compared with 1986, 2010, Taiyuan basin and

Table 2 The water resources utilization in Fen River basin (10^8 m^3)

Year	River segment	Runoff			Groundwater		
		Water resource	Output of supplying	Use ratio (%)	Water resource	Output of supplying	Use ratio (%)
2007	Whole	15.3199	9.7062	63.4	24.0365	15.9466	66.3
	Mid-upper	9.7679	6.2232	63.7	15.9439	10.593	66.4
	Down stream	5.552	3.483	62.7	8.0926	5.3536	66.2
2008	Whole	12.8737	10.0685	78.2	22.0819	15.7927	71.5
	Mid-upper	8.1561	6.877	84.3	14.8955	10.9407	73.4
	Down stream	4.7176	3.1915	67.7	7.1864	4.852	67.5
1956–2000	Whole	20.67	7.93*	38.4	24.09	17.02*	70.7
	Mid-upper	13.27	–	–	14.76	–	–
	Down stream	7.4	–	–	9.33	–	–

Note *the data was for 2000

Linfen basin in plain groundwater level still fell cumulatively by 8.15 m and 3.81 m, respectively [12].

2.3 *Water Quality of Fen River Basin*

The environmental quality information posted on the website of China Public Environmental Research Center [13] showed that Fen River is one of the most polluted tributaries of the Yellow River. According to the report of the water resources in Shanxi Province in 2006–2008 [2], down from Taiyuan Shop Bridge the Fenhe River quality was all worse that can just be used as agriculture use water and general landscape water. Actually, back by 10 years, from 1998 to 2007, it had always been like this [14]. Upstream Lan Village and Jingle water quality was relatively good, except for individual year individual items slightly exceeding the standard, and could basically achieve II and III class standard. However, on the village cross section, which is in Gujiao city, the water quality was always worse. Regarding Fen River pollution, its mainly exceeded standard items are ammonia nitrogen, petroleum, chemical oxygen demand (COD) and volatile phenol, which are mainly from effluent sewage of coal gas, coking, petroleum, chemical, steel and paper industry, synthetic materials, food, livestock and poultry feed and other factories. According to the statistics, from 1998 to 2006 Shanxi Province annually discharged effluent sewage by an average of 8.45 × 108 tons, of which industrial effluent accounted for 77.5% and Fen River basin accounted for 40.1% [14].

3 Review of the Water Resources Development and Ecological Economic Harnessing

3.1 *Traditional Development Processes*

As one of the most important forefronts of industrial and agricultural economic development in Shanxi Province, Fenhe basin reflected the characteristics of a different era of water resources development pattern and intensity with its economic and social development process. Before the metaphase of the 1960s, it mainly held the characteristics of retaining surface water, and had a large number of medium and small reservoirs and irrigation areas. Although the water resources development growth is larger, it was overall on a smaller scale, and water resources utilization rate was less than 30%, which belongs to a low-level water consuming pattern. In the mid-1960s to mid-1980s, priority was given to large-scale development of groundwater with drilling supporting, and different degrees of water shortage areas began to appear. That limited the further development and potential of regional water resources. Therefore, industrial and agricultural production water consuming began to transition to a water-saving style, which started the attention to the integrated management of water resources. Since the mid-1980s, it became a water resources comprehensive development period to meet the energy-based construction and the economic and social development in Shanxi Province. The water resources utilization rate had reached 70%, and the groundwater resources development degree was higher. It was close to a threshold by the end of the 1990s; the whole Fenhe basin had none or small further development and utilization potential [7]. This period development approach gave priority to deep development and optimizing configuration, and the integrated water resources management reached quite a high level. The deep-water resources development lasted until the early 21st century, as the results of Zhang and Feng [15]. Using the method of ecological footprint to research Shanxi Province water ecological carrying capacity, in 2001–2010 Fenhe basin showed ecological footprints that were greater than the water resources carrying capacity and presented a water deficit over the years [15].

3.2 Clear Water Reoccurring Engineering of Fenhe Water Streams

Continuous development caused Fenhe basin to face water resources shortage and poor water quality, as well as ecological environment deterioration problems, which had become a bottleneck restricting the further development of the local economy. Since 2008, a systematic engineering involving many departments and multi-disciplines, Fenhe Water Reflowing Engineering, officially started the ecological environment governance with repair and protection work in the new century. It also was an important part of the Large Water Project in Shanxi Province. Shanxi large water network aimed at constructing an engineering skeleton covering the entire province's six big basins and major economic centers, taking the north natural trunk of Yellow River and Fen River as the two main longitudinal lines and plus the ten key traverse water supply system, to focus on the uneven distribution of water resources and realize sustainable utilization efficiently. As one of the two main lines, Fenhe streams in engineering played a vital role. According to the serial dissertation reported [16], NingWu of Xinzhou, where Fenhe source lies, was focusing on soil and water conservation and rational utilization. It was proposed by implementing in stages of recent (2 years), medium (5 years) and forward future (10 years) repairing, immigration reclamation and water conservation projects, to reduce soil and water loss. In the middle stream Taiyuan and Jinzhong have expanded and constructed new sewage treatment plants for sewage purification, shut-in wells to compress groundwater consumption, and constructed artificial wetlands to realize water body restoration. All these actions obtained certain effects; for example, the underground water level of two water sources of LAN Village and Xizhang had stopped decreasing and realized an increase again for the first time since 1987. Linfen City, the downstream segment that was facing serious water pollution, started unprecedented governance within 3 km of Fenhe range by closing down or relocating polluting enterprises that did not meet industrial policy or affect Fen River water quality. In 2012, Linfen City began constructing "Fen River Ecological Economic Zone in half hundred kilometers". It took ecological restoration as the starting point and implemented construction of roads, industrial parks, cultural tourism and urban areas.

3.3 Further Ecological Restoration Plan in Fen River Basin

The new century's Fen water reflow project has achieved initial progresses: surface water regulation and storage functions have been significantly improved; the histories of the drying of the Fenhe have ended; part of the local groundwater table also achieved a cease in the decrease and a rebounding. However, the whole Fen River Basin is still faced with falling water tables, reducing surface runoff, decrease and degradation of vegetation, soil erosion and other outstanding issues. The watershed's own "blood" production and flow function have not been restored. In July 2015, Shanxi provincial government formulated the "Fen River Basin Ecological Restoration Plan (2015 to 2030)". It can be said that this is the fifth large-scale governance program of the Fen River in Shanxi Province. Its overall goal is with multi-investment of ¥130 billion, after five years of construction and 10 years of natural restoration, that is after a total of 15 years of effort, to make a substantial increase in groundwater reserves and the area of vegetation in Fen River basin, to effectively control erosion and water pollution and make the groundwater level rise, so as to have the reappearance of the splendid scenery of Fen River. There are six specific measures included: 1. adhere to the principle of water-saving priority through scientific configuration of land and water resources, vigorously promote a water-saving society and encourage efficient use of water resources; 2. rely on the built programs of Wanjia Village Yellow River Diversion, Yumenkou Water Lifting, Citing Qin Water to Fen River, and the under-constructed Guiding Yellow River in Central, Dongshan Water Supply Project, implement the "Giving assistance to Fenhe with five waters", increase the amount of surface water resources and to protect the watershed healthy economic and social development through the water transfer network in Fen River basin; 3. restore and build a number of regulating water storage projects such as "string of pearls", "bunch of grapes" in low-lying areas on both sides of the Fen River, to make full use of flood resources and make efforts to increase groundwater recharge; 4. define according to law the Fen River and its major 9 tributaries' source protection areas, closing off to restoring vegetation, so as to achieve water conservation and increasing streams and greenery; 5. strictly control the exploitation of groundwater in the basin, shut down coal mines within protected spring fields and the nine tributary source protected areas according to the law, strengthen the protection of eight large karst fields and groundwater systems; and 6. in mountainous areas vigorously implement the building of clean small watersheds; in plain areas control sewage discharge, strengthen pollution prevention through comprehensive treatment and sewage interception at the 10 tributaries in Taiyuan City to utilize water resources [17, 18].

4 Thinking and Proposing

From the point of historical exploitation of water resources and ecological environment management view, Fen River Basin watershed has also experienced the old route of resource over-exploitation, environmental degradation and then ecological restoration. More than 50 years of development after the founding of the nation almost depleted its groundwater resources and caused it to face the problems of surface runoff which is shrinking year by year, vegetation cover reduction, exasperation of water quality and environmental degradation. Fortunately, at the dawn of the new century, with the rise of the concept of sustainable development, people began to awaken. Starting in 2008 the Water Flow Reoccurring up until introduced in 2015 "Fen River Basin Ecological Restoration Plan (2015–2030)" [18], the Fen River Basin governance and economic restructuring and development organically combined, makes the future completely effective ecological recovery of Fen River basin to be just around the corner. However, looking at the several developing governance actions in Fen River Basin, there are two major issues to note: first, it pays more attention to the development of resources but lacks effective water pollution prevention and controlling; second, too much rely on the government-leading and less popular participation.

4.1 Resolving the Importance to Water Pollution Problem

Water scarcity is a big problem throughout Shanxi Province, so it is natural that several decades of governance of the Fen River basin was mainly centered on how to get more water. The design of "Five Waters Assisting Fen River" in Shanxi Large Water Network Engineering is mainly aimed at solving the uneven distribution and shortage of water resources in the basin. In the "Fen River Basin Ecological Restoration Plan (2015–2030)" the treatment of sewage discharge on 10 tributaries in Taiyuan was mentioned, but obviously the water pollution problem was not the topical subject. It cannot be neglected that Fen River water quality has not been resolved, especially, so far most of the time the middle and lower reaches of the river were inferior class V, while the "Fen River Basin Water Pollution Prevention Act" [19] has been issued and began to be implemented in 2005. It is considered that in the early stages of the management of the water reflow project, implementation of flood control, environmental protection, afforestation, landscaping, leisure and entertainment as one integrated work in Taiyuan, Linfen, etc. basically avoided the sewage governance [20]. It is time for Fen River to take unified action on trunk and tributaries for pollution control, learning from the experiences of River Thames and others.

4.2 Integrating Departments to Comprehensive Management Within the Basin

To improve the antifouling and pollution control, it is necessary to pay much attention to it and take action from the top to the bottom. The Government should not just focus on GDP development overlooking and winking at contaminant enterprise; those who created the pollution are responsible for treatment and it should not be just a slogan as well. To urge its real action, it is rational to proceed supporting subsidies to enterprise that have the ability to perform pre-sewage treatment and to charge those with reasonable fees based on the number and nature of emissions for sewage treatment plants' operation and maintenance. Comprehensive management of the whole basin requires breaking the boundaries within administrating departments and industry sectors, having good communication and coordination on technology, policy, action and benefits, moving from divide and conquer to basin cohabitation. A comprehensive management committee of Fen River basin could be established on the basis of the original relevant departments, such as the Upstream and Downstream Water Management Sites, Fenhe Reservoir Authority and the Fen River Irrigation Authority, to discuss and decide the allocation and utilization of water resources. Environmental Protection Departments should be given sufficient regulatory powers for illegal sewage investigating and treating. The Water Information Department should organize and release information dynamically for the convenience of water use, management and conservation.

4.3 Monitor and Open Dynamic Information to Common Participation

Automatic monitoring stations built on municipal boundaries of Fen River basin have been put into use. Relevant departments should make full use of this new dynamic monitoring platforms to keep abreast of the changes in water quality and to provide information supporting water pollution control, not only to the water departments, but also to the public. Basin water governance requires the top-down driven auspices, but also the bottom-up supporting and participation of the popular. In terms of water pollution monitoring, if the public could immediately report to the environmental protection department as soon as an abnormal condition occurred in the river or they found some secretly sewage discharging, it would greatly improve the efficiency of supervision. To increase public participation in the initiative, in addition to a good publicity and education program, more open and transparent information is necessary. Bulletining on schedule the changes of water quality and quantity, associated with development and utilization planning and implementation progress basin-wide, so that all interested people can learn readily and conveniently, thus to actively participate in governance of the basin.

5 Summary

As the largest river in Shanxi Province and number two of Yellow River's first-grade tributaries, Fen River is confronting serious ecological environment problems; increasingly serious shortage of water resources and pollution trouble; and it is bringing about the bottleneck of ecological and economic development in Shanxi Province, and affected the water quality and flow of Yellow River. A new round of river basin ecological management measures are imperative, which must include: ecological—resources—environmental—society—economy. The five elements should be balanced in consideration; none are dispensable; from divided ruling to shared governance within the basin, from passive decontamination to active anti-fouling, from focusing on engineering construction at the master stream to balanced controlling of the whole basin, all relevant departments and personnel need to create new ideas and actions.

References

1. Wang, H.-J., Xiao, W., et al.: The impact of climate change on the duration and division of flood season in the Fenhe River basin, China. Water **8**(3), 105 (2016). https://doi.org/10.3390/w8030105
2. Shanxi Provincial Department of Water Resources. http://www.sxwater.gov.cn/
3. Zhao, X.-M., Hu, C.-H., et al.: Climatic analysis of precipitation change trend over the Fenhe basin. J. Arid Land Geogr. **3**(1), 53–59 (2007). https://doi.org/10.3321/j.issn:1000-6060.2007.01.009
4. Yang, P.-G, Zheng, F.-Y.: Temporal and spatial change features of 50 years in Fen River basin. J. Arid Land Resour. Environ. **12**, 108–111 (2008)
5. Li, W.-H.: Precipitation variation feature analysis within 58 years in Fenhe River basin. J. Shanxi Hydrotech. **1**, 58–62 (2015). https://doi.org/10.3969/j.issn.1006-8139.2015.01.019
6. Ren, S.-F., Wang, S.-Y.: Research on the Fenhe River valley environment. J. Taiyuan Normal Univ. Nat. Sci. Ed. **9**(2), 99–104 (2010)
7. Xiao, M.-Y., Dong, Y.-L.: Assessment of water resources carrying capacity in Shanxi Province. J. Shanxi Hydrotech. **4**, 5–12 (1998)
8. Liu, C.-M.: China's water resources supply and demand trends and key issues of the 21st century. New Problems, New Technologies and New Methods of Hydrological Science Research in the 21st Century of China, pp. 3–10. Science Press, Beijing (2001)
9. Li, L.-J., Zheng, H.-X.: A preliminary study on the ecological environment water demand of river system. New Problems, New Technologies and New Methods of Hydrological Science Research in the 21st Century of China, pp. 74–80. Science Press, Beijing (2001)

10. Sun, Y.-F., Liang, X.-J.: Protection practice and thinking of groundwater in Fenhe River basin. J. Shanxi Water Resour. (Water Manag.) **4**, 7–8(2013). https://doi.org/10.3969/j.issn.1004-7042.2013.04.007
11. Zhang, J.: Some thoughts about the Shanxi water resources bulletin. J. Shanxi Water Resour., 24–25 (2005)
12. Institute of Public and Environmental Affairs. http://www.ipe.org.cn/MapWater/water.aspx?q=2
13. Wang, H., Deng, A.-L.: Re-check and evaluation of groundwater overexploitation areas in Shanxi Province. J. Water Resour. Prot. **28.5**, 80–82 (2012). https://doi.org/10.3969/j.issn.1004-6933.2012.05.017
14. Liang, X.-Y.: On analysis of Fen River's water quality pollution in recent ten years and suggestion of its improvement and protection. J. SHANXI Hydrotech. **1**, 76–78 (2009). https://doi.org/10.3969/j.issn.1006-8139.2009.01.029
15. Zhang, J.-L., Feng, H.-M.: Water resources carrying capacity in Shanxi Province based on ecological footprint method. J. Eng. Heilongjiang Univ. **4**(1), 61–65 (2013)
16. Serial special interviews on "Thousands of miles Fenhe water reflows", Shanxi news. http://www.sxrb.com/
17. Shanxi Institute of Water Resources and Hydropower Research: "Three steps" crack water problem. http://www.shxsky.com.cn/viewarticle.asp?c=SLZS&n=1458260804
18. Pan, J.-F.: The Deputy of National People's Congress and the Director of Shanxi Provincial Water Resources Bureau—Fen River ecological restoration should be 15 years, People's Daily, the livelihood of the people weekly. http://www.shxsky.com.cn/viewarticle.asp?c=SLZS&n=1459828548
19. Fen River Basin Water Pollution Prevention Act. http://www.tyswzw.gov.cn/news_detail.aspx?titleID=806
20. Chen, Z-T.: Suggestion on pollution-improving in whole basin of Fen River. J. Shanxi Hydrotech. **1**, 11–13 (2010). https://doi.org/10.3969/j.issn.1006-8139.20, https://doi.org/10.01.005

Current Situation and Analysis of Water Saving Irrigation Project in Hebei

Wenbin Chen

Abstract Hebei is a water resource shortage area, which relies on groundwater alone. Massive and unreasonable exploitation of groundwater caused a series of ecological and geological environmental problems, which pose a threat to the economic and social development. In recent decades, Hebei has pioneered some water-saving irrigation projects, aimed at increasing water and land productivity. Analyzing the present water-saving irrigation situation in this province, benefits and areas for improvement are proposed in this paper and can have a certain value in this fields.

Keywords Groundwater · Water-saving irrigation · Water-saving Hebei · China

1 Introduction

In China, the problem of water resources shortage is outstanding. Hebei is one of the provinces with the largest amount of groundwater exploitation in China. According to the Bulletin of Water Resources in Hebei, the total water consumption is 193.1 billion m^3 of which agricultural water consumption accounted for the largest share of 72% [1]. Meanwhile, the overexploitation area of the plain has reached 92%. Overexploitation of groundwater contributed to the groundwater level decline, land subsidence, etc. The ecological and geological environment has gradually deteriorated.

From 2003 to 2011, the settlement rate in the eastern part of Beijing exceeded 10 cm/year [2]. The deep groundwater in the north China plain has formed 13 subsidence centers, such as Cangzhou and Hengshui, and even the underground water level of nearly 70,000 km^2 is lower than the sea level. According to the China

W. Chen (✉)
Hebei Engineering University, Zhonghua Street, Hanshan District, Handan, Hebei, China
e-mail: 1192757756@qq.com

W. Dong et al. (eds.), *Sustainable Development of Water Resources and Hydraulic Engineering in China*, Environmental Earth Sciences,
https://doi.org/10.1007/978-3-319-61630-8_17

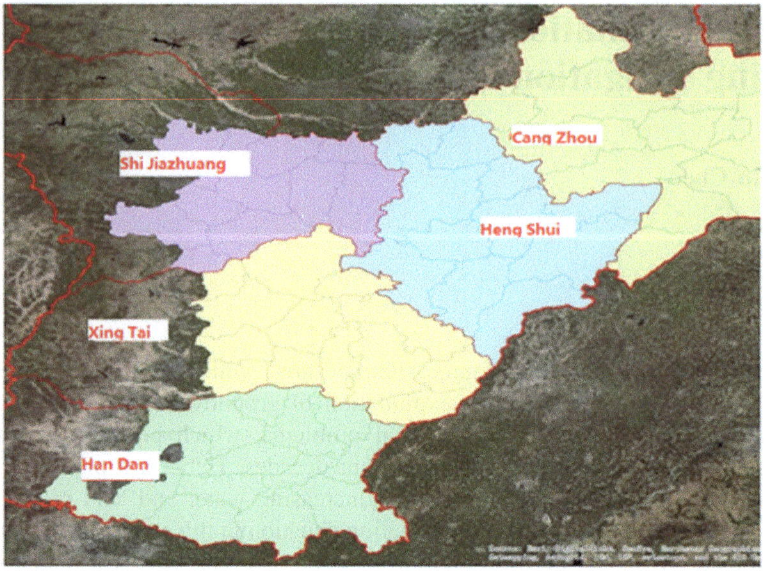

Fig. 1 Distribution of water-saving irrigation test points in Hebei Province

Geological Survey and according to data released since 1959, the north China plain ground settlement accumulated over 200 mm, which km^2 is more than 60,000 km^2, more than 40% of the north China plain area of 14 km^2.

Hebei has less than 300 m^3 of water per person and groundwater levels are falling. Therefore, in 2015, we have selected five regions in Hebei province. These five areas are the most serious land subsidence in Hebei province. Hebei province was selected as the test site, where we implemented the underground water pressure mining project and carried out water-saving irrigation measures. See Fig. 1 for the test point area. The colored areas in the figure are the test points for the implementation of water-saving irrigation measures.

2 Agronomic Water-Saving Irrigation

2.1 The Technology of Water-Saving and Stable Production of Winter Wheat

Wheat fields in overexploitation of groundwater were selected as these fields have a greater ability to store water. In these areas, new seed varieties are promoted as they have the characteristics of water-saving drought resistant. Meanwhile, some new techniques were included, for instance, deep loosening of the soil and adding reused straw to the fields, etc. During the fertility period of wheat, water should be watered down 1–2 times. After the implementation of the project, field irrigation water can save 50 m^3/μ (μ is an area unit; 1 μ is 666.7 m^2.).

2.2 *Adjustment of Planting Structure*

Adjustment of planting structure is one of the main ways of saving and efficiently utilizing water resources in the region. The principle of planting structure adjustment is to consider the law of crop water demand, water supply and water shortage under the premise of ensuring food security [3]. According to market demand, water consumption should be increased, and production value should be increased. Establishment of a water-saving and efficient planting structure by the proportion of good economic crops should be put forward.

This measure is to change the original high-water consumption of field crops into low water crops, trees, or fallow ground to aim at reducing the water consumption. In the area of serious overexploitation of groundwater, the area of winter wheat cultivated by groundwater irrigation should be reduced properly. The crops of winter wheat and summer maize are changed into crops such as corn, cotton, peanuts, anemone, and grains. To fully explore the yield potential of autumn crops in the same period, in combination with the development of animal husbandry, the development of corn, alfalfa and other crops is supported.

Due to the sharp decrease of irrigation water diversion, infiltration and replenishment of underground water decreased, groundwater level decreased, groundwater depth increased, and soil evaporation decreased in the field [4]. Through the adjustment of the planting structure, the planting area of the crops with large water consumption is reduced, the transpiration water consumption of the field crops is greatly reduced, the water demand of the crops is reduced, and the irrigation water is reduced.

3 Engineering Water-Saving Irrigation

In Hebei, governments often focus on two aspects in engineering water-saving irrigation: surface water replacement of groundwater technique, and improvement of groundwater utilization. Choosing a reasonable engineering technique could improve the benefit of water-saving in agriculture.

3.1 Determine the Irrigation Quota Under Different Irrigation Methods

In order to ensure the accuracy and practicality of the data in this paper, the area with different irrigation methods was selected as the experimental area and the irrigation experiment was carried out. The experiment of determining irrigation quota is divided into volume method and ultrasonic flow method.

(1) Volume method:

Firstly, on the water outlet, we should connect to a large bucket of 175L should be connected and the time recorded when a bucket of water was filled. By formula: $q = v_{(bucket)}/t$, the flow "q" is calculated. (In this formula, "q" stands for flow; "t" is the time it takes to fill a bucket of water; and "v" is the volume of the barrel for the test.)

Secondly, the flow is not always stable throughout the irrigation. Therefore, the volume method was used to measure three times and then calculate the average of its flow. The formula is: $Q_{(average)} = (q_1 + q_2 + q_3)/3$.

Thirdly, the time required for irrigation completion of the selected pilot area is recorded. By the formula: $V = Q_{(average)} * T$, the amount of water needed at the end of the irrigation pilot zone could be calculated. (In this formula, "V" is for irrigation water consumption; "$Q_{(average)}$" stands for average flow; and "T" stands for the time required for irrigation to complete the pilot zone.) See Fig. 2 for the field experiment.

(2) Ultrasonic flow monitor method:

The ultrasonic flow monitor was connected with the pipe by a coupling agent. After the flow was stable, it was measured once every minute. Finally, the flow value "q" in the steady process is obtained. By formula: $V = q * t$, the water consumption—"V" could be calculated. Figure 3 is the field diagram of the ultrasonic flow monitor.

Fig. 2 Volume method

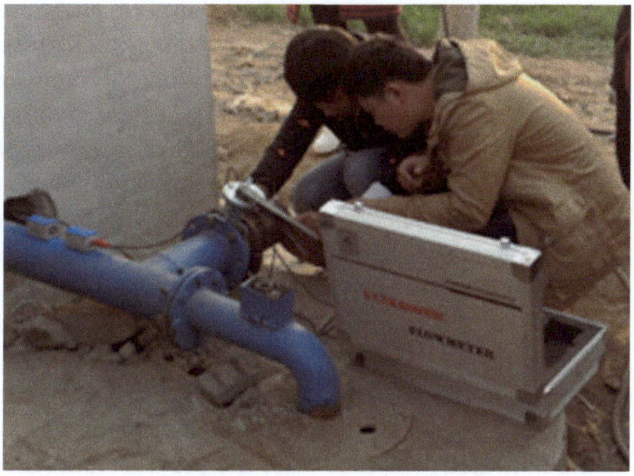

Fig. 3 Ultrasonic flow monitor method

Table 1 Water saving in different engineering water-saving irrigation (m^3/μ)

Category	Field crops[a]	Field crops	Vegetable	Chinese herbal medicine
Irrigation method	Micro sprinkler	Sprinkling irrigation	Drip irrigation	Micro sprinkler, drip irrigation, etc.
Groundwater saving	85	80–95	100–150	100

[a]The field crops here refer to wheat and corn and μ is an area unit, 1 μ is equal to 667 m^2

After the experiment, if the above two methods are basically consistent, then it can be considered as an effective test. Finally, the irrigation quota under different irrigation methods according to the formula: irrigation quota = water consumption/ area can be calculated.

Through the above two methods, the water consumption can be measured before and after the installation of water-saving measures, and then the water-saving amount of different irrigation methods determined, as shown in Table 1.

3.2 *Improve Groundwater Utilization*

To improve the efficiency of groundwater use sprinkler irrigation, implementation water and fertilizer integration technology, etc. can be installed.

Water and fertilizer integration technology is the use of the water pressure irrigation system, soluble solids or liquid fertilizers dissolved in irrigation water. Then, through the controlled pipeline system, fertilizer is placed directly into the soil near the root of the crop, which can accurately control the amount of irrigation and fertilization, significantly improve the utilization of water and fertilizer, and reduce water consumption. Meanwhile, water-saving irrigation equipment plays an important role in improving agricultural water use efficiency and increase in the income of farmers and construction of ecological environment [5]. It also has an important strategic position to ease the shortage in water resources, ensure the national food security, and promote the rural economy.

In recent years, although the speed of micro-irrigation in Hebei has developed rapidly, the products of micro-irrigation equipment are basically matched; and the model specifications are complete; however, the high-end products are less.

At present, this technology is applied in three aspects in Hebei: are field crops, vegetables, Chinese herbal medicine. Although these three aspects of water-saving principle are the same, the irrigation methods are different (see Table 1).

From Table 1, it can be seen that it is also a field crop, which is more water-saving than micro-spraying. After the implementation of water-saving measures, the vegetable water-saving effect is the largest, which can save 100–150 m^3/μ of groundwater compared with the previous border irrigation.

3.3 *Surface Water Replacement of Groundwater Technique*

Overexploitation of groundwater causes ecological problems such as subsidence of the ground. Through the construction of channels and ponds, the part of irrigation district irrigation using surface water, reduced the consumption of groundwater.

Because the water demand in each project area and the amount of river water allocated by the government are different, the calculation method of groundwater saving is different. If the water requirement of the surface water in the region is greater than the amount of water allocated to the region by the government, the surface water can displace part of the groundwater. If the water requirement for the surface water in the area is less than the amount of water allocated to the region by the government, all the groundwater will be replaced by the surface water.

Through the investigation of the engineering test site, it is concluded that the implementation area of surface water replaces the groundwater project and the underground water saving in the urban. This is shown in Table 2.

Table 2 In the surface water displacement groundwater project, the underground water saving in each area

	Area (m^2)	Groundwater saving (m^3)
Shijiazhuang	4.85	365.40
Cangzhou	51.13	6170.51
Hengshui	69.35	5459.91
Xingtai	24.85	2245.88
Handan	47.14	5194.92
Total	197.32	19,436.62

4 Management Water-Saving Irrigation

Agriculture is a big water user in all aspects, and it is an important work to strengthen the use of management in agricultural. Management of irrigation water is an important link in the application of water-saving irrigation.

First, with the increasing shortage of water resources and the increase of water demand in the development of agricultural production, the importance of planned water and water volume work is becoming more and more important. Hebei carried out a program to strengthen the management of water supply quotas.

Second, agricultural water price reform: development of reform programs to estimate the price standards and funds in different regions in Hebei. With the new situation of the agricultural economy structure adjustment, the scale of the water-saving irrigation project was expanded; and the state of the implementation of the policy to encourage grain crops, accelerate the pace of price reform, the establishment of a reasonable price formation mechanism and effective water fee collecting methods is imperative.

Third, finish water right on the basis of the right to register: (water right refers to the ownership of water and the general term for the right to use water.) Hebei selected pilot areas for the trading of the city's agricultural water rights. In recent years, water rights, water prices, water market transactions and other water rights management problems have been applied and practiced in Hebei, which show significant water-saving effects. The initial water rights, the water market and the transfer of water rights are the basic contents of the water right management system.

5 Benefits and Suggestions

5.1 Benefits

Water-saving irrigation could reduce the water lost in the irrigation process and increase the yield and output value. As Hebei is paying close attention to the development and research of water-saving irrigation technology, remarkable success has been achieved. Refer to Table 3 on different measures of water saving.

From Table 3, it can be seen that surface water displacement groundwater project saves the most groundwater. However, that does not negate the effect of other water-saving facilities because the area of different water-saving measures has a great impact on saving the underground water. The area of planting structure is small, so the underground water conservation is relatively small. Generally speaking, water-saving facilities play an active role in saving groundwater.

Table 3 Water saving under the different measures (million m^3)

Categories	Agronomic water saving irrigation		Engineering water saving irrigation		Sum
	The technology of water-saving and stable production of winter wheat	Adjustment of planting structure	Improve groundwater utilization	Surface water replacement of groundwater	
Groundwater saving	34,294.82	4302.32	8862.71	19,436.62	66,896.47

5.2 *Suggestions*

Although Hebei is developing water-saving irrigation systems in a wide scope, the majority of farmers cannot reasonably use water-saving irrigation facilities. Thus, water saving cannot be maximized and waste of resources. It is necessary to do a good job of education guidance to farmers ahead of time, execute the standards strictly, ensure famers can select equipment and improve management. However, there are many difficulties in managing these concerns. Government and authorities should strengthen supervision and administration of their work in water-saving irrigation.

To sum up, following are suggestions:

1. Governments at all levels, especially those below the county level, must establish special organizations in the formulation of water-saving plans. Formulating a unified water-saving development plan. A file should be set up and strictly enforced.
2. In the implementation of water-saving irrigation on flexibility, attempt to meet the masses need to avoid the contradiction of farming. To do a thorough and careful investigation into the household, it is necessary to determine the actual outlet of the peasant household, because farmers are the performers of water-saving irrigation.
3. The state should give different support in accordance with the degree of local water shortage and local financial capacity. Local governments should make a correct assessment of local water resources. The state should increase support for areas with severe water shortage and overdependence of agricultural production on groundwater. Before putting water resources into various economic work, water-saving construction funds should be greatly increased. Most of the money invested should be used to solve the engineering aging and reconstruction and improve the publicity effect of the water law.
4. Increase external surface water volume. Ways should be found to increase the amount of external surface water, and even refer to the water diversion in some parts of the south-to-north water. (The "south-to-north water diversion project" is a strategic project of the People's Republic of China, which mainly solves the

water shortage in northern China. Through the connection of three water diversion lines with the Yangtze River, the Yellow River, the Huaihe River and the Haihe River, the rational allocation pattern of the north and south allocation of water resources in China is realized.) To ensure the water availability of the project areas, the comprehensive benefit of surface water replacement of the groundwater project is given. Construction should be speeded up to meet the needs of supporting projects in the south-to-north water diversion project and realize the replacement of external water sources as soon as possible.

6 Conclusion

In Hebei, low use efficiency of water now is supposed to be the important constraints of national economic development and social sustainable development. Therefore, it has a special significance to intensely develop and spread advanced water-saving irrigation techniques. In this paper, through field experiment, the irrigation quota under a water-saving facility was studied. At the same time, by visiting and investigating the operation of different water-saving facilities in the county stage, some suggestions were put forward. These suggestions are of great significance for future research on the efficient use of water resources and water conservancy construction.

Acknowledgements This work has benefited from the National Natural Science Foundation of Hebei province underground water pressure mining project in 2015. Thank you to all the researchers working on this topic, who offered me the confidence and discussed with me about my paper.

References

1. Hou, Z., Zhang, H., Hao, N., et al.: The present situation of water resources in Hebei province and countermeasures for agricultural water conservation. Build. Mater. Decor. **19** (2017)
2. Zhao, H., Zhang, R., Qi, Y., et al.: Analysis on the dynamic changes of the water body in the underground water pressure mining area in Hebei Province. Water Saving Irrig. **7**(7), 66–69 (2017)
3. Zhang, J., Pei, Y., Guo, B., et al.: Effects of planting structure adjustment on regional water circulation. Geogr. Arid Reg. (Chin. version) **34**(1), 26–33 (2011)
4. Cui, Y.: An Example of Agricultural Water-Saving Irrigation Technology and Application. Chemical Industry Press (2005)
5. Zheng, Y.M., Hao, B., Huang, Z.B., Tian, X.L., Nie, F.Q., Zhao, Y., Zhai, J., Jiang, L.: Directional water collection on wetted spider silk. Nature **463**, 640–643 (2010)

Numerical Simulation for Impacts of Mountainous Tunnel Drainage on Groundwater Environment

Yong Xiao, Qichen Hao, Jingli Shao, Yali Cui and Qiulan Zhang

Abstract Mountainous tunnel drainage can cause various negative impacts on the groundwater environment and human life; as a result, it is necessary that the drainage is quantitatively estimated and minimized during the construction period. In this study, a numerical model was conducted to predict the influences of mountainous tunnel drainage on the groundwater environment in northern China. The results show that the drainage would change the groundwater flow field and form drawdown funnels; however, it would not cause regional groundwater drawdown. Besides, the discharge amount of springs was also affected by the tunnel drainage, and the maximum reducing amount was up to 25%. The storage resources of the aquifers were decreased under the effect of tunnel drainage. All negative influences could be gradually eliminated after the strong drainage. This research can provide effective methods to measure and decrease the impacts of tunnel drainage.

Keywords Numerical modelling · Groundwater · Tunnel drainage
Water inflow · Environment

Y. Xiao · J. Shao (✉) · Y. Cui · Q. Zhang
School of Water Resources and Environment,
China University of Geosciences (Beijing), Beijing 100083, China
e-mail: jshao@cugb.edu.cn

Y. Xiao
e-mail: xiaoyong@cugb.edu.cn

Y. Cui
e-mail: cuiyl@cugb.edu.cn

Q. Zhang
e-mail: qiulanzhang919@cugb.edu.cn

Y. Xiao
Faculty of Geosciences and Environmental Engineering,
Southwest Jiaotong University, Chengdu 610031, China

Q. Hao
Institute of Hydrogeology and Environmental Geology,
Chinese Academy of Geological Sciences, Shijiazhuang 050061, China
e-mail: haoqichen1986@163.com

W. Dong et al. (eds.), *Sustainable Development of Water Resources and Hydraulic Engineering in China*, Environmental Earth Sciences,
https://doi.org/10.1007/978-3-319-61630-8_18

1 Introduction

As one of the most important parts of communication construction, a tunnel can significantly reduce the distance of transportation and improve traffic efficiency. However, the construction of a tunnel would cause a series of hydrogeological and environmental issues on groundwater dependent ecosystems [1]. Due to the increasing concern about environmental issues of large engineering projects, it is agreed that the effects of the tunnel on the quantitative and qualitative status of the water masses excavation should be avoided [2]. To weaken the impacts of tunnel excavation, measures should be taken appropriately so that the potential impacts can be correctly identified [3].

Currently, qualitative methods and quantitative methods are the main methods for identifying tunnel excavation impacts. The qualitative methods provide a basis for establishing objective conceptual hydrogeological models and determining the evolution trend of the groundwater environment under the condition of tunnel excavation. However, qualitative methods could not illustrate the degree of the impact on the water environment caused by tunnel excavation. While, quantitative methods can provide convincing information for estimating the effects. Numerical models are widely used to conduct quantitative methods. It could take all related variables into consideration, which is suitable for such complex hydrological conditions.

In this study, a 3-D numerical flow model was conducted to analyze the impacts of a mountainous tunnel construction on groundwater flow, private wells and springs. The results can provide reference to evaluate the influences of mountainous tunnel construction on the groundwater environment.

2 The Study Area

The tunnel is in a mountainous area in northern China (Fig. 1). The main stratums outcropping in this area include: Quaternary, Tertiary, Cambrian and Sinian strata (Fig. 2). The Quaternary deposits get thicker from north to south. The upper Quaternary deposits are Aeolian loess with vertical bedding joints, and the lower Quaternary deposits are solid clay with very poor permeability, leading to weak hydraulic connection between Quaternary strata and Tertiary strata. The upper Tertiary strata is mainly comprised by weakly cemented conglomerate, mudstone and sandy mudstone, while the lower Tertiary strata is mainly comprised of conglomerate, sand, sandstone and sandy mudstone. The lower Tertiary strata is the main groundwater aquifer. The Cambrian and Sinian strata develop fractures differentially; the Cambrian develop more fractures than Sinian strata.

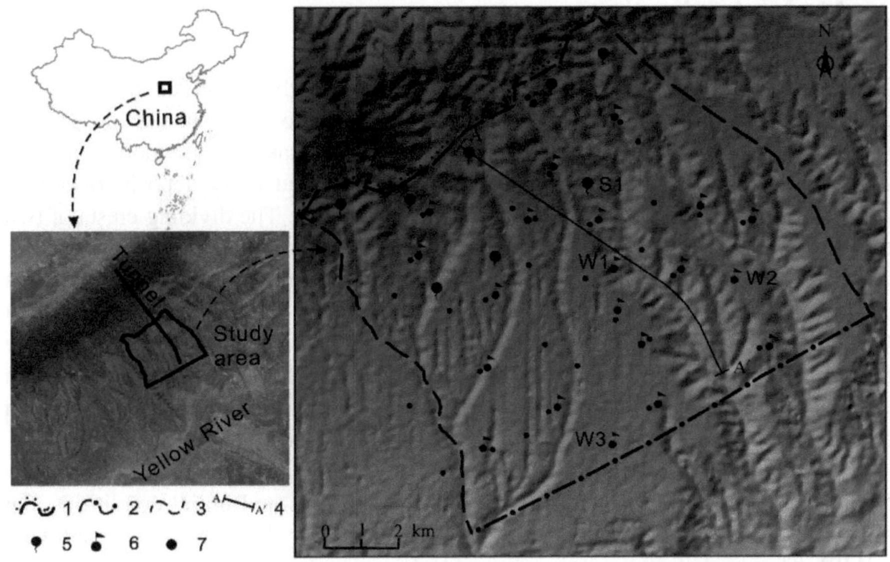

Fig. 1 Schematic map of model domain and location of observation wells. 1-Given flux boundary, 2-Given head boundary, 3-General head boundary, 4-Cross section, 5-Spring, 6-Well, 7-Village

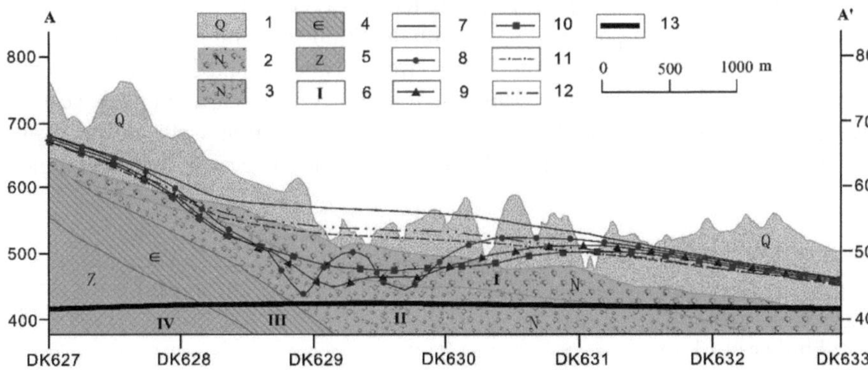

Fig. 2 Section view of lithology and groundwater level drawdown along the A-A'. 1-Quaternary deposits, 2-Upper Tertiary strata, 3-Lower Tertiary strata, 4-Cambrian strata, 5-Sinian strata, 6-Aquifer label, 7-Initial water level, 8-Water level in 6th month, 9-Water level in 12th month, 10-Water level in 18th month, 11-Water level in the 1st year after construction, 12-Water level in the 3rd year after construction, 13-Design tunnel

Under natural condition, groundwater is recharged by precipitation in the northern area, and flows from north to south. Groundwater discharges include springs, lateral outflow and evaporation and exploitation. However, the quantity of exploitation is less than 2000 m^3/d and is mainly exploited in the Quaternary aquifer and Tertiary aquifer.

3 Model Application

The conceptual model of the study area was established based on identification and characterization of the hydrogeological conditions. Due to the weak specific yield of rock in the northern area, the southern deposits area was taken as the modelling area. The northern boundary is between the bedrock and Quaternary deposits occurring area. The southern boundary is the Yellow River terraces. The dividing crests of two rivers in the east and west were set as general head boundaries. The northern boundary was defined as the given flux boundary, and the southern boundary was defined as the given (constant) head boundary. As the poor permeability of Quaternary deposits in the lower part, the strata below Quaternary deposits were taken as the modelling strata and divided into four aquifers. The upper Tertiary strata was defined as aquifer I, and the lower Tertiary strata was aquifer II. The Cambrian and Sinian strata are defined as aquifers III and IV, respectively.

The regional groundwater flow can be expressed as the equation described by Wang et al. [4]. Combing with the boundary conditions and initial conditions, the numerical model was established using Groundwater Model System software (GMS). The modelling area was discretized as a matrix of 205 rows × 210 columns × 4 layers. The total number of valid cells was 26,403 and each cell was 100 m × 100 m.

The numerical model was manually adjusted and calibrated using trial and error method. As shown in Fig. 3, there are ten observed wells with the fitting error less than 1 m, and seven observed wells with the fitting error ranging between 1 and 2 m. The number of wells with the fitting error between 2 and 3 m is three. Only one well has the fitting error over 3 m. The four kinds of observation wells account for 47.6%, 33.3%, 14.3% and 4.8% of the total observation wells, respectively. The results suggest that the model can reflect the objective hydrogeological

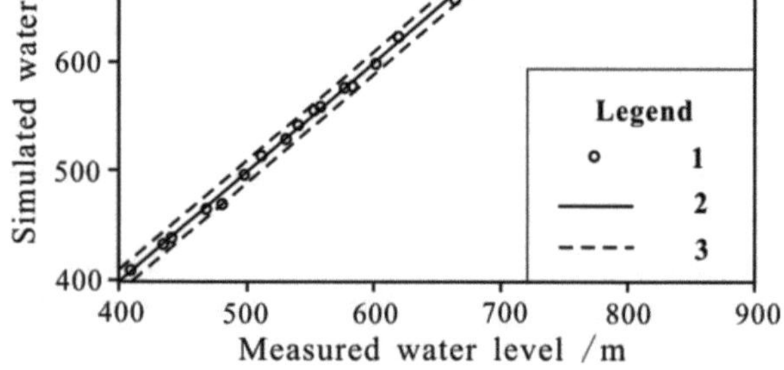

Fig. 3 Comparison chart of simulated and measured water level. 1-Wells, 2-Diagonal line, 3-Error line (±3 m)

condition of the study area and can also be used to simulate the impacts of tunnel drainage.

4 Tunnel Impact Evaluation

4.1 Tunnel Drainage Scenarios

According to the tunnel design, the construction period of the tunnel is five years. The predicting period of the simulation ranged from 2016 to 2045. The water quantity of tunnel drainage during the construction period is estimated using Eq. (1), and that after the construction is calculated using the equation described by Yang et al. [5]:

$$Q = \frac{4pmLK(H - r)}{\ln(2(H - r)/r)} \tag{1}$$

where,

K is the hydraulic conductivity (m/d),
H is the vertical distance of the tunnel below the water level (m),
r is the radius of the tunnel (m),
L is the distance from the tunnel face to secondary lining (m),
m is a conversion coefficient, which usually equals to 0.86.

4.2 The Impacts on Groundwater Level

Figure 2 shows the variation of groundwater level during the construction period. In the initial period of construction, the discharge amount is small and has little effects on the groundwater flow system. After six months, with the increase of water drainage amount, the groundwater flow pattern begins to change, and two groundwater drawdown funnels were formed in the area with high specific yield. At DK629 + 900, the groundwater level starts to increase due to the reducing of water drainage amount; however, the groundwater drawdown area keeps expanding. After 12 months, the two groundwater drawdown funnels expand to one funnel. As the lag of groundwater system response, the groundwater drawdown area keeps expanding until one year after the construction completion. The maximum drawdown area covers 23.95 km^2 (Fig. 4). Groundwater level was observed to increase rapidly after strong drainage; and about one year later, groundwater flow pattern recovered to a state similar to the initial natural state.

The groundwater level generally decreases with the increase of the distance from the center of the drawdown funnel. The largest groundwater drawdown depth is

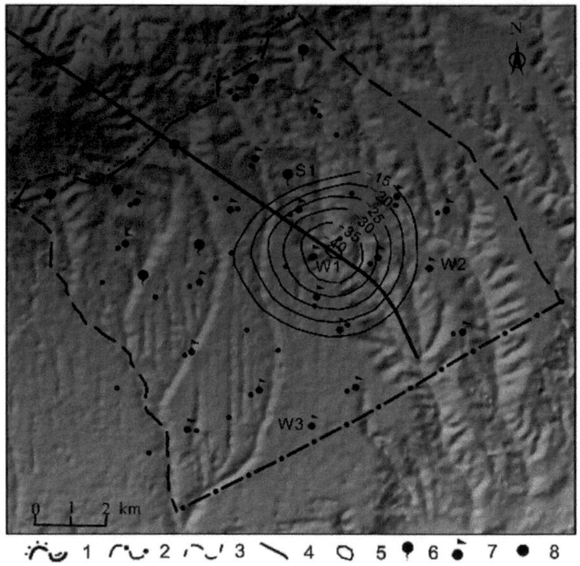

Fig. 4 The maximum impact range of groundwater funnel. 1-Given flux boundary, 2-Given head boundary, 3-General head boundary, 4-Tunnel, 5-Contours of drawdown depth (m), 6-Spring, 7-Well, 8-Village

175 m in the sixth month, and the maximum radius of the drawdown funnel reached 400 m in the twelfth month. Groundwater level fluctuation shows an increasing trend and has a short recovery time in the high specific yield area. Meanwhile, the groundwater level drawdown and the recovery time have little negative effects on the private wells.

4.3 The Impacts on Springs and Groundwater Storage Resource

There are two kinds of springs distributed in the study area, one is the erosion springs located in the loess area, and the other is the overflow springs in the mountainous area. Most of the springs are distributed away from the area of high specific yield; the discharge amount of the springs is less affected by the tunnel drainage. Only one erosion spring (S1) located in the east of DK 627 + 500 is close to the area of high specific yield and is strongly affected by the tunnel drainage (Fig. 5). The initial discharge of S1 is 841 m^3/d and maximally decreases to 629 m^3/d during the tunnel drainage, which is only 75% of the initial discharge. After 3.5 years, with the decrease of the tunnel drainage amount, the spring discharge gradually recovers to 90%.

Surface water reservoirs are distributed in the study area, which have weak hydrological connection with groundwater due to the existence of the aquitards. Under current conditions, the recharge amount is 4320 m^3/d; while, half a year after the construction, the recharge amount is 3866 m^3/d, which is the 89% of the initial

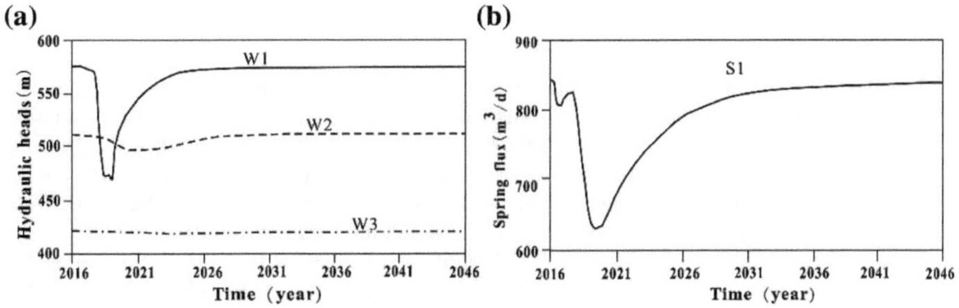

Fig. 5 **a** Process curve of water level fluctuation for typical observation wells, **b** Process curves of spring flux fluctuation

amount. The recharge amount can recover to 4093 m^3/d after 5.5 years, accounting 95% of the initial amount. Furthermore, the tunnel drainage has temporal influences on the reservoir resource and can quickly recover after the construction. As a result, the tunnel construction has little effects on the groundwater system in respect to the long term.

5 Summary

In this study, the impacts of mountainous tunnel drainage on groundwater were quantitatively assessed using numerical modelling approach. The results indicate that numerical simulation is a useful and effective tool to identify potential impacts on the groundwater environment. The tunnel drainage can result in fluctuations of groundwater level, and forming groundwater drawdown funnels. However, the serious drawdown area is no more than 400 m away from the tunnel and can be recovered rapidly after the tunnel construction. Tunnel drainage also reduces the discharge amount of springs and the aquifer reservoir quantity; however, the influences can be rapidly eliminated after the strong drainage. The private wells close to the tunnel are affected by tunnel drainage, but the influence is limited and temporal. In general, mountainous tunnel drainage would have negative influences on the groundwater environment as well as human life, but these influences can be eliminated rapidly after the tunnel construction due to the rich rainfall in the mountainous area.

Acknowledgements This research was financially supported by the Fundamental Research Funds for Central Public Research Institutes (YYWF201626), the Fundamental Research Funds for the Central Universities (No. 2652016022), the Public Welfare Industry Special Funds for Scientific Research from Ministry of Land and Resources of China (201211079-4) and China Railway Tunnel Survey & Design Institute CO., LTD.

References

1. Vincenzi, V., Gargini, A., Goldscheider, N.: Using tracer tests and hydrological observations to evaluate effects of tunnel drainage on groundwater and surface waters in the northern apennines (Italy). Hydrogeol. J. **17**, 135–150 (2009). https://doi.org/10.1007/s10040-008-0371-5
2. Kazagli, E., Bierlaire, M.: Directive 2000/60/ec of the European parliament and of the council establishing a framework for the community action in the field of water policy (EU Water Framework Directive). In: International Choice Modelling Conference (2015)
3. Raposo, J.R., Molinero, J., Dafonte, J.: Quantitative evaluation of hydrogeological impact produced by tunnel construction using water balance models. Eng. Geol. **116**, 323–332 (2010). https://doi.org/10.1016/j.enggeo.2010.09.014
4. Wang, S., Shao, J., Song, X., Zhang, Y., Huo, Z., Zhou, X.: Application of modflow and geographic information system to groundwater flow simulation in north China plain, China. Environ. Geol. **55**, 1449–1462 (2008). https://doi.org/10.1007/s00254-007-1095-x
5. Yang, Y.N., Zhang, Q., Xu, M.: Numerical simulation method utilization in water gushing yield forecasting of paoziling tunnel in Hunan Province, China. Appl. Mech. Mater. **580–583**, 1392–1397 (2014). https://doi.org/10.4028/www.scientific.net/AMM.580-583.1392

The Influence of Straw Mulch on the Transport and Distribution of Water and Salt in Indirect Drip Irrigation with Saline Water

Ce Zhang, Lihua Jia, Yingying Shen and Chao Ma

Abstract Straw mulch could effectively inhibit the ineffective evaporation of soil water, which could significantly improve the water use efficiency of crops and inhibit the return of soil salt. The research and popularization of straw mulch technology is beneficial to the sustainable development of agriculture. An indoor, indirect drip irrigation experiment with saline water was done under different straw thicknesses and straw comminution degrees, analyzing the soil water content and soil electrical conductivity. The results showed that: the hindering effect of straw mulch on water and salinity increased with the increase of straw mulch thickness; soil water content decreased with the distance from the water-conducting device increasing; soil electrical conductivity reached the minimum in the water outlet of a water-conducting device; and it reached a maximum in the straw mulching edge and wetting front boundary. The hindering effect of straw mulch on water and salinity decreased with the increase of the straw comminution degree.

Keywords Straw mulch · Straw thickness · Straw comminution degree
Soil water content · Soil electrical conductivity

C. Zhang (✉) · C. Ma
Hydropower College, Hebei University of Engineering, Handan 056000, China
e-mail: 1325411917@qq.com

C. Ma
e-mail: 18231086956@163.com

L. Jia
Shaanxi Agricultural Engineering Survey and Design Institute, Xian 710068, China
e-mail: 2375256152@qq.com

Y. Shen
China Irrigation and Drainage Development Center, Beijing 100000, China
e-mail: shenyingying_cugb@163.com

W. Dong et al. (eds.), *Sustainable Development of Water Resources and Hydraulic Engineering in China*, Environmental Earth Sciences,
https://doi.org/10.1007/978-3-319-61630-8_19

1 Introduction

Water resource is an important strategic resource in the development of a regional social economy [1]. At present, drought and water shortage has become an important restriction factor of national economic and social sustainable development, especially the shortage of freshwater resource has become increasingly prominent. Therefore, water resource has become an important limiting factor in agricultural irrigation, a serious impediment to sustainable agriculture development [2]. Drip irrigation and micro-irrigation with brackish water are two types of water-saving irrigation. In addition, straw mulching and water saving technology are also widely used. About the effect of straw mulch, many domestic and foreign experts and scholars have done a great deal of research work. The test results showed that the coverage measures have the advantages of lowering cost, inhibiting soil water evaporation, improving crop water use efficiency and preventing soil salinization; and the problem of “white pollution” does not exist. The research and popularization of straw mulching technology is beneficial to the sustainable development of agriculture [3, 4]. Using drip irrigation with saline water can maintain high crop yield [5, 6]. The results showed that using drip irrigation of saline water irrigation can obtain a higher yield than traditional ground irrigation, greatly reducing the consumption of water resources. In recent years, domestic and foreign scholars have done many researches on brackish water irrigation, drip irrigation, straw covering and returning technology; however, the combination of the three above is less. There will be enormous potential, if the straw mulching, drip irrigation and brackish water irrigation are combined. There were influences of straw mulching under the saline water on water and salt transport in soil [7, 8]. Therefore, this paper will discuss the impact on the distribution of soil water and salt movement under the condition of straw mulching on micro-saltwater indirect drip irrigation, which will provide a theoretical basis for micro-salty water and straw resources for agricultural irrigation and water-saving agriculture.

2 Materials and Methods

2.1 Experimented Soil

The experiment was conducted in the laboratory of soil and water analysis, Hebei University of Engineering. Soil samples for the test were separated through a 2 mm sieve after being dried and crushed. The prepared soil with specific moisture content were rammed within 5 cm thick according to the specific layered bulk density of the soil, and then placed into an experimental soil box. To make the close contact between each layer, the lower layer soil should not be compacted before loading the upper layer of soil. The soil should be naturally settled one day to obtain a uniform initial water content profile of the soil.

2.2 Experimented Device

The test device was composed of two parts, the test soil box and the water supply device, which are shown in the Fig. 1. The soil box was made of column and organic glass with the length, width and height of 20 cm × 15 cm × 25 cm. The water supply bottle was a Markov bottle, which can provide a constant water head. The Markov bottle was also made from organic glass and the bottle body had scales and the cross-sectional area which was 30.5 cm^2. The drip emitter was modeled with the medical syringe and the needles were connected with PE pipe. During the experiment, the Markov bottle was connected with PE pipe, adjusting the injector control valve to control dripper discharge. Dripper discharge was measured with a stopwatch and cylinder, maintaining the same flow rate of effluent in different test.

2.3 Experiment Method

According to the layered loading soil method above, there were four layers, 5 cm per layer; there were 2100 g soil in each layer, compacted and polished between each upper and lower layers. Finish loading the last layer of soil, surfacing with different thickness of straw or shredded, according to the different level of the test, was finally compacted.

Fig. 1 Device of experiment

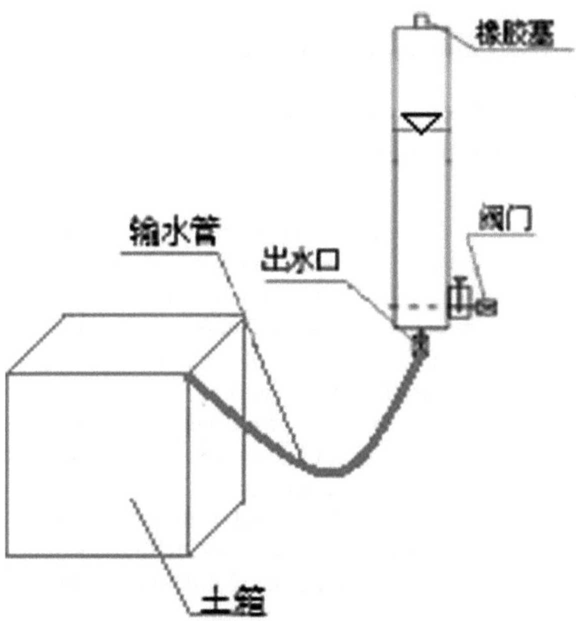

The solution was prepared with mixed sodium sulphate, calcium chloride, and magnesium chloride, together with a certain proportion. The mineralization degree of saline water was 4 g/L, sodium adsorption ratio (SAR) was 10, and volume of solution was 2 L. Then the prepared solution was poured into the Markov bottle.

The infiltration time and cumulative infiltration were recorded by the Markov bottle marks. During the test, the dynamic change process of wetting front on both sides of the soil box was depicted. After the end of the trial, the wetting front shape was described on tracing paper; and the exact coordinates of the wetting front were read on coordinate paper, and the horizontal and vertical maximum wetting distance were recorded. When water was dropped out, the soil levelled, and quickly borrowed soil. Soil samples lay on where vertical and horizontal spacing were also 2.5 cm, and the depth reached the top of the wetting front after separating the moist soil from the dry soil. The moisture content of the soil sample was measured by the drying method.

2.4 Experiment Design

This experiment was a two-factors test; the test factors were straw mulching thickness and straw degree. Both factors set two levels; straw mulching thicknesses were 2, 4, and 6 cm; and straw degrees were 0.5, 1, and 1.5 cm. Soil moisture content, electrical conductivity and infiltration rate were measured. The wetting front was observed and described. Finally, a regular pattern was observed.

3 Results and Discussion

3.1 The Effect of Different Mulching Straw Thickness on Wetting Front Migration

The curve of wetting front migration is shown in Figs. 2, 3, and 4, under the condition of different straw mulching thickness (2, 4, and 6 cm), the depth of water conducting device (4 cm) and the straw crushing degree (1 cm). The effect of different mulching straw thickness on the horizontal wetting front migration is shown in Fig. 5. The effect of different straw mulching thickness on the vertical wetting front is shown in Fig. 6.

Under the same conditions, except different thickness of straw mulching, horizontal and vertical wetting front migration distance both were increasing along with the increase of the infiltration time; and the vertical wetting front migration rate was faster than that of the horizontal wetting front. In the initial stage, the wetting front migration rate was larger; however, with the extension of the infiltration time, the wetting front migration rate of different thickness of straw mulching decreased

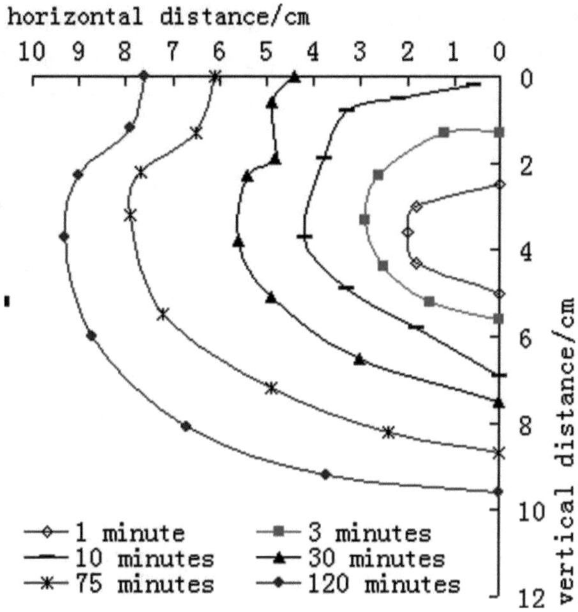

Fig. 2 Straw mulching thickness was 2 cm

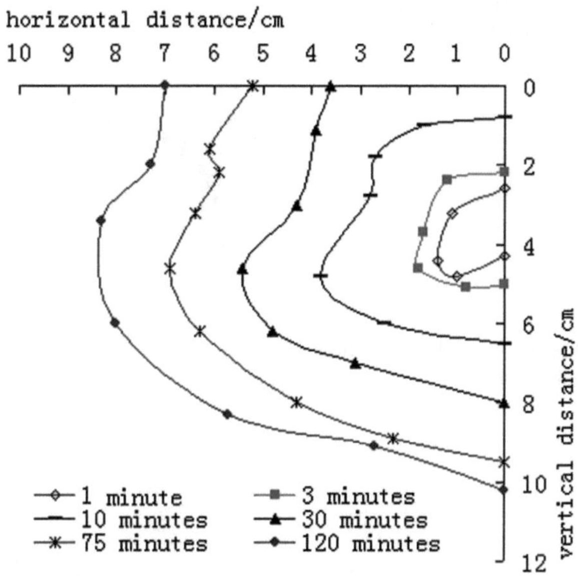

Fig. 3 Straw mulching thickness was 4 cm

gradually. The thicker the surface of straw, blocking infiltration of straw was more obvious in the horizontal direction and the level of horizontal wetting front was slowing; however, the vertical wetting front migration was relatively faster.

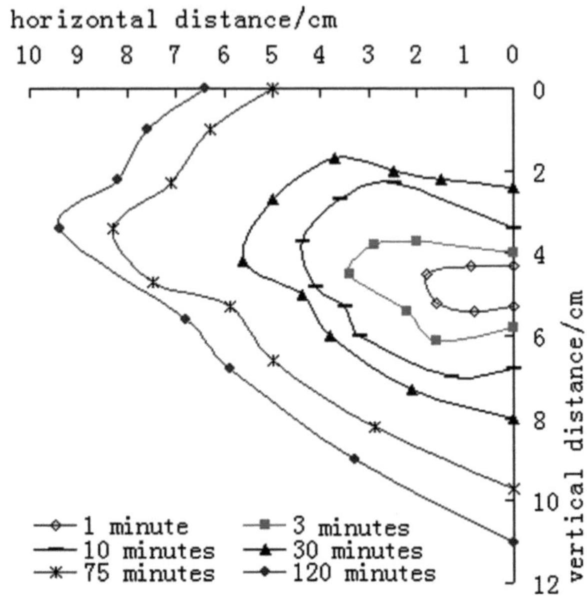

Fig. 4 Straw mulching thickness was 6 cm

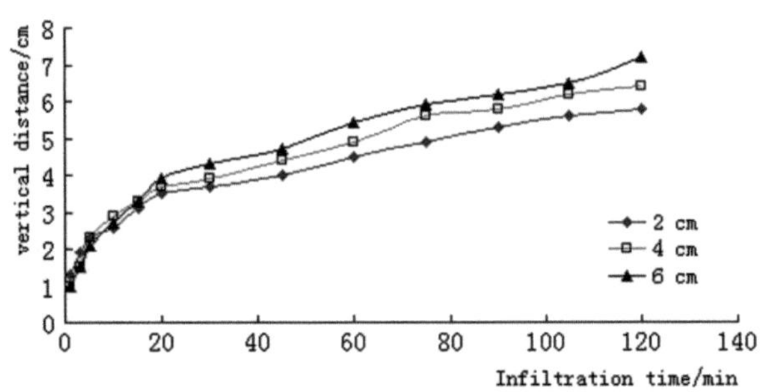

Fig. 5 Horizontal wetting front migration

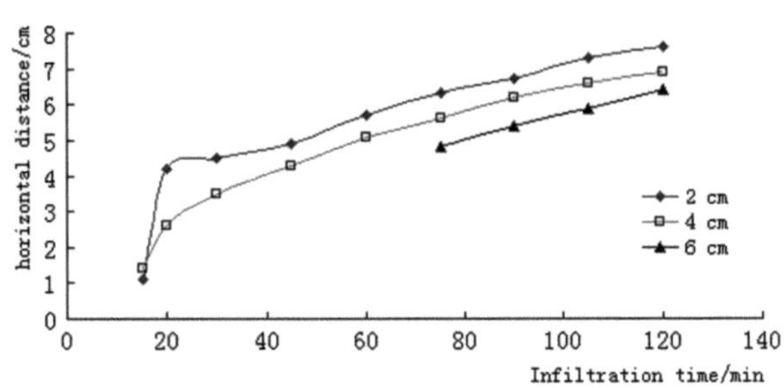

Fig. 6 Vertical wetting front migration

3.2 The Effect of Different Thickness of Straw Mulching on Soil Water Content and the Soil Electrical Conductivity Distribution

Changes of soil water content and soil electrical conductivity is shown in Tables 1, 2, and 3, under the condition of different straw mulching thickness (2, 4, and 6 cm), the depth of water conducting device (4 cm), and the straw crushing degree (1 cm).

When the other conditions were the same, the soil water content of the water-conducting device outlet was the highest under different straw mulching thickness. The final soil water content showed a decreasing trend with the increase in the distance from the water outlet center. The greater the center distance from the water outlet, the smaller the soil water content was.

There was a small part that the soil electrical conductivity was small at the water outlet of the water-conducting device, and an increasing salt area was formed outside that part. Conductivity gradually decreased first and then gradually became larger from the inside to the outside with the water outlet as the center, and there was a performance of salt leaching. The thicker the straw mulching layer, the greater the conductivity there was.

3.3 The Effect of Different Straw Comminution Degree on Wetting Front Migration

The curve of wetting front migration was shown in Figs. 7, 8, and 9, under the condition of different straw comminution degree (0.5, 1, and 1.5 cm), the depth of water conducting device (4 cm), and the straw mulching thickness (4 cm). The effect of different straw comminution degree on the horizontal wetting front migration is shown in Fig. 10. The effect of different straw comminution degree on the vertical wetting front is showed in Fig. 11.

Under the same conditions, except different straw comminution degree, horizontal and the vertical wetting front migration distance both were increasing along with the increase of the infiltration time; and the vertical wetting front migration

Table 1 Soil water content and soil electrical conductivity under the condition that straw mulching thickness was 2 cm

Soil water content					Soil electrical conductivity (μs/cm)				
10 cm (%)	7.5 cm (%)	5 cm (%)	2.5 cm (%)	0 cm	10 cm	7.5 cm	5 cm	2.5 cm	0 cm
22.2	27.3	29.7	28.0	2.5	180	176	134	168	2.5
23.0	27.5	28.6	29.7	5	120	107	116	125	5
20.0	25.0	25.7	25.0	7.5	118	100	108	118	7.5
	20.5	20.9	22.2	10		126	115	120	10

Table 2 Soil water content and soil electrical conductivity under the condition that straw mulching thickness was 4 cm

Soil water content					Soil electrical conductivity (μs/cm)				
10 cm (%)	7.5 cm (%)	5 cm (%)	2.5 cm (%)	0 cm	10 cm	7.5 cm	5 cm	2.5 cm	0 cm
26.3	23.8	35.3	35.7	2.5	411	325	316	289	2.5
18.2	35.7	38.5	41.7	5	342	463	300	280	5
18.2	23.3	24.2	24.1	7.5	253	267	221	296	7.5
	19.4	20.7	22.2	10		253	255	302	10

Table 3 Soil water content and soil electrical conductivity under the condition that straw mulching thickness was 6 cm

Soil water content					Soil electrical conductivity (μs/cm)				
10 cm (%)	7.5 cm (%)	5 cm (%)	2.5 cm (%)	0 cm	10 cm	7.5 cm	5 cm	2.5 cm	0 cm
23.0	25.0	30.3	29.0	2.5	215	214	166	200	2.5
22.8	32.1	30.8	34.8	5	221	185	136	156	5
22.2	29.0	29.6	36.0	7.5	212	176	168	152	7.5
	19.0	21.4	22.0	10		174	134	130	10

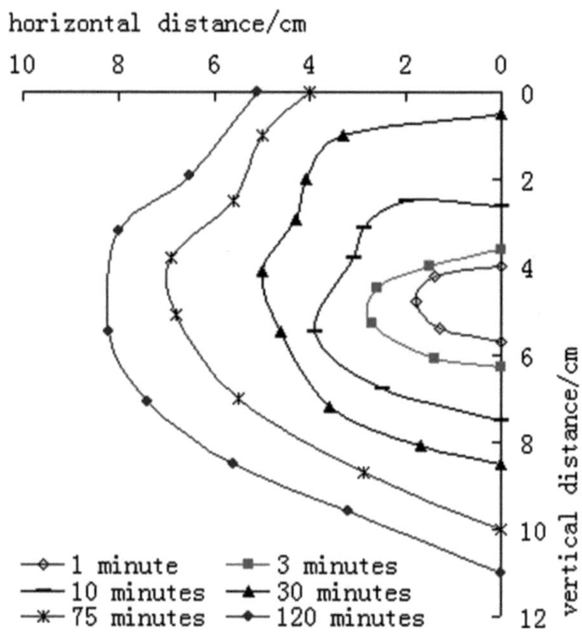

Fig. 7 Straw comminution degree was 0.5 cm

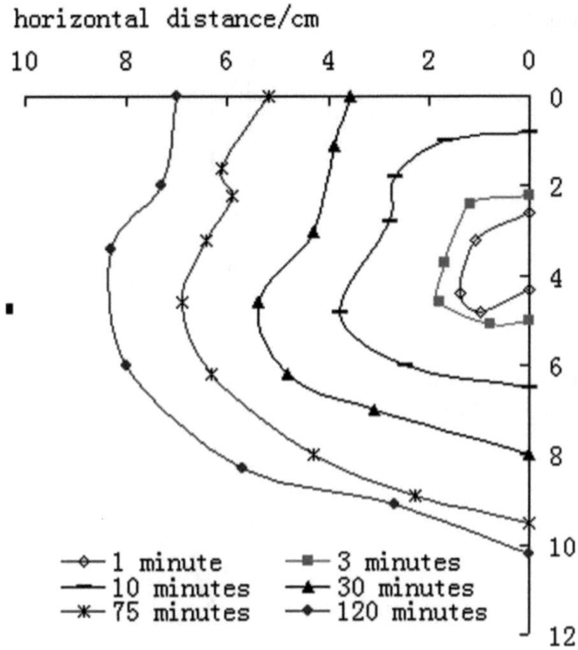

Fig. 8 Straw comminution degree was 1 cm

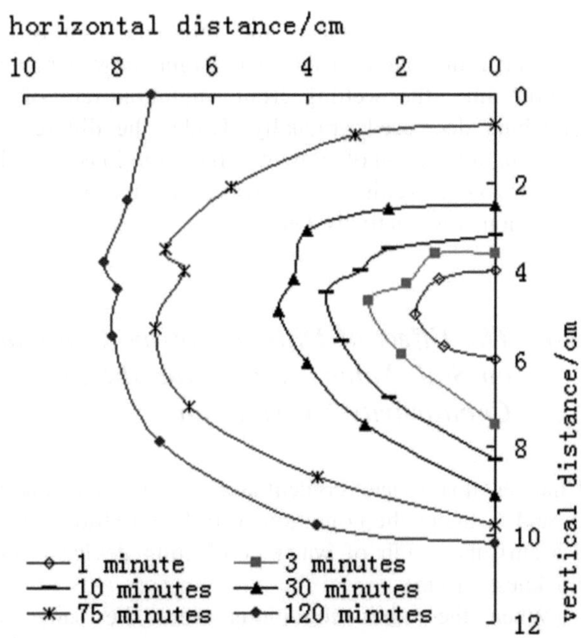

Fig. 9 Straw comminution degree was 1.5 cm

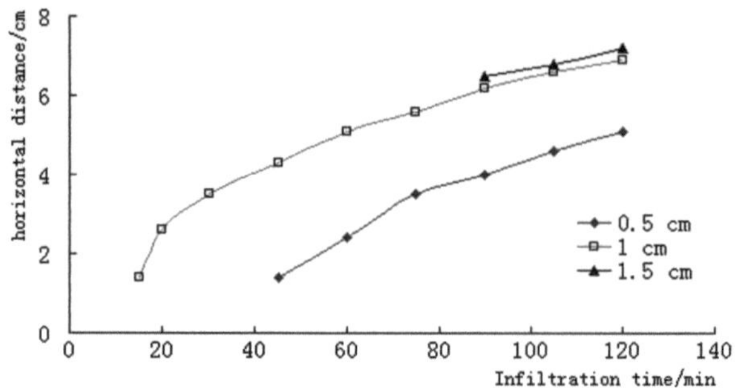

Fig. 10 Horizontal wetting front migration

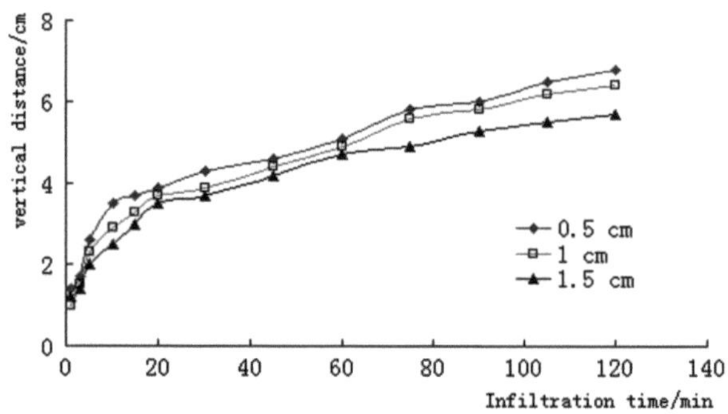

Fig. 11 Vertical wetting front migration

rate was faster than that of the horizontal wetting front. In the initial stage, the wetting front migration rate was larger; however, with the extension of the infiltration time, the wetting front migration rate of different thickness of straw mulching decreased gradually. Under the different straw comminution degree, blocking infiltration of straw was more obvious in the horizontal direction; the level of horizontal wetting front was slowing; however, the vertical wetting front migration was relatively faster.

3.4 The Effect of Different Straw Comminution Degree on Soil Water Content and the Soil Electrical Conductivity Distribution

Changes of soil water content and soil electrical conductivity is shown in Tables 4, 5, and 6, under the condition of different straw comminution degree (0.5, 1, and 1.5 cm), the depth of water conducting device (4 cm), and the straw mulching thickness (4 cm).

When the other conditions were the same, the soil water content of water-conducting device outlet was the highest under different straw comminution degree. The final soil water content showed a decreasing trend with the distance

Table 4 Soil water content and soil electrical conductivity under the condition that straw comminution degree was 0.5 cm

Soil water content					Soil electrical conductivity (μs/cm)				
10 cm (%)	7.5 cm (%)	5 cm (%)	2.5 cm (%)	0 cm	10 cm	7.5 cm	5 cm	2.5 cm	0 cm
25.0	25.8	29.0	36.7	2.5	248	192	208	212	2.5
20.5	24.1	23.7	29.4	5	167	162	176	175	5
24.3	21.9	25.0	25.8	7.5	155	151	150	155	7.5
	21.4	20.5	22.6	10		148	149	145	10

Table 5 Soil water content and soil electrical conductivity under the condition that straw comminution degree was 1 cm

Soil water content					Soil electrical conductivity (μs/cm)				
10 cm (%)	7.5 cm (%)	5 cm (%)	2.5 cm (%)	0 cm	10 cm	7.5 cm	5 cm	2.5 cm	0 cm
26.3	23.8	35.3	35.7	2.5	411	325	316	289	2.5
18.2	35.7	38.5	41.7	5	342	463	300	280	5
18.2	23.3	24.2	24.1	7.5	253	267	221	296	7.5
	19.4	20.7	22.2	10		253	255	302	10

Table 6 Soil water content and soil electrical conductivity under the condition that straw comminution degree was 1.5 cm

Soil water content					Soil electrical conductivity (μs/cm)				
10 cm (%)	7.5 cm (%)	5 cm (%)	2.5 cm (%)	0 cm	10 cm	7.5 cm	5 cm	2.5 cm	0 cm
	17.2	17.5	23.1	2.5		273	397	418	2.5
22.2	23.8	26.9	31.0	5	460	240	330	300	5
21.2	23.8	25.0	26.5	7.5	270	226	194	216	7.5
	21.7	22.9	23.3	10		233	238	228	10

from water outlet center increasing. The greater the center distance from the water outlet, the smaller the soil water content was.

There was a small part that the soil electrical conductivity is small at the water outlet of the water-conducting device, and an increasing salt area was formed outside that part. Conductivity gradually decreased first and then gradually became larger from the inside to the outside with water outlet as the center, and there was a performance of salt leaching. The greater the straw comminution degree, the smaller the water infiltration resistance was, and the faster the salt migration. That means conductivity was increasing with the straw comminution degree. The salt content was relatively larger on the straw layer and the wetting front edge region.

4 Conclusions

In this paper, the effect of straw mulching indirect drip irrigation on soil water content and electrical conductivity was studied, and the water and salt distribution in the soil under the condition of straw mulching indirect drip irrigation were discussed. The main conclusions are as follows:

In the test conditions of straw mulching thickness and straw comminution degree, horizontal and vertical wetting front migration distance both were increasing along with the increase of the infiltration time; and the vertical wetting front migration rate was faster than that of the horizontal wetting front. In the initial stage, wetting front migration rate was larger; however, with the extension of the infiltration time, the wetting front migration rate all decreased gradually. The soil water content of water-conducting device outlet was the highest. The soil water content showed a decreasing trend with the increase in the distance from water outlet center.

Conductivity gradually decreased first and then gradually became larger from the inside to the outside in water outlet as the center, and there was a performance of salt leaching. The salt content was relatively larger on the straw layer and the wetting front edge region. Soil moisture distribution opposed the distribution of salt, and that was favorable to the crop root growth and crop yield. The straw layer conductivity was relatively larger, indicating that the straw has some hindrance to salt and water movement. With the comprehensive effect of salt leaching and straw mulching, salt migrated and accumulated at the straw layer and wetting front edge region.

Utilizing brackish water to drip irrigation, under the other same conditions, the thicker of the straw mulching, the more obvious the hindering effect of straw on horizontal moisture permeation was; that was to say the horizontal wetting front transported more slowly, and vertical wetting front transport was relatively faster. Straw in the horizontal water movement was slower with decrease in the straw comminution degree; however, the vertical water movement was relatively faster. The larger the straw comminution degree was, the faster the salt migrated. Salinity content was higher at straw layer and wetting front edge region.

Acknowledgements This work was supported by the Natural Science Foundation program of Hebei Province (No. E2016402169) and Hebei Province Science and Technology Research Project (No. 14227002D.

References

1. Danilov-Danil'yan, V.I.: Water resources: a strategic factor in the long-term development of the Russian economy. J. Her. Russ. Acad. Sci. **79**(5), 420–428 (2009)
2. Zhao, L., Ma, J.: Agricultural sustainable development in China-restriction of water resources and its countermeasures. J. Guizhou Agric. Sci. **06**, 84–85 (2004). (in Chinese)

3. Zhao, Y., Li, Y., Jing, W., et al.: Buried straw layer plus plastic mulching reduces soil salinity and increases sunflower yield in saline soils. J. Soil Tillage Res. **155**, 363–370 (2016). https://doi.org/10.1016/j.still.2015.08.019
4. Wu, J., Zhu, Z, Zheng, J., et al.: Effects of straw mulching on soil physical and chemical properties and crop yield. Southwest China J. Agric. Sci. **02**, 192–195 (2006). (in Chinese). https://doi.org/10.16213/j.cnki.scjas.2006.02.0065
5. Li, S., He, X., Wang, Z., et al.: Experimental research on the effects of drip irrigation with brackish water on soil salinity and cotton output. China Rural Water Hydropower **07**, 16–20 (2011)
6. Zheng, J.: Studies on Utilization of Saltish Water Under Straw Mulching. D. Shandong: Shandong Agricultural University, pp. 34–35 (2002). (in Chinese)
7. Sun, S.M., An, Q.X., Cai, H.J., Yao, B.L., Qiao, Y.: Research on salt movement law in jujube root zone under indirect subsurface drip irrigation. Trans. Chin. Soc. Agric. Mach. **45**, 160–169 (2015)
8. Moreno, F., Cabera, F., Andreu, L., et al.: Water-movement and salt leaching in drained and irrigated marsh soils of southwest Spain. Agric. Water Manag. **27**(1), 25–44 (1995). https://doi.org/10.1016/0378-3774(95)01128-6

The Design of a Check Gate in Wu Village's Branch Canal

Si Li, Defeng Yang and Zhouxiang Yuan

Abstract In engineering construction, the design of sluice often appears; the check gate (also known as regulating sluice) is one of a common kinds of sluice. This article will concentrate on the check gate in Wu Village and give an example to introduce its layout, downstream energy dissipation and anti-scour and stability of locker chamber, which provides the basis and reasons for the design of small- and medium-sized check gates.

Keywords Check gate · Energy dissipation and erosion control
Stability of sluice chamber

1 Introduction

Wu village's branch canal is in northeastern Guangping County. Several counties lie across the branch canal; there are seven counties in all: Nanyangbu Xiang and Pinggudian Zhen, which include Houdazhai, Damiao, Beiwu, Linhebao, Cuizhuang, Xiwangfeng and Nanwangfeng. Guangping County has a typical temperate and monsoonal climate with four clearly distinct seasons, with a characteristic of rain in summer and aridness in autumn. Drought is the most fluent weather disaster. The annual average precipitation is 514.3 mm and average evaporation is 1038.6 mm in this area. The yearly frost-free time adds up to 201.8 days. Guangping County is in Handan, Hebei Province; it belongs to alluvial-proluvial plain, which also has a flat landscape. At the same time, it has

S. Li (✉) · D. Yang · Z. Yuan
College of Water Conservancy and Hydropower, Hebei University of Engineering,
Hebei Zhonghuanan Street, Hanshan District, Handan, Hebei, China
e-mail: 649667809@qq.com

D. Yang
e-mail: 740105676@qq.com

Z. Yuan
e-mail: 1131524172@qq.com

W. Dong et al. (eds.), *Sustainable Development of Water Resources and Hydraulic Engineering in China*, Environmental Earth Sciences,
https://doi.org/10.1007/978-3-319-61630-8_20

many old canals that are partially covered or filled up. On both sides of the canal, there are many trees and fields; according to research, the maximum depth of exploration is less than 5.0 mm. The loam is engendered by the pluvial-alluvial force in the Quaternary strata. The stratum from up to down is sandy loam, loam, sandy loam in a proper disorder.

The foundation surface of a check gate is planned for construction in the loam, because loam has some degree of strength and engineering geological conditions are good. The admissible value of subgrade-bearing capacity of the loam is 110 kp_a.

Guangping County's ground motion peak acceleration is 0.10 g. The corresponding basic earthquake intensity is VII [1]; characteristic period of response spectra is 0.40 s. To make it much more convenient for Wu village branch canal irrigation and storage water, designers should raise the water level for obtaining water; to drain away water in time after the irrigation period, workers constructed a set of check gates away from the crossing of Wu village branch canal and Wangfeng main canal. The design flow is 4 m^3/s. The node gate is open gate, which can be divided into entry section, gate section and exit section.

Based on the above background, this paper designed the control gate of Wucun branch canal to supplement and improve the comprehensive underground water overrun project in 2014. Firstly, the layout of the control branch gate was discussed. Then the gate width and downstream energy dissipation. Finally, the stability of the gate checking, water conservancy designers for the design of control gates provide a reference.

2 Check Gate Design

The design of the control gate mainly includes the arrangement form, gate width, downstream energy dissipation and stability check; and the calculation is made one by one to ensure the design rationality.

2.1 Engineering Layout

A check gate, as an open gate, can mainly be divided into entry section, gate section and exit section. The entry section employs the C30 plain concrete slope protection, with a thickness of 0.2 m; C30 reinforced concrete is used for bottom protection, with a thickness of 0.4 m. At the bottom it also has a plain concrete cushion of 0.1 m. The height of the channel should flush with the channel bottom; the height of the end of the entry should fit the height of the sluice bottom; the slope of both sides should connect perfectly with the slope of the channel. To connect better, both slopes should connect to the slope of the canal. The proportion of length of the boundary and the slope is 1:2; the width of the check gate needs to correspond to the width of the canal upstream and downstream. When adopting alternation irrigation, the design flow of the check gate differs between downstream and upstream; when adopting continued irrigation, the water level needs to take the corresponding water level according to the sluice; however, the level downstream should take the influence of the next check gate's damming into consideration.

The gate section is a C30 reinforced concrete structure, with a length of 4.0 m; the thickness of the bottom is 0.6 m, with a plain concrete cushion layer of 0.1 m.

2.2 The Calculation of Sluice Hole

The gate width affects the overall flow of the inlet and outlet as well as the structural safety and stability and should be calculated using Eqs. (1), (2) and (3):

$$B_0 = \frac{Q}{\sigma \varepsilon m \sqrt{2g} H_0^{3/2}} \tag{1}$$

$$\varepsilon = 1 - 0.171\left(1 - \frac{b_0}{b_s}\right)\sqrt{4}\,\frac{b_0}{b_s} \tag{2}$$

$$\sigma = 2.31\frac{h_s}{H_0}\left(1 - \frac{h_s}{H_0}\right)^{0.4} \tag{3}$$

In this formula: B_0 equals sluice hole's total net width (m); Q equals diversion flow; H_0: weir head, which includes the water head of approach velocity (m); h_s depth of water of weir crest downstream (m); g: acceleration of gravity, can adopt 9.8 (m^2/s); m: weir flow coefficient, can adopt 0.38; ε weir flow-lateral contraction coefficient; h_0 gate net width (m); b_s half width of the water deep upstream (m); σ weir coefficient of submergence.

The barrier mainly involves the following indicators: design, flow, depth, calculated hole width, design of a single hole width, and the number of holes. Improvement of the above several indicators can design the gate hole. Due to the simplicity and clear structure of the control sluice involved in this paper, detailed parameters of the orifice size of the irrigation sluice are shown in Table 1.

Table 1 Check gate technical indicators

Channel name	Check gate name	Design flow (m)	Canal water depth (m)	Calculated sluice hole width (m)	Designed single hole width (m)	Hole number (individual)
Wu village canal	Check gate of Wu village canal	4	1.43	1.94	3.0	1

2.3 Calculation of Energy Dissipation and Scouring Protection Downstream

The downstream energy dissipation and erosion calculation is an inevitable process in the design of a control gate to reduce the energy generated by the head impact. Most check gates are open gates; the height of the lock sill should equal the bottom of the channel; the flat broad-crested weir adopted makes it easier for energy dissipation and scouring protection. The beginning flow can depend on the apron upward stilling pier to distribute the water flow and crash and scour the energy. To make the wing walls downstream and upstream smooth, warp surface as a transition is usually used, thus causing the loss of water head [2–4].

① The water flow within a short period, significantly increased the water depth; the corresponding reduction in flow rate will have a water jump phenomenon; the water depth should be calculated after the jump.
The calculating formula is as follows:

$$h_c^{''} = \frac{h_c}{2}\left(\sqrt{1 + \frac{8aq^2}{gh_c^3}} - 1\right)\left(\frac{b_1}{b_2}\right)^{0.25} \tag{4}$$

h_c should be solved as the following formula:

$$h_c^3 - T_0 h_c^2 + \frac{aq^2}{2g\varphi^2} = 0 \tag{5}$$

In the formula: $h_c^{''}$: conjugate water depth of the water depth of vane contraction (m); h_c: water depth of vena contraction (m); q unit discharge (m^3/s); b_1: head end width of the stilling basin (m); b_2 end width of the stilling basin (m); a: water flow energy correction coefficient; T_0: total potential energy calculated from the bottom of the stilling basin; φ: coefficient of velocity.
In the downstream of the discharge structure, the phenomenon of water jump occurs in the downstream of the building, which is unfavorable to the structural stability. The increase of the stilling pool can effectively reduce the head energy. To ensure that the stilling pool is effective and economical, the depth and length of the stilling pool need to be calculated. The flow of water out of the pool should be calculated to ensure that it is within reasonable limits.

② Depth of stilling basin: the calculation formula is as follows:

$$d = \sigma_0 h_c^{''} - h_s^{'} - \Delta Z \tag{6}$$

Table 2 Calculation results table of sluice energy dissipation

Canal name	Check gate	Rate of flow (m^3/s)	Depth of stilling basin (m)	Length of stilling basin (m)
Wu village branch canal	Wu village canal check gate	4	0.37	4.6

Drop out of the pool:

$$\Delta Z = \frac{aq^2}{2g\varphi^2 h_s'^2} - \frac{aq^2}{2gh_c''^2} \tag{7}$$

In the formula: depth of stilling basin (m); σ_0 submerging coefficient of water jump; h_c'' depth after water jump (m); h_c contracted depth (m); q locking unit discharge (m^2/s); h_s' water depth of riverbed out of pool (m).

③ Length calculation of stilling basin: Length of stilling basin L_{sj} can be calculated as follows:

$$L_{sj} = L_s + \beta L_j \tag{8}$$

$$L_j = 6.9(h_c'' - h_c) \tag{9}$$

In the formula: L_S: slope hole deviation of stilling basin (m); β: correction coefficient of water jump; L_j length of water jump.

For the data query, the required data parameters were selected; with the use of the above formula, the energy-saving effect of the control gate to calculate the data is shown in Table 2.

2.4 Calculation of Stability of Lock Chamber

Loading Combination of Lock Chamber Stability Calculation. Lock chamber loads include self-weight, water pressure, uplift pressure, earthquake loading, etc. This working condition involves the construction of two basic combinations of anhydrous and normal retaining and the special combination of normal retaining during earthquakes. Each combination has a different load composition, and the load effect combination is shown in Table 3.

Lock Chamber Stability Calculation. Lock chamber stability calculation applies the formula in *Design specification for sluice* SL265—2001 [5].

Table 3 Loading combination of lock chamber stability calculation

Loading combination	Calculation cases and explanation		Load				
			Self-weight	Water weight	Hydrostatic pressure	Uplift pressure	Earthquake load
Combination	Built without water	No water before or after gate	√				
	Normal back astern	Highest level of back astern, no water downstream	√	√	√	√	
Special combination	Normal back astern with earthquake	7° earthquake	√	√	√	√	√

① Brake gate to resist the horizontal force caused by the slip, that is, the impact of water pressure, the structure needs to be stable against sliding; the formula is shown in Eq. 10:

$$k_c = \frac{f\sum G}{\sum H} \tag{10}$$

In the formula: k_c: safety factor against sliding which along the bottom of the gate; f: frictional factor between the bottom of the gate and foundation; $\sum G$: all vertical load which acts on the foundation lock chamber(KN); $\sum H$: all horizontal load acts on retaining wall(KN); $[K_c]$: permissible value of safety factor against sliding: fundamental combination $[K_c] = 1.35$, accidental combination first class $[k_c] = 1.20$, accidental combination second class $[K_c] = 1.10$.

② The pressure, size and distribution of the structural foundation on the surface contact of the foundation involve many aspects such as the structure of the superstructure; and the stress on the foundation should be calculated by the formula 11:

$$P_{\min}^{\max} = \frac{\sum G}{A} + \frac{\sum M}{W} \tag{11}$$

In the formula: $p_{\min}^{\max}$ the maximum or the minimum of the base stress (kp_a); $\sum G$: all vertical load which acts on the gate floor (KN); $\sum M$: vertical load and horizontal load which act on the gate floor producing torque to the geometrical cross section's axis ($KN \cdot m$); W moment of area between foundation and that surface's geometrical cross section's axis (m^3).

③ Checking the calculation to stress nonuniformity coefficient of foundation: Permissible value of stress nonuniformity coefficient of foundation, according to the formula in *Design specification for sluice* SL265—2001 [5].

The formula of stress non-uniformity coefficient of foundation is:

$$\begin{gathered} \eta \le [\eta] \\ \eta = \frac{\sigma_{\max}}{\sigma_{\min}} \end{gathered} \tag{12}$$

In the formula: η calculating stress nonuniformity coefficient of foundation $\sigma_{\max}$; $\sigma_{\min}$ maximum and minimum stress pressure (kp_a); $[\eta]$ normalizing permitting stress pressure nonuniformity coefficient of foundation, fundamental combination applies for $[\eta] = 2.0$, special combination applies for $[\eta] = 2.5$.

Table 4 Stability calculation of sluice

Calculation cases	Safety factor against sliding		Foundation stress (kp_a)			Stress nonuniformity coefficient of foundation η	
	K_c	$[K_c]$	σ_{up}	σ_{down}	σ_{av}	Calculated value	Allowable value
Base I	∞	1.2	75.01	59.75	67.38	1.26	2
Base II	1.67	1.2	54.44	62.02	58.23	1.14	2
Special I	1.28	1.0	42.81	73.65	58.23	1.72	2.5

Based on the two basics and a special three conditions [6–8], the need for all situations are calculated; for the results of the design of anti-sliding stability and safety factor, the foundation stress and stress non-uniform coefficient, the stability of the results is shown in Table 4.

According to the report of geological exploration, the bearing stratum of the gate foundation is the loam layer; the sub-grade capacity is 110 kp_a; and the calculation result shows that the sub-grade capacity and the gate both meet the requirements.

3 Conclusion

This check gate is mainly a supplement and perfection to the project of groundwater overdraft and management in 2014. Because Wu village branch canal does not set the check gate and with a result of the bad influence dispatching management in the irrigation period, setting the check gate in Wu village can adjust the upstream water level and control the downstream discharge volume. The canal check gate can use the open gate; the height of the gate should equal the foundation of the canal, adapting the flat broad crested weir. This is easy for sluice to dissipate energy and control erosion. When it begins to let down stream, the settled baffle pier comes into use. The baffle pier can crash and make energy dissipation. The mini-type check gate is widely used in China; this check gate design is to design the Wu village check gate and give an example for water conservancy designers to check gate designation.

References

1. Li, S.Y.: China Earthquake Intensity Table (2008)
2. Li, J., Zhao, Z.: Hydraulics. Hohai University Press, Nanjing (2001)
3. Liu, J.: A discussion on design essentials of check gate. Inner Mongolia Sci. Technol. Econ. **05** (2011)
4. Wang, Z.: Analysis and solution on erosion reason of check gate. Anhui Vocat. Tech. Coll. Water Conserv. Hydropower J. **03** (2007)

5. China Water Power Press: Design Specification for Sluice, Beijing (2001)
6. Wang, H.: A discussion on methods of sluice design. Sci. Technol. Enterp. **22**(12), 195–19 (2013)
7. Tang, T.: A discussion on key methods of sluice design. New Mater. New Declar. **12**(34), 24–25 (2013)
8. The Second Exploitation and Design Institute of Hebei Province of Water Conservancy and Hydropower: Feasibility report about the project of Guangping County's comprehensive treatment groundwater overdraft and project of surface water irrigation (2007)

Application Research of SWMM in the Simulation of Large-Scale Urban Rain Flood Process—A Case Study of Yizhuang District, China

Xiaoran Fu, Qinghua Luan, Haichao Wang, Jiahong Liu and Xuerui Gao

Abstract Storm Water Management Model (SWMM) is established based on the data of the Yizhuang Beijing Economic Development Zone in the core region of China, including the information of the terrain variation, the underlying surface conditions and the storm sewer system. The data of two typical rainstorm events in Beijing were selected; the model parameters were calibrated; and the simulation accuracy results were validated. The simulation regional average depth in different return periods designed precipitation scenarios were calculated. From a comparison of these results, potent measures to deal with the calibration and validation in a condition of the development zone lack of measured rainfall flow data have been suggested. The green space and square land is much better than the others in rain flood digestion. It is important and instructive for the application and popularization of SWMM in large scale-urban areas in China.

X. Fu (✉) · J. Liu
Beijing University of Technology, Chaoyang, Beijing 100124, China
e-mail: fuxiaoran@emails.bjut.edu.cn

J. Liu
e-mail: liujh@iwhr.com

Q. Luan
Hebei University of Engineering, Handan 056001, Hebei, China

Q. Luan
Research Center for Water Ecological Civilization & Social Governance of Hebei Province, Handan 056001, Hebei, China

H. Wang (✉)
China International Engineering Consulting Corporation, Haidian, Beijing 100048, China
e-mail: wanghaichao@ciecc.com.cn

J. Liu
China Institute of Water Resources and Hydropower Research, Haidian, Beijing 100038, China

X. Gao
Northwest A&F University, Yangling 712100, Shaanxi, China

W. Dong et al. (eds.), *Sustainable Development of Water Resources and Hydraulic Engineering in China*, Environmental Earth Sciences,
https://doi.org/10.1007/978-3-319-61630-8_21

Keywords SWMM · Storm flood · City waterlogging · Yizhuang economic development zone · Large scale urban area

1 Introduction

Waterlogging refers to heavy rainfall or continuous rainfall over urban drainage ability, leading to groundwater disasters [1]. Under the situation of rapid urbanization, frequent and wide waterlogging occurs in China at present, seriously threatening the urban public infrastructure and the lives and property safety of the residents [2]. Combined with larger spatiotemporal difference in urban construction scale and waterlogging forecast level between the north and south of China, the aid of the model in accurate simulation research on urban storm runoff is an important approach and a prerequisite condition to solve the problem of waterlogging. Storm Water Management Model (SWMM) is relatively mature in the aspect of urban rain flood runoff simulation technology [3, 4]. In 1974, SWMM was first proposed to apply in simulation stimulation of a wetland with an area of 3.37 km^2 by Meinholz et al. [5] to discuss and validate the simulation results. Since then, SWMM is more and more applied in this aspect, and the study areas ranged from 0.02 to 6.47 km^2 [6–9]. Since the end of the 20th century, domestic application research based on SWMM is booming. In 1994, SWMM was first introduced and put into application in Jizhuangzi test area of Tianjin by Liu et al. [10], creating the precedent of domestic application of the multi-functional general urban hydrology, hydraulic and water quality model. In 2011, Beijing Xiangshan area was selected for simulation by Wang et al. [11], and GIS and RS technology were applied to identify key parameters of the model. Good simulation effect has been achieved in the mountainous cities of northern China. Domestic application fields of SWMM for research are mainly divided into two types. One type refers to the research object derived from SWMM's own instance with the scale often confined to the small and medium-sized areas with the area less than 10 km^2; the other type is mainly used in single land use type, such as the recessed overpass, residential community and school, while rare studies exist in the big scale cities of mixed land use types [12–15].

2 Materials and Methods

2.1 Study District

Beijing Yizhuang economic and technological development zone (hereinafter referred to as development zone) is in the southeast of downtown Beijing at the junction between Daxing District and Tongzhou District, which is the important development base of the high-tech industry in Beijing. Development zone has clear and accessible boundary conditions, and the southwestern study area is bounded by Liangshui River from Dayangfang Bridge in the north to Dayangfang Gully in the east, where various types of land use are distributed interlocked. Among them, the residential land area accounts for 13%; the land use area of public service and commercial service facilities accounts for 21%; industrial land area accounts for 46%; the land use area of road and transportation facilities accounts for 9%; and green space and square land area accounts for 11%. The land use type distribution of development zone is shown in Fig. 1, and the parameters of each land use type are shown in Table 1.

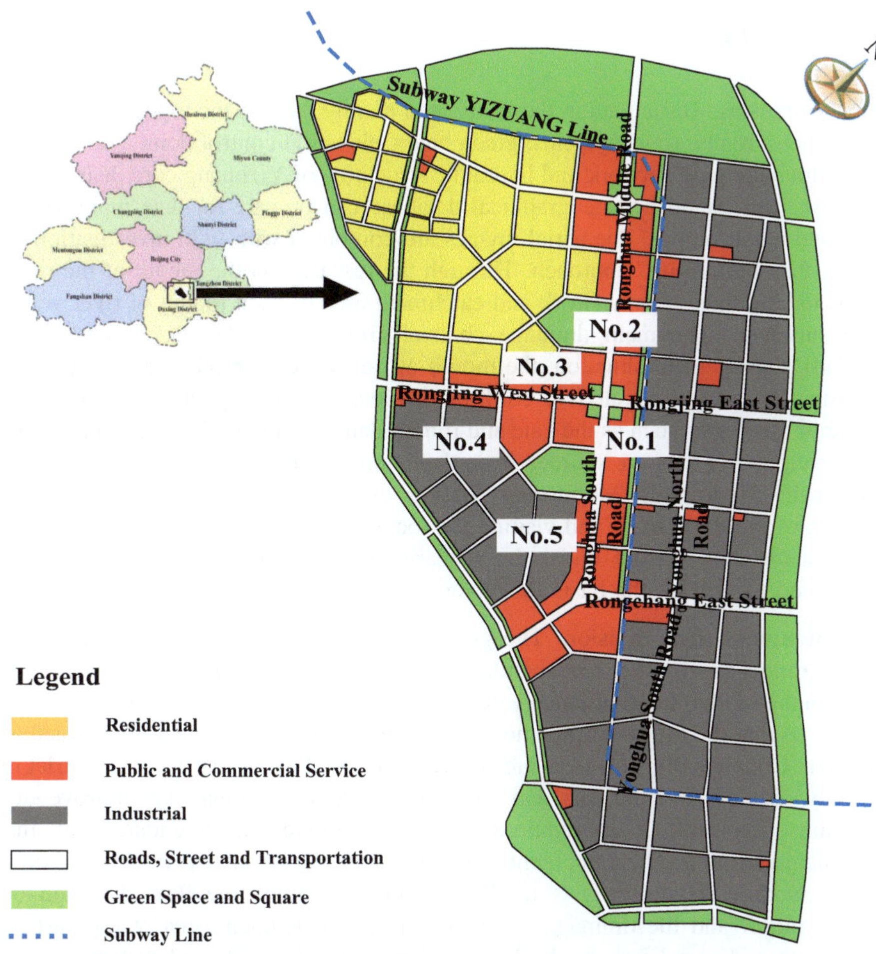

Fig. 1 Location and distribution of land use types

Table 1 Parameters of land use types

Land use types	Typical plots	Generalization runoff coefficient	Percentage of area (%)	Position
Roads, street and transportation	No. 1	0.85–0.95	9	Rongjing East Street Subway
Green space and square	No. 2	0.10–0.20	11	Broad park east gate
Residential	No. 3	0.60–0.70	13	Nobita tulip homes
Public and commercial service	No. 4	0.45–0.60	21	Beijing Shengshiyali fashion company
Industrial	No. 5	0.20–0.45	46	Pullman Beijing South

2.2 Modeling

Data processing. Basic data in the study area mainly include the topographic map, aerial photo, roads, water systems, green lands, elevation control points, rain check wells, drainage pipe network and hydraulic structures of Yizhuang core district. On this basis, the established geographical database can prepare for data input, operation and result output. The aerial images are combined with field investigation to choose the "bottom-up" approach. Through the research on the mouth of the river water; the drainage pipe network and catchment area of each mouth of river water were surveyed. The main drainage channel cross-section layout, flood control standard, management situation, the mouth of rain water and other information of core district are investigated to confirm the location and operation situation of hydraulic structures such as the gate and rubber dam running in the river channel on site. Precipitation data are sourced from the 1 min precipitation process data at the Songlinzha Hydrologic Monitoring Station on June 23, 2011. Through field research of the flood mark and visiting the local people, the waterlogging situation of Yizhuang core district under the scenario of "6.23" torrential rain is deeply understood to provide the basis for model parameter calibration.

Sub-catchment area division. The goal of catchment area division is to assign surface runoff flow to the corresponding drainage pipe network nodes (check well nodes in the model) in accordance with runoff generation and confluence in the core district, so that the inflow of pipe network system is more conformant to the actual situation. Concrete division steps are shown as below. First of all, the core district is coarsely divided into the east and west large catchment areas. To improve the modeling precision, it is required to further refine the boundaries of the sub-catchment area. The catchment area tool (Watershed) in ArcGIS 10.1 hydrological module (Hydrology) is used to further determine the flow direction of surface runoff and the drainage boundary of each catchment area in contrast to DEM data and aerial images. Fully considering the diversity and distribution of

Fig. 2 The result of sub-catchment generalization

land use types in large-scale areas as well as the mutual contact and influence between adjacent sub-catchment areas, 1072 sub-catchment areas were divided eventually with a total catchment area of 17.84 km^2. Among them, the maximum sub-catchment area is 286,207 m^2, while the minimum area is 578.13 m^2. In the study area, the outlet is set as free flow. Division results of the sub-catchment area are shown in Fig. 2.

Drainage system generalization. The whole core district is divided into the east and west major catchment areas by Ronghua Middle Road—Ronghua South Road, where surface water flows along the boundary across the roads, ditches and drainage pipelines eastwards and southwards to Dayangfang Gully and Liangshui River, respectively, which discharges from the study area at the junction of the two river channels eventually. Considering the size of the study area, data accuracy, model calculation speed and other factors, on the basis of old and new drainage pipe network data, the rivers, ditches, partial secondary trunk roads and branches were unified to generalize into drainage pipeline, and there were 3621 water wells finally generalized with 3274 rainwater pipelines and total pipeline length of 107.1 km. The generalized situation of drainage pipe network in core district is shown in Fig. 3.

2.3 *Parameter Empirical Calibration and Validation*

Initial parameter setting and empirical calibration. Model parameters are involved in the rainfall infiltration process, and SWMM is simulated by the Horton

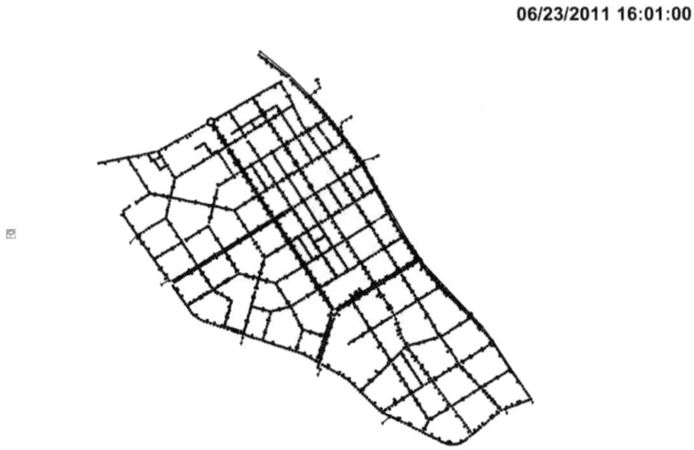

Fig. 3 The result of drainage system generalization

model principle [16, 17]. Initial values of average surface slope, manning coefficients of the sub-catchment area, pipe or open channel are set according to the typical values of SWMM user manual [17]. In view of the development zone lack of measured rainfall flow data, it is difficult to calibrate and validate model parameters. Through field investigation in search of the building walls or traffic facility bridge piers adjacent to observation points, additional surveys and records were made on the left natural traces indicating the maximum water level of waterlogging. Taking the lowest point of the sub-catchment area as the datum, the measured maximum water depth is set as the main indicator of parameter empirical calibration and validation. The advantage of such a reference index is that maximum water depth data can not only completely and clearly reveal the real situation of field flooding, but also have the operability of field monitoring or a subsequent additional survey. The waterlogging and flooding situation of Yizhuang core district in the scenario of "7.21" torrential rains in Beijing in 2012 are deeply understood. For the calibration point with large deviation between analog value and survey value, it is required to further debug model parameters until the analog values and measured values of maximum water depth at all validation points are close to the most accurate value.

The water depth process line and rainfall process line at key nodes simulated by SWMM were compared to analyze the relevance, tendency, recession time, occurrence time of maximum water depth and other parameters to determine the reliability and accuracy of simulation results. The distribution of all validation points is shown in Fig. 1, and the main properties are shown in Table 2.

Model validation. In the case of no measured flow data, all validation points are in the construction, transportation and population concentration areas of the core zone. Urban waterlogging caused by heavy rain will inevitably bring economic losses to the residents, leading to inconvenient life. The maximum surface water depth

Table 2 Main parameters of SWMM

Types	Reference value	Remark
Average slope (%)	0.05	–
Overland flow concentration manning coefficient	0.012	Roads
	0.013	Roof and square
	0.15	Green space
	0.4	Forest
Closed conduits manning coefficient	0.013	–
Open conduits manning coefficient	0.05	–
Sub-catchment width (m)	5.57	Min
	174.81	Max

Table 3 "6.23" precipitation maximum depth analog error (Unit: m)

Typical plots	Testing position	Measured value	Simulation value	Error
No. 1	Bridge piers	0.24	0.262	0.022
No. 2	Square sculpture	0.37	0.418	0.048
No. 3	Cell wall	0.70	0.683	−0.017
No. 4	Shopping mall wall	0.95	0.931	−0.019
No. 5	Hotel stone pillars	1.15	1.162	0.012

induced by the "6.23" heavy rain in Beijing in 2011 is taken as the main validation indicator. According to the waterlogging maximum water depth data reported in the press and site records, the maximum water depth and measured water depth at each validation point simulated by SWMM were compared. Verification results are shown in Table 3, and the error values are within the allowable scope of runoff depth simulation forecast, indicating better overall simulation accuracy.

3 Results and Discussion

There are over 20 years of self-recording precipitation data in Yizhuang core district. To improve the simulation accuracy of SWMM on design storm flood, according to the preparation method of rainstorm intensity formula in Outdoor Drainage Design Specification in contrast to return period, rainfall intensity and rainfall duration contour maps of Beijing Hydrologic Manual (first volume of rainstorm atlas), the storm duration is determined and designed as 6 h from seven standard durations [18]. The present flood control standard of two drainage systems in Liangsuhi River and Dayangfang Gully is for a 20 year return period; the checking standard is 50-year return period; and the design of storm return period is derived as 5, 10, 20 and 50 year statistics [19]. Through SWMM simulation

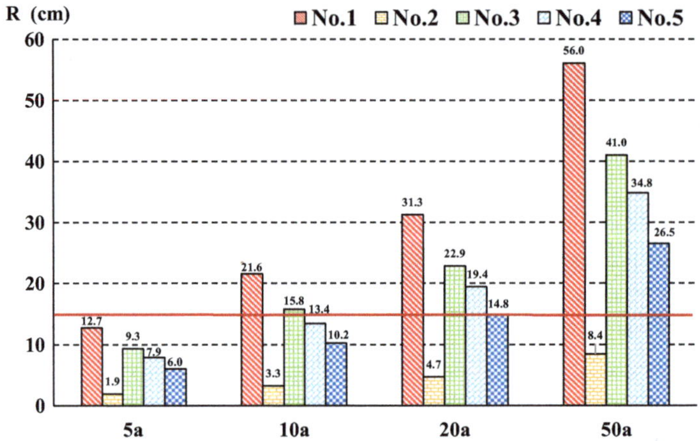

Fig. 4 Correlation section of maximum runoff depth

calculation, the maximum water depth of each validation point in different scenes of design storm is shown in Fig. 4.

As viewed from the figure, in terms of the designed heavy rains with the same return period, for the designed heavy rains with the return period of 5 years, only No. 1 of each land use type will produce the waterlogging with the depth of more than 10 cm; except No. 2, the waterlogging with the depth of 10–15 cm is generated in other land use types, i.e., the rain is easy over the road, which needs to take appropriate protective measures to prevent rain from entering the interior. The designed heavy rain with the return period of 50 years and above will lead to the waterlogging with the depth of over 20 cm on No. 1–No. 5 surface, extremely easy to cause the infestation of waterlogging, and rainfall will overflow to subway, resulting in endangering public safety. For the plot of the same land use type, the current rainfall flood elimination and acceptance ability is lower at No. 1 and No. 3, and the storm with the return period of 10 years is easy to cause flooding in the large scale. No. 4 and No. 5 have a certain waterlogging resistance. The Roads and Traffic Facilities Land represented by Boda Park has outstanding rain flood elimination and acceptance ability, which can resist in severe storms with the return period of 50 years.

4 Conclusion

Through empirical calibration and test of model parameters, in the absence of observed runoff data, the aid of the model refined input technology and the introduction of maximum torrential rain runoff depth were taken as the important basis of empirical calibration. Through error analysis, SWMM can be considered to have a certain simulation precision and the credibility of rain flood runoff simulation in the large scale urban areas similar to Yizhuang, which can be applied to analyze and evaluate drainage pipe network as well as study urban flood control in a designed rainstorm.

The above conditions of designed storm bring no hazards to waterlogging at No. 2 area. On the one hand, because the selected typical plot is relatively independent with higher terrain, the location is the upstream of the drainage system. The other important reason is due to abundant vegetation in No. 2 area, the permeable rate is high with a certain rain flooding elimination and acceptance ability.

In future applications and practices, the model validation method with no measured data mentioned in this study can be taken as a reference, and more scene reconnaissance and observations are recommended. This method plays a positive role in improving precision of model simulation. In the meantime, the rain flood elimination and acceptance abilities of designed storms in different return periods at the plots of different land use types were compared to provide practical experience and technical support of cavernous urban design concept with certain reference significance to urban waterlogging control in China.

Acknowledgements The researchers would like to extend thanks to the Chinese National Natural Science Foundation (No. 51739011, No. 51409275, No. 51522907). The study was also supported by the Research Fund of the China Institute of Water Resources and Hydropower Research (No. WR0145B502016, No. 2017ZY02).

References

1. Ministry of Housing and Urban-Rural Development of the People's Republic of China: Code for Design of Outdoor Wastewater Engineering: GB 50014-2006 (2016 Version). China Planning Press, Beijing (2016)
2. Ministry of Water Resources of the People's Republic of China: The Flood and Drought Disaster China (2006–2014). http://www.mwr.gov.cn/zwzc/hygb/zgshzhgb/
3. Wang, H., Chen, J., Zhang, S., et al.: Application status and comparative analysis of urban storm flood. Water Resour. Hydropower Eng. **42**, 10–13 (2011). https://doi.org/10.3969/j.issn.1000-0860.2011.11.003
4. Liu, J., Chen, G., Wang, H., Chen, X., Lu, L.: Storm Water Management Model Theory and Application—Take SWMM as an Example. Science Press, Beijing (2015)
5. Meinholz, T.L., Hansen, C.A., Novotny, V.: Application of the storm water management model. Ky. Univ. Off. Res. Eng. Serv. Bull. **106**, 109–113 (1974)
6. Marsalek, J.: A technique for flow measurements in urban runoff studies. In: International Seminar and Exposition on Water Resources Instrumentation, pp. 508–516. Ann Arbor Science Publishers, Chicago (1975)
7. Cole, G.D., Shutt, J.W.: SWMM as a predictive model for runoff. In: Proceedings of the National Symposium on Urban Hydrology, Hydraulics, and Sediment Control, pp. 193–201. Kentucky (1976)
8. Wang, C.H., Williams, S.D.: SWMM application in Indian river county, Florida. In: Proceedings of the 1989 National Conference on Hydraulic Engineering, pp. 454–459. ASCE, New York (1989)

9. Williams, W.M., Cave, K.A., Hoang, L.T.: Design storm selection for stormwater management master plans. In: Proceedings of the 1987 National Conference on Engineering Hydrology, pp. 85–90. ASCE, New York (1987)
10. Liu, J., Zhao, H., Li, D.: Influence of urbanization on the change of rain and flood in Tianjin city. Haihe Water Resour. **05**, 48–52 (1994)
11. Wang, H., Chen, J., Kong, G., Wang, Y., Liu, D.: SWMM construction based on GIS and RS technology. Beijing Water **03**, 46–49 (2011). https://doi.org/10.3969/j.issn.1673-4637.2011.03.014
12. Zhou, Y., Yu, M., Chen, Y.: Estimation of sub-catchment width in SWMM. China Water Wastewater **30**, 61–64 (2014)
13. Huang, G., Huang, J., Yu, H., Yang, S.: Secondary development of storm water management model SWMM based GIS. Water Resour. Power **04**, 43–46 (2011). https://doi.org/10.3969/j.issn.1000-7709.2011.04.014
14. Jia, H., Yao, H., Tang, Y., Yu, S., Richard, F., Anthony, N.: LID-BMPs planning for urban runoff control and case study. Adv. Water Sci. **02**, 260–266 (2014). https://doi.org/10.14042/j.cnki.32.1309.2014.02.019
15. Zhang, Q., Su, B., Yuan, J.: Simulating rainfall-runoff in urban residential area based on SWMM. J. BNU (Natural Science) **03**, 276–281 (2012)
16. Rui, X.: The discovery and development of runoff formation models. Adv. Sci. Tech. Water Resour. **33**, 1–6, 26 (2013). https://doi.org/10.3880/j.issn.1006-7647.2013.01.001
17. Rossman, L.A.: Storm Water Management Model User's Manual (Version 5.0). US EPA, Washington DC (2009)
18. Beijing Water Authority: Beijing Hydrological Handbook—The First Section of the Rainstorm Atlas. Beijing Water Authority, Beijing (1999)
19. Beijing Planning Commission, Beijing Municipal Bureau of Quality and Technical Supervision: Urban Storm Water System Planning and Design Storm Runoff Calculation Standard: DB11/T 969-2016 (2016)

The Response Relationship Between Flushing Time and Water Quality in Xiangxi Bay of the Three Gorges Reservoir

HongQing Zhang and ZhongJin Bao

Abstract For the problem of the spring blooms in the tributary embayment of the Three Gorges Reservoir, 3D hydrodynamic and water quality mode based on the background of Xiangxi River analyzed the response relationship between flushing time and water quality of Xiangxi Bay during the discharge period. The results show that in the upper reaches and near the estuary of the bay, chlorophyll and dissolved silicon concentration are smaller. The chlorophyll concentration in the waters of 12.5 km (in spring, 2005) and 7.5 km (in spring, 2007) away from the estuary remained at a relatively high value, while the dissolved silicon concentration is the opposite. In general, there was a positive correlation between flushing time and chlorophyll concentration, but a negative correlation between flushing time and nutrient salt for the growth of algae. The results can provide a reference for preventing the water environment problems of the tributary embayment.

Keywords Tributary embayment · Flushing time · Water quality
3D-hydrodynamic and water quality model

1 Introduction

The Three Gorges Project is a large water conservancy project, which attracts worldwide attention. It has a great benefit in flood control, power generation, shipping and so on. However, the water body before the dam is changed from the original natural river to the river channel type reservoir after the construction of the dam; then significant changes have occurred in the structure of the water flow in the reservoir area, and the water dynamic conditions and water quality factors in the reservoir area have obvious spatial and temporal characteristics. At the end of 2012,

H. Zhang (✉) · Z. Bao
Zhejiang Institute of Hydraulics & Estuary, Hangzhou 310020, Zhejiang, China
e-mail: zhanghangqing8@126.com

Z. Bao
e-mail: bzjin@tom.com

W. Dong et al. (eds.), *Sustainable Development of Water Resources and Hydraulic Engineering in China*, Environmental Earth Sciences,
https://doi.org/10.1007/978-3-319-61630-8_22

the outbreak of "water bloom" in the Yangtze River was still increasing year by year. The Yangtze River is in the non-flood period every spring; the water level in Three Gorges Reservoir is higher, making the flow velocity in the upstream of the reservoir tributaries slower. At the same time, the temperature has risen gradually, and all these conditions promoted the growth of algae. However, according to the statistical data of water bloom in the upper reaches of the Three Gorges Reservoir during 2003–2008 [1–6], spring has become the main season of the bloom.

Phytoplankton plays a key role in circulation of materials in the lake ecological system and energy transfer. Chlorophyll is the vital component of algae; its concentration level can be used to evaluate phytoplankton species and quantity and can represent the environmental quality in some degree. If the chlorophyll concentration in a certain area maintains at a high level, it must have a greater degree of water quality deterioration, a lesser degree conversely [7]. Prior to 2008, the dominant algae of Xiangxi Bay were mainly diatoms. The water quality monitoring data of Xiangxi Bay show that silicate can affect the extent and intensity of outbreaks of diatoms in the bay more than phosphate.

In recent years, the scholars often used the flushing time [8] to study the convection diffusion transfer process in an estuary, bay or lake. Flushing time not only reflects the hydrodynamic system, but also reflects the exchange characteristics of waters and the outside world water and internal transport rules of the water on their own [9]. Therefore, for the flow which is very complex, the existing thermal density flow phenomenon of tributaries, the calculation of flushing time of tributaries bay can provide a new way to solve water quality problems. In this paper, combined with the impoundment process of Three Gorges Reservoir, and considering that the dominant algae in Xiangxi Bay before 2008 is consistent basically, choosing 2005-2-22 to 2005-4-28 and 2007-3-1 to 2007-5-31 as the research period, then the 3D hydrodynamic and water quality model is established to study the response relationship between flushing time distribution and chlorophyll and dissolved silicon concentration.

2 Research Method

Based on Delft3D, the hydrodynamic and water quality of Xiangxi Bay will be researched. Considering hydrostatic pressure assumption, Boussinnesq assumption and buoyancy effect, the hydrodynamic model is established on a basis of the N-S equation and vertically discrete by using σ coordinate. The hydrodynamic model is solved by ADI algorithm in the category of finite difference method, and the wind stress on free surface, shear stress at the bottom of the river as well as Coriolis force were considered to form its model boundaries when the vertical acceleration is considered far less than gravitational acceleration. Based on the results of the hydrodynamic model, the water quality model, shown as Eq. 1, is used to solve material transportation diffusion equation further and get dissolved water quality in σ coordinate:

$$\frac{\partial C}{\partial t}+\frac{\partial(uC)}{\partial x}+\frac{\partial(vC)}{\partial y}+\frac{1}{H}\frac{\partial(wC)}{\partial\sigma}=\frac{\partial}{\partial x}(\Gamma_H\frac{\partial C}{\partial x})+\frac{\partial}{\partial y}(\Gamma_H\frac{\partial C}{\partial y})+\frac{1}{H^2}\frac{\partial}{\partial\sigma}(\Gamma_V\frac{\partial C}{\partial\sigma})+S+D \tag{1}$$

In Eq. (1), u, v, w are velocity components in every direction; Γ_H, Γ_V are horizontal and vertical diffusion coefficient, respectively; S, D are represent source sink term and biochemical reaction, respectively.

2.1 Computation Region and Meshing

The computational domain consisted of Xiangxi River (about 32.2 km) and the section of the Yangtze River (5 km upstream and downstream of Xiangxi River); the total area is 2.175×10^7 m^2. The calculation area is divided into 50 m × 50 m orthogonal grids, and the number of the planar mesh is 640 × 160, while the vertical is divided into 10 layers. The Xiangxi River-Yangtze River region and the calculation mesh are shown in Fig. 1. Considering the problem of graphic display, the partial section of the grid is enlarged.

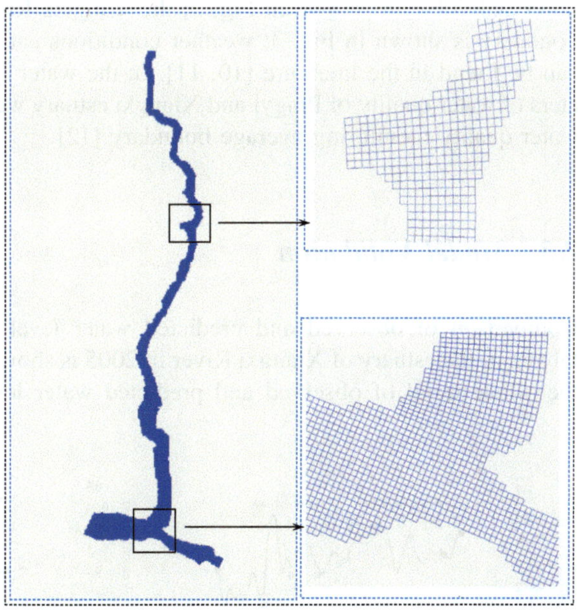

Fig. 1 Schematic diagram of domain and mesh

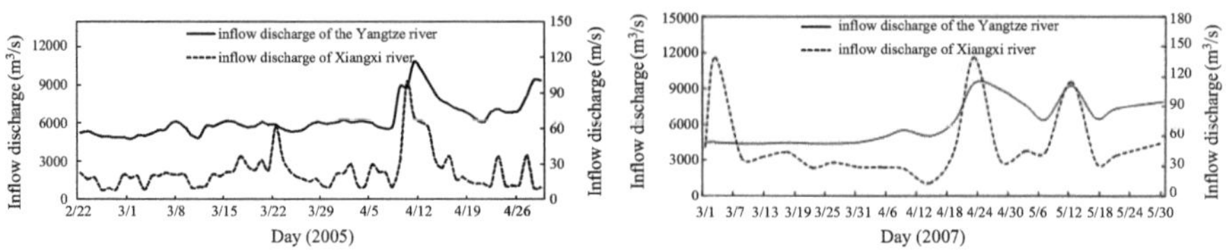

Fig. 2 Distribution of import flow boundary of the Yangtze River and the Xiangxi River in spring, 2005 and 2007

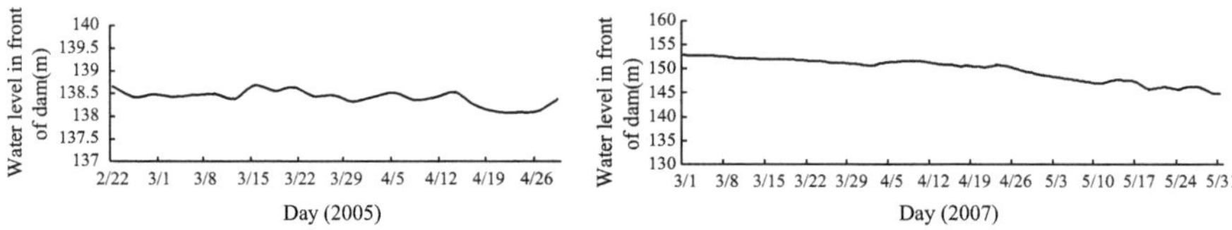

Fig. 3 Distribution of water level in front of dam

2.2 Boundary Condition

In the hydrodynamic model, Xiangxi Bay and Yangtze River inlet flow boundary in 2005 and 2007 are shown in Fig. 2; the Yangtze River downstream export water boundary is shown in Fig. 3; weather conditions and water temperature boundary can be found in the literature [10, 11]. In the water quality model, related parameters of water quality of Pingyi and Xiangxi estuary were selected as inlet and outlet water quality monitoring average boundary [12].

2.3 Model Validation

Comparison of observed and predicted water level and average velocity at the talweg of the estuary of Xiangxi River in 2005 is shown in Fig. 4, which shows that the whole trend of observed and predicted water level and average velocity are

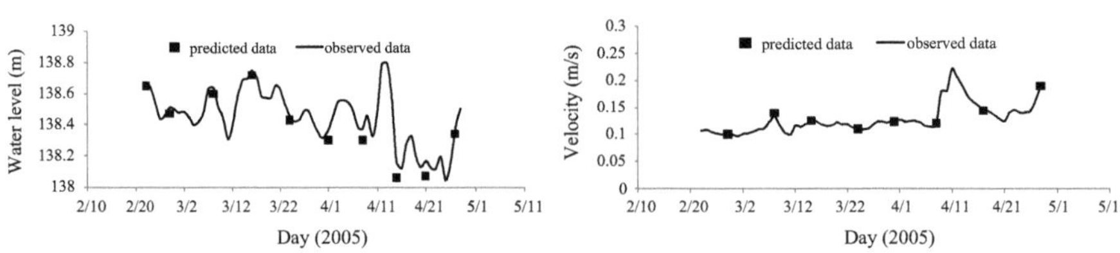

Fig. 4 Comparison of observed and predicted water level and average velocity at the talweg of Xiangxi River estuary in 2005

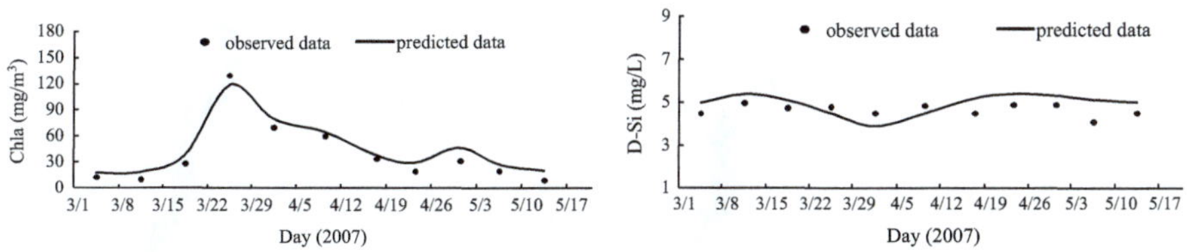

Fig. 5 Comparison of observed and predicted water quality indexes at an observation station about 3 km away from the estuary of Xiangxi River in spring, 2007

consistent, and the errors are small. In the spring of 2007, average chlorophyll and dissolved silicon concentration changes of the observation point about 3 km away from the downstream estuary of Xiangxi Bay are shown in Fig. 5 which shows that the calculated value of total chlorophyll and dissolved silicon concentration trend conform to monitoring value, although there exists an error, but it can be ignored. Therefore, the 3D hydrodynamic-water quality model is established to simulate the change of the water quality process of Xiangxi Bay.

3 Results and Analysis

The temporal and spatial distribution of chlorophyll concentration of Xiangxi Bay in the spring of 2005 and 2007 is shown in Fig. 6, and the spatial distribution of the averaged flushing time of each layer in the spring of 2005 and 2007 is shown in Fig. 7.

As seen in Fig. 6a, in the spring of 2005, there is a smaller value in the upper reaches and the estuary nearby of Xiangxi Bay, a larger value in the middle reaches.

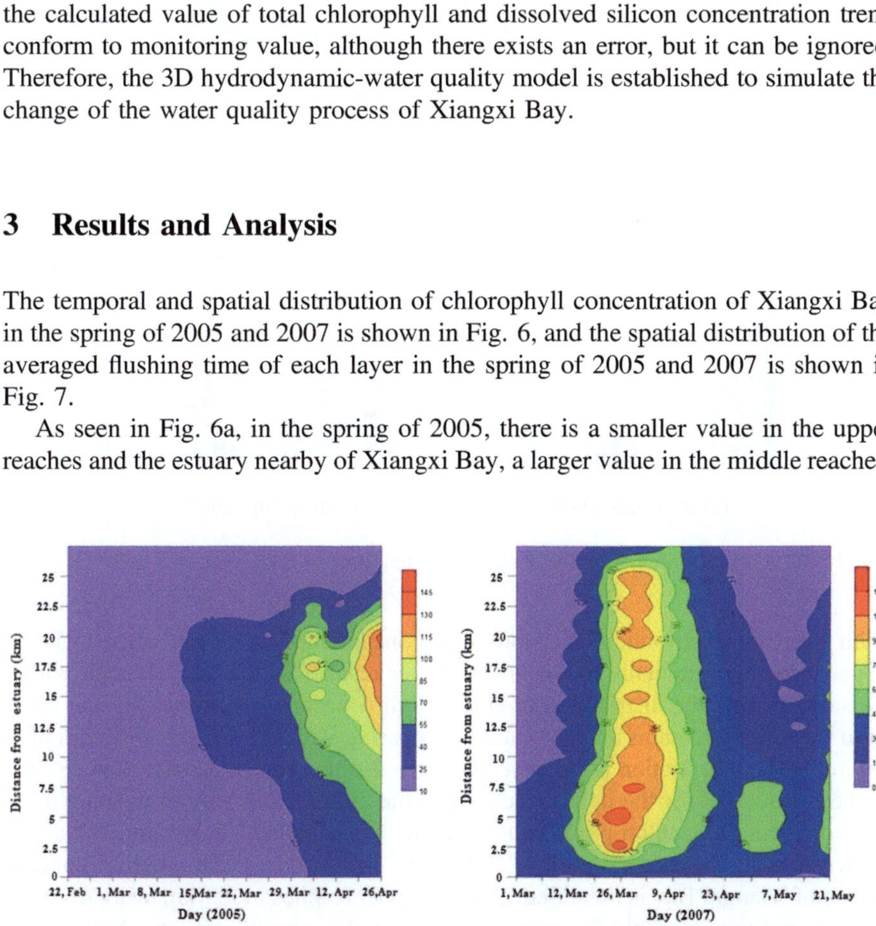

(a) in spring, 2005 (b) in spring, 2007

Fig. 6 Spatio-temporal distribution of chlorophyll concentration in 2005 and 2007

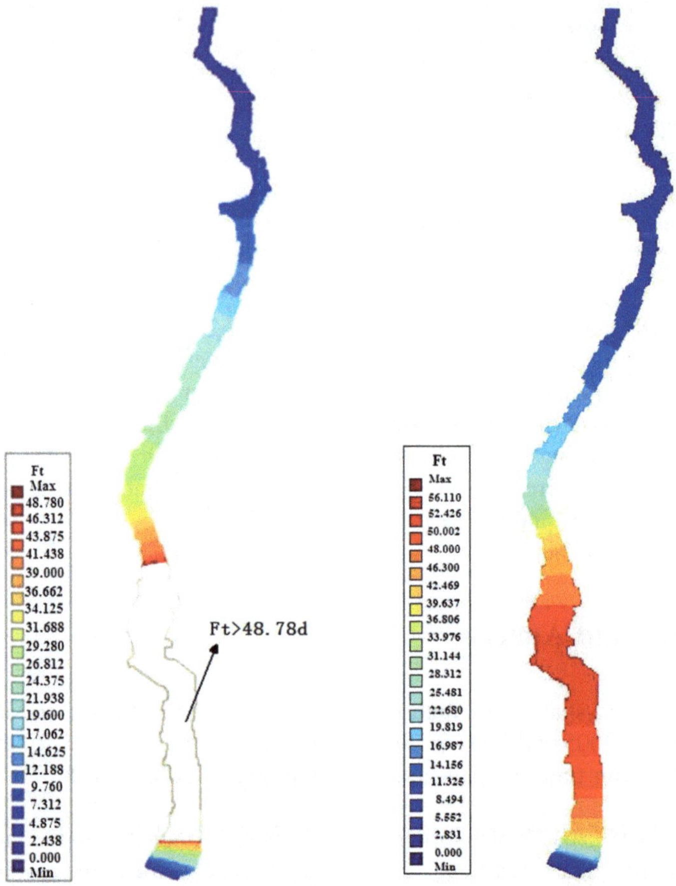

(a) in spring, 2005 (b) in spring, 2007

Fig. 7 The spatial distribution of the averaged flushing time of each layer of Xiangxi Bay

With time, the distribution of chlorophyll concentration increases where it has a larger value and gradually moves to the lower reaches of the bay. Overall, within the study period, the distribution of chlorophyll concentration about 12.5 km away from the estuary remains a larger value. Comparing with Fig. 7a, the area with high chlorophyll concentration is the area where the flushing time is greater than 48.78 d in the period, and for the upper reaches and the estuary with shorter flushing time, the corresponding chlorophyll concentration has remained a smaller value. As shown in Fig. 6b, in the spring of 2007, the distribution of chlorophyll concentration of Xiangxi Bay is increased from the upper reaches to 2.5 km away from the estuary, and the chlorophyll concentration is smaller near the estuary. At the same time, it remains a larger value on the chlorophyll concentration from 2.5 km to 10 km away from the estuary. Comparing with Fig. 7b, the area with the high

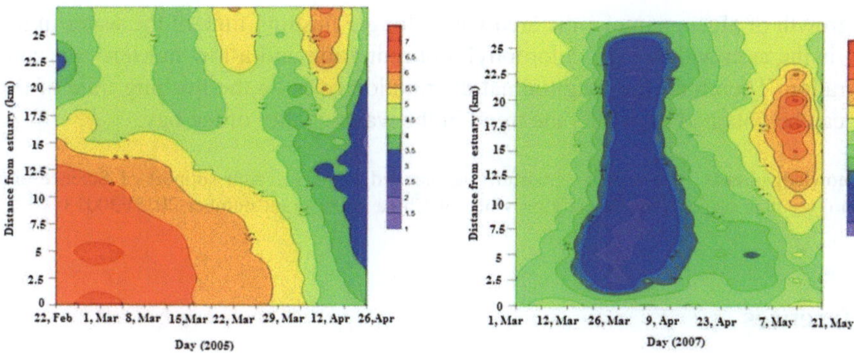

(a) in spring, 2005 (b) in spring, 2007

Fig. 8 Spatio-temporal distribution of dissolved silicon concentration

chlorophyll concentration is the area with the longer flushing time and the estuary where the chlorophyll concentration remains a small value has a shorter flushing time.

The temporal and spatial distribution of dissolved silicon concentration of Xiangxi Bay in the spring of 2005 and 2007 is shown in Fig. 8. As seen in Fig. 8a, in the spring of 2005, there is a larger value in the upper reaches and the estuary near Xiangxi Bay, a small value in the middle reaches. With time, the distribution of dissolved silicon concentration increases where it has a smaller value and gradually moves to the lower reaches of the bay. Overall, within the study period, the distribution of dissolved silicon concentration where it is 12.5 km away from the estuary remains a smaller value. As shown in Fig. 8b, in the spring of 2007, the distribution of dissolved silicon concentration of Xiangxi Bay is decreased from the upper reaches to 2.5 km away from the estuary, and larger near the estuary. At the same time, the dissolved silicon concentration remains a smaller value from 2.5 to 10 km away from the estuary.

4 Conclusions

The 3D hydrodynamic and water quality model based on the background of Xiangxi River was used to analyze the response relationship between flushing time and water quality of Xiangxi Bay during the discharge period. The results show that in the upper reaches and near the estuary of the bay, chlorophyll and dissolved silicon concentration are smaller. The chlorophyll concentration in the waters of 12.5 km (in the spring of 2005) and 7.5 km (in the spring of 2007) away from the estuary remained at a relatively high value, while the dissolved silicon concentration is the opposite. In general, there was a positive correlation between flushing time and chlorophyll concentration, but a negative correlation between flushing time and nutrient salt for the growth of algae, namely, the area with a longer flushing time of the water in the bay, is the area with a high chlorophyll concentration and a low nutrient salt concentration. Therefore, the longitudinal distribution rules on the flushing time of the bay can be obtained to predict the trend in the water quality of the bay.

Acknowledgements The authors gratefully acknowledge the research support of Science and technology plan projects in Zhejiang province of China under grant number 2017F30013.

References

1. Tang, H.-B., Liu, G.-X., Hu, Z.-Y.: Preliminary research on the algal bloom of peridiniopsis sp.in Gaolan River of the three Gorges Reservoir. Acta Hydrobiol. Sin. **1**, 47–51 (2006). https://doi.org/10.3321/j.issn:1000-3207.2006.01.009
2. Zhou, G., Kuang, Q., Hu, Z., Cai, Q.: Assessment of algal diversity and water blooms prevention in four tributaries of Three Gorges Reservoir. China Environ. Sci. **3**, 337–341 (2006). https://doi.org/10.3321/j.issn:1000-6923.2006.03.019
3. Zhou, X., Luo, G., Yang, Q., Chen, G., Li, Y., Gong, L.: Effects of different environmental factors on algal growth in the backwater reaches of the tributary to the Three-Gorges Reservoir: a simulation. Acta Sci. Circum. **3**, 558–562 (2008). https://doi.org/10.3321/j.issn:0253-2468.2008.03.023
4. Cai, Q.-H., Hu, Z.-Y.: Studies on eutrophication problem and control strategy in the Three Gorges Reservoir. Acta Hydrobiol. Sin. **1**, 7–11 (2006). https://doi.org/10.3321/j.issn:1000-3207.2006.01.002
5. Ye, L., Han, X.-Q., Cai, Q.-H.: Kinetic study of the dissolved organic carbon in Xiangxi Bay of Three Gorges Reservoir region during the spring bloom period. Acta Hydrobiol. Sin. **1**, 80–83 (2006). https://doi.org/10.3321/j.issn:1000-3207.2006.01.015
6. Wang, L., Cai, Q., Zhang, M., Tan, L., Xu, Y., Kong, L.-H.: Spatiotem poral dynamics and related affecting factors of summer algal blooms in Xiangxi Bay of Three Gorges Reservoir. Chin. J. Appl. Ecol. **8**, 1940–1946 (2009). https://doi.org/10.13287/j.1001-9332.2009.0274
7. Ye, L.: Studies on the Eutrophication and Spring Phytoplankton Bloom in Xiangxi Bay of Three Gorges Reservoir. Chinese Academy of Sciences, Beijing (2006)
8. Dyer, K.R.: Estuaries: A Physical Introduction. Wiley, London (1973)
9. Zhang, M.: Numerical study on water environment spatial distribution characteristics and transport time scales of shallow lakes. Hohai University, Nanjing (2014)
10. Yu, Z.: Numerical study on internal wave characteristics of Xiangxi Bay. Hohai University, Nanjing (2011)
11. Huang, Y.: Study on the Formation and Disappearance Mechanism of Algal Bloom in the Xiangxi Bay at Three Gorges Reservoir. North West Agriculture and Forestry University, XiAn (2007)
12. Yang, Z.J.: The mechanisms of algal blooms and its operation method through water level fluctuation under the situation of the bidirectional density currents in tributaries of the Three Gorges Reservoir. Wuhan University, Wuhan (2014)

Removal of Phthalate Esters by Combination of Activated Carbon with Nanofiltration

Long Wang, Qiaoling Wan, Junjie Wu, Ming Guo, Shuang Mao and Jiaqi Lin

Abstract Nanofiltration (NF), combined with activated carbon (AC) for phthalate esters (PAEs) treatment, were studied. Three kinds of typical PAEs dimethyl phthalate (DMP), di-(2-ethylhexyl) phthalate (DEHP) and dioctyl phthalate (DOP) were selected as target contaminant. Fifty μg/L of each PAEs liquid match was prepared with natural river water and treated by NF combined with 10 mg/L or 50 mg/L powdered AC. The test was processed at normal temperature, 0.4 MPa pressure, and pH value as 7. The results indicated that AC can be used as the pretreatment of NF to remove the majority of dissolved organic matter and inorganic particles and can be a safeguard for NF membrane. The removal rates of three PAEs by AC-NF were all above 99%, which suggests that the AC-NF process is feasible for PAEs treatment.

Keywords Removal · Phthalate esters · Activated carbon · Nanofiltration

L. Wang · S. Mao
Chongqing Academy of Metrology and Quality Inspection,
Chongqing 400020, China
e-mail: cquwang@163.com

S. Mao
e-mail: maoshuang@163.com

Q. Wan
Chongqing Monitoring Station, Water Quality Monitoring Network
of National Urban Water Supply, Chongqing, China
e-mail: qiaolingwan@163.com

J. Wu · M. Guo · J. Lin (✉)
Hebei University of Engineering, Handan 056038, China
e-mail: jackstarfly@sohu.com

J. Wu
e-mail: junjiewu@163.com

M. Guo
e-mail: mingguo@163.com

W. Dong et al. (eds.), *Sustainable Development of Water Resources and Hydraulic Engineering in China*, Environmental Earth Sciences,
https://doi.org/10.1007/978-3-319-61630-8_23

1 Introduction

Phthalate esters (PAEs) are synthetic compounds that are used as polymer additives in the production of plastics, rubber, cellulose and styrene. PAEs are categorized as environmental hormones. They are present in many consumer products, such as children's toys, cosmetics, personal care products, blood bags, organic solvents, packaging materials, paper coatings and insecticides. Because of their widespread use, PAEs have become ubiquitous environmental pollutants [1]. PAEs adversely affect human health and are suspected to be estrogenic and reproductive toxins [2]. PAEs can be accumulated through the food chain because of their lipophilicity and stability [3]. Consequently, they are recognized globally as hazardous substances that should be controlled.

Many techniques have been developed for the removal of PAEs, including biodegradation [2], hydrolysis [4], and oxidation technologies [5]. These techniques have limitations, such as high cost, low removal efficiency, lack of availability of bacteria for biodegradation, even some degradation products may be more harmful than the reactants.

AC is known as the most effective adsorbent that is used for the removal of contaminants from polluted water. AC adsorption is recommended by the Environmental Protection Agency (EPA) as a treatment technique suitable for the elimination of NOM from water to attain the established regulations [6], and it has been shown that the AC is efficient in practice for PAEs [7]. Membrane technique is regarded as the most promising water treatment technology in the 21st century; among them, NF membrane can remove the organic pollutants with relative molecular mass of 150–100, and most of the relative molecular mass of PAEs are in the range. In this study, batch experiments were conducted to investigate the removal of PAEs in water by the NF or AC-NF process.

2 Experimental

2.1 Reagents and Materials

Three representative PAEs, DMP, DEHP, and DOP were selected as targets for the test in this work mainly due to their molecular polarities and molecular sizes. Three PAEs, analytically pure, DMP, DEHP, and DOP (Sinopharm Chemical Reagent Co., Shanghai, China) were of analytical grade. Domestic coconut shell based activated carbon (Tangshan United Carbon Technology Co., Ltd. Tangshan, China) with granularity of 200 mesh. NF membrane (American Membrane Corporation, Michigan, USA) was of roll type polyamide composite membrane. Demineralized water was obtained from a Milli–Q Water System (Millipore, Temecula, CA, USA).

2.2 *Effects of Different Water Quality Conditions on the Removal of PAEs by NF*

In an NF membrane test, the influencing factors of NF on PAEs treatment were studied, such as pressure, temperature, pH, ionic strength, dissolved organic matter (DOM) and inorganic particles. Fifty (50) μg/L of each PAEs liquid match was prepared with pure water at different simulated water quality conditions.

The results showed that the main influencing factors of NF on PAEs treatment includes DOM, ionic strength and inorganic particles. Retentate rates of PAEs enhanced as DOM, ionic strength and inorganic particles increased. The retentate rates of DMP enhanced as the inorganic particles increased, and the retentate rates of DEHP and DOP enhanced as the DOM increased.

2.3 *Stability of NF on PAEs Treatment for Long Running*

The stability of NF on PAEs treatment during the long time running was also studied. Fifty (50) μg/L of each PAEs liquid match was prepared with natural river water (with a turbidity of 40 NTU, UV254 0.1 and conductivity of 0.23 ms/cm). The water sample was treated with NF system at room temperature, 0.4 MPa operating pressure, and pH 7. The effluent was taken out after 10 min, 20 min, 30 min, 1 h, 3 h, 6 h, 24 h, 48 h, respectively. The results are shown in Fig. 1.

From Fig. 1 it could be seen that the retentate rates of PAEs by NF membrane were high at the beginning followed by a drop and the equilibrium achieved at 30 min later. During the next 48 h, flux and retentate rate of NF did not produce apparent change. The retentate rates of PAEs by NF membrane were positively

Fig. 1 The removal of PAEs by NF during the long time running

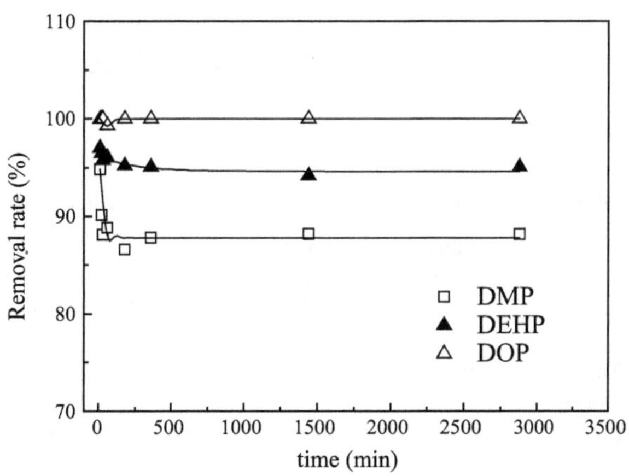

correlated with the molecular weight and size of PAEs because of their difference on molecular weight and molecular structure, and the rejection rate was above 99.31% (DOP), followed by 94.16% (DEHP) and 87.78% (DMP).

2.4 *Removal of PAEs by NF Combined with AC*

Fifty (50) μg/L of each PAEs liquid match was prepared with natural river water. Removal of PAEs by NF, combined with 10 mg/L or 50 mg/L powdered AC, was studied at room temperature, 0.4 MPa operating pressure, and pH 7. The concentration of PAEs in effluent treated by AC or by NF combined with AC was detected, and the results were shown in Fig. 2 and Table 1.

From Fig. 2 and Table 1, it could be learned that the turbidity and UV254 of the effluent treated by AC decreased above 98 and 90% than that of natural water, and the removal rates of three PAEs by AC-NF were all above 99%.

Fig. 2 The AC-NF for PAEs treatment. Notes: 1. Treated by NF; 2. Treated with 10 mg/L powdered AC; 3. Treated with 50 mg/L powdered AC; 4. Treated by NF combined with 10 mg/L AC; 5. Treated by NF combined with 50 mg/L AC

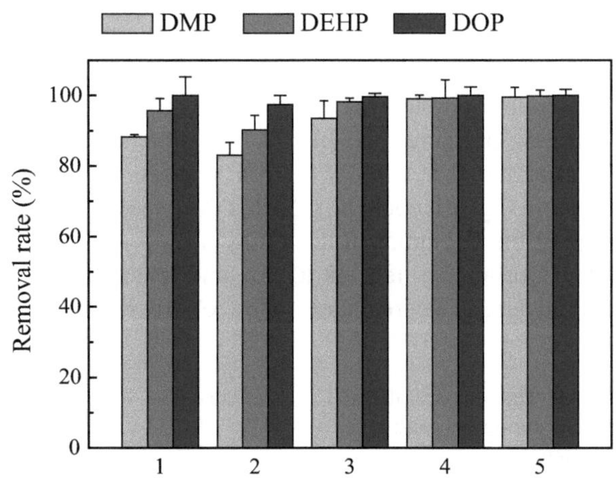

Table 1 The turbidity and UV_{254} of the effluent treated in different processes

	Natural water	Effluent treated by 10 mg/L AC	Effluent treated by 50 mg/L AC	Effluent treated by NF combined with 10 mg/L AC	Effluent treated by NF combined with 50 mg/L AC
Turbidity (NTU)	40	0.65	0.40	0.32	0.30
UV_{254}	0.10	0.01	0.01	0	0

3 Discussion and Conclusion

In the removal of PAEs by NF, the retentate rates of PAEs enhanced as the DOM or inorganic particles increased. The cause may be that DOM or inorganic particles were adsorbed and stayed on the surface of the NF membrane to form a layer and blocked the PAEs flow.

The water quality of effluent treated by AC could meet the Standards for Drinking Water Quality of our country (GB5749-2006). However, the concentration of PAEs in effluent was still at an elevated level; the highest concentration of DMP could be reached 7.86 μg/L and that of DEHP could be reached 3.65 μg/L. Thus, in the actual application of water treatment, NF can be combined with other advanced treatment processes such as AC to achieve more effective removal of PAEs in water. AC can be used as the pretreatment of NF to remove the majority of dissolved organic matter and inorganic particles and can be a safeguard for NF membrane. Therefore, the AC-NF is feasible for PAEs treatment.

References

1. Arvaniti, O.S., Stasinakis, A.S.: Review on the occurrence, fate and removal of perfluorinated compounds during wastewater treatment. Sci. Total Environ. **524–525**, 81–92 (2015). https://doi.org/10.1016/j.scitotenv.2015.04.023
2. Monsalvo, V.M., Mcdonald, J.A., Khan, S.J., Le-Clech, P.: Removal of trace organics by anaerobic membrane bioreactors. Water Res. **49**, 103–112 (2014). https://doi.org/10.1016/j.watres.2013.11.026
3. Wijekoon, K.C., Mcdonald, J.A., Khan, S.J., Hai, F.I., Price, W.E., Nghiem, L.D.: Development of a predictive framework to assess the removal of trace organic chemicals by anaerobic membrane bioreactor. Bioresour. Technol. **189**, 391–398 (2015). https://doi.org/10.1016/j.biortech.2015.04.034
4. Yim, B., Nagata, Y., Maeda, Y.: Sonolytic degradation of phthalic acid esters in aqueous solutions. Acceleration of hydrolysis by sonochemical action. J. Phys. Chem. A **106**, 104–107 (2001). https://doi.org/10.1021/jp011896c
5. Liu, J., Li, Y., Wang, J., Chen, K.: Removal of phthalic acid esters from water by advanced catalytic oxidation process. Ecol. Environ. Sci. **23**, 904–910 (2014). https://doi.org/10.3969/j.issn.1674-5906.2014.05.027
6. Hua, J., Martinb, A., Shang, R.: Anionic exchange for NOM removal and the effects on micropollutant adsorption competition on activated carbon. Sep. Purif. Technol. **129**, 25–31 (2014). https://doi.org/10.1016/j.seppur.2014.03.019
7. Venkata, M.S., Shailaja, S., Rama, K.M., Sarma, P.N.: Adsorptive removal of phthalate ester (di-ethyl phthalate) from aqueous phase by activated carbon: a kinetic study. J. Hazard. Mater. **146**, 278–282 (2007). https://doi.org/10.1016/j.jhazmat.2006.12.020

Startup of Formatting Biological Membrane in Denitrifying Filter at Low Temperature

Long Wang, Yanyan Dou, Qiaoling Wan, Shuang Mao, Wen Zhang and Jiaqi Lin

Abstract Pollution and treatment of the wastewater have been the focus of environmental protection. In the microbial nitrogen removal processes, a denitrification filter has the advantages of low investment, less land occupied, high biomass and high treatment efficiency. Therefore, it is one of the most widely used processing technologies in advanced treatment. This study uses secondary effluent of the East Sewage Treatment Plant as the research object. The effluent of the waste water treatment plant (WWTP) has a low temperature about 10–20 °C. The inner diameter of the filter column is 150 mm, and the height of the column is 2.3 m. The denitrification filter used boring exposure and continuous water to select the advantage bacterium group to form biological membrane attached on the surface of the filter material. This experiment studies the different treatment effect when forming biological membrane using two different packing processes (ceramsite packing and polyethylene polyhedral hollow ring filter packing) in the

L. Wang · S. Mao · W. Zhang
Chongqing Academy of Metrology and Quality Inspection,
Chongqing 401123, China
e-mail: cquwang@163.com

S. Mao
e-mail: maoshuang@163.com

W. Zhang
e-mail: zhangwen@163.com

Y. Dou
College of Energy and Environmental,
Zhongyuan University of Technology, Zhengzhou 450000, China
e-mail: douyanyan@163.com

Q. Wan
Chongqing Monitoring Station, Water Quality Monitoring
Network of National Urban Water Supply, Chongqing, China
e-mail: qiaolingwan@163.com

J. Lin (✉)
Hebei University of Engineering, Handan 056038, China
e-mail: jackstarfly@sohu.com

W. Dong et al. (eds.), *Sustainable Development of Water Resources and Hydraulic Engineering in China*, Environmental Earth Sciences,
https://doi.org/10.1007/978-3-319-61630-8_24

denitrification filter. During the experiment, TN, NO_3^-–N, NO_2^-–N, NH_4^+–N, COD_{Cr} were constantly measured as well as other regular indicators. The above regular indicators show the biological membrane attachment condition during the formation of the biological membrane. To confirm this procedure successfully, the mark was that the COD_{cr} removal rate reached 50% and the NO_3^-–N removal rate reached 60%. It was shown that the ceramsite packing needed 25 d, and the polyethylene polyhedral hollow ring packing needed 30 d. This experiment indicates that the startup time, membrane growth and removal efficiency of the ceramsite filter are better than the polyethylene polyhedral hollow ring filter.

Keywords Low temperature · Denitrification filter · Removal efficiency

1 Introduction

The startup of formatting a biological membrane in a denitrification filter is formatting a biofilm on the surface of packing. It is a process of interaction between packing and microorganisms. Related studies have shown [1, 2] whether microorganisms can be attached on the surface of packing which is determined by three aspects: the first one depends on the characteristics of the bacteria; the second one depends on the physical and chemical properties of the packing surface; and the third one depends on the filter's environment conditions; and the nitrogen removal of the denitrification filter is determined by the stability of the microbial in biofilm. After a series of complex processes—physical, chemical, biological, etc.—the biofilm is formed. Therefore, the startup of formatting biological membrane is extremely important for the stable operation of the denitrification filter.

At present, the startup of formatting biological membrane in the denitrification filter usually chooses two ways: natural formatting biological membrane and inoculation formatting biological membrane [3]. Natural formatting biological membrane is the process of gradually forming mature biofilm. In this process, the microorganisms use the nutrients in raw water to grow. The inoculation formatting biological membrane is artificially added with a certain volume of sludge in the reactor to speed up formatting the biological membrane through the combination method of stuffy exposure and exchange water to improve the microorganism's adaptive velocity and shorten the time of the formatting biological membrane.

The denitrification filter depends on the biofilm attached to the surface of the packing for denitrification, and microorganisms give priority to the denitrifying bacteria in the biofilm, the mechanism of nitrogen as described below. The biological denitrification method is an effective and exhaustive elimination of nitrogen pollution. Under the action of nitrifying bacteria and denitrifying bacteria, the nitrogen pollutant in the polluted wastewater translates into gaseous nitrogen by ammonification, nitrification and denitrification reaction [4]. Ammonification processes are that ammonifiers decompose and transform organic nitrogen compounds into ammonia nitrogen. Nitrification processes are to further decompose ammonia nitrogen by denitrifiers. Firstly, ammonia nitrogen is transformed into nitrous nitrogen by nitrosobacteria; secondly, nitrite nitrogen is transformed into nitrate nitrogen by nitrate bacteria; thirdly, nitrate nitrogen and nitrous nitrogen are reduced into N_2 by denitrifiers [5]. In the biological filter, denitrifying bacteria is a type of heterotrophic facultative anaerobic bacteria; it proceeds denitrification reaction. The biological denitrification process is divided into assimilative denitrification process and dissimilative denitrification. The assimilative denitrification is also called anabolism. Under the metabolism of denitrifying bacteria, inorganic nitrate is reduced into organic compounds. The organic compounds become a part of the cell for their own use. The dissimilative denitrification is also called catabolism. Nitrogen was converted into N_2 by denitrification bacteria, then returns to the atmosphere. Specifics are shown in Eq. 1 [6].

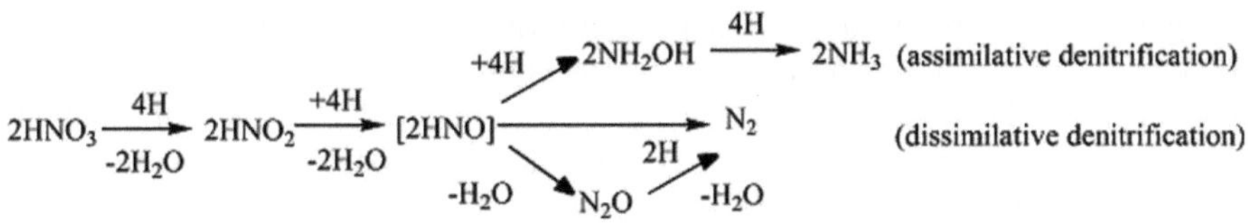

(1)

Equation 1 Nitrate nitrogen transformation pathways and results in the biological digestion.

Denitrifying bacteria in the process of dissimilative denitrification uses organic carbon as an electron donor and nitrate as an electron acceptor. As with the methanol, the dissimilative denitrifying process can be finished according to the following two steps:

$$6NO_3^- + 2CH_3OH \rightarrow 6NO_2^- + 2CO_2 + 4H_2O \tag{2}$$

$$6NO_2^- + 3CH_3OH \rightarrow 3N_2 + 3CO_2 + 3H_2O + 6OH^- \tag{3}$$

$$6NO_3^- + 5CH_3OH \rightarrow 3N_2 + 5CO_2 + 7H_2O + 6OH^- \tag{4}$$

By the above formula, the denitrifying bacteria for denitrification reaction can effectively remove nitrogen compounds and achieve the goal of denitrification.

Denitrification biofilter has several characteristics of low investment, less land occupation, high biomass, etc.; and it is one of the most widely used technologies in advanced wastewater nitrogen removal treatment.

Dahab [7] used the synthetic wastewater as raw sewage and used the static up flow reactor. Adding acetic acid as a carbon source, potassium nitrate as nitrate, they research and analyze the denitrification effect. The result shows that the nitrate removal rate of the reactor is more than 90% when the nitrate concentration of influent is 100 mg/L, C/N is 1.5, HRT is 9 h.

In 1988, Soares [8] designed a kind of sand column, which uses sucrose as a carbon source. The study shows that denitrification can fully process with sufficient carbon source and improve the nitrogen removal efficiency. It also shows that the osmotic pressure in the sand column will gradually reduce as time continues. It is because of the existence of gas in the filter material. France built denitrification biofilter in 1981, which chose Biolite (a kind of filter material which has an average grain diameter of 1.7 mm) as the filter material and set it at 8 m/h, and the depth of the material is 3 m. The result of the experiment shows that this kind of filter is applicable for high concentration nitrate wastewater treatment. Nowadays, denitrification biofilters have already caused the widespread attention of scholars, especially focused on different influence of factors like types of filter media, biofilm inoculation methods, temperature, etc.

Huang et al. [9] used a kind of Acinetobacter Y16, which has a strong ammonia nitrogen removal ability in 2 °C, and it can turn ammonia to nitrogen. Under the condition of using sodium acetate as a carbon source, it can remove 66% ammonia nitrogen from the wastewater, and it has a high removal efficiency of ammonia nitrogen in low temperature and low C/N so that Acinetobacter Y16 can be used in ammonia nitrogen removal in low temperature water in winter.

Sahinkaya and Kilic [10] evaluated the difference between the dissimilative denitrification and SCAD for the surface water including nitrate with chromate. The results showed that elemental sulfur and nitrate were used as an electron donor and acceptor in SCAD process; denitrification is reacted thoroughly; and the concentration of Chromate is lower than 0.5 mg/L. Although the denitrification rate of SCAD is lower than the one of dissimilative denitrification, SCAD can avoid the secondary pollution caused by the organic added.

Tang et al. [11] used a denitrification biofilter in advanced treatment of secondary effluent of a WWTP. During biofilm colonization, dose Sodium Acetate external as a carbon source, which only costed 15 d to complete the biofilm attachment. After stable operation, the COD removal efficiency of denitrification biofilter was established at 72–83%; the COD average concentration of final effluent is 11.71 mg/L, which is lower than secondary effluent of WWTP.

The total nitrogen of raw water in this study is mainly composed of water nitrate nitrogen, so that reducing the TN concentration of effluent needs to reduce the concentration of nitrate nitrogen. Aiming at the characteristics of secondary effluent of WWTP, denitrification biofilter is used in advanced treatment in this study.

2 Materials and Methods

2.1 Water Quality

This experiment works in the East Sewage Treatment Plant; it chooses the combination process "denitrifying filter-coagulation sedimentation-O_3-biological sand

Table 1 The main specification of secondary effluent from WWTP

Specification	COD_{Cr} (mg/L)	TN (mg/L)	NH_4^+–N (mg/L)	NO_3^-–N (mg/L)	NO_2^-–N (mg/L)	TP (mg/L)	pH	Temperature (°C)	Turbidity (NTU)
Maximum	45.3	21.87	2.42	17.58	0.79	1.54	8.0	28	12.02
Minimum	20	16.44	1.86	12.48	0.51	0.62	7.4	10	0.94
Average	32.65	19.16	2.14	15.03	0.65	1.08	7.7	–	6.48

filter" for deep processing to reduce the nitrogen. This experiment influent is the WWTP's secondary effluent; the main specification of secondary effluent from WWTP is shown in Table 1.

2.2 *The Experimental Materials*

The secondary effluent of WWTP is pressed to the original water tank by a submersible sewage pump. The original tank water is pushed by peristaltic pump into the two denitrifying filters. The effluent of denitrifying filter flow is pushed into coagulation sedimentation tank by gravity. The gas that washes the denitrification filter is provided by the air compressor. The experimental schematic diagram is shown in Fig. 1.

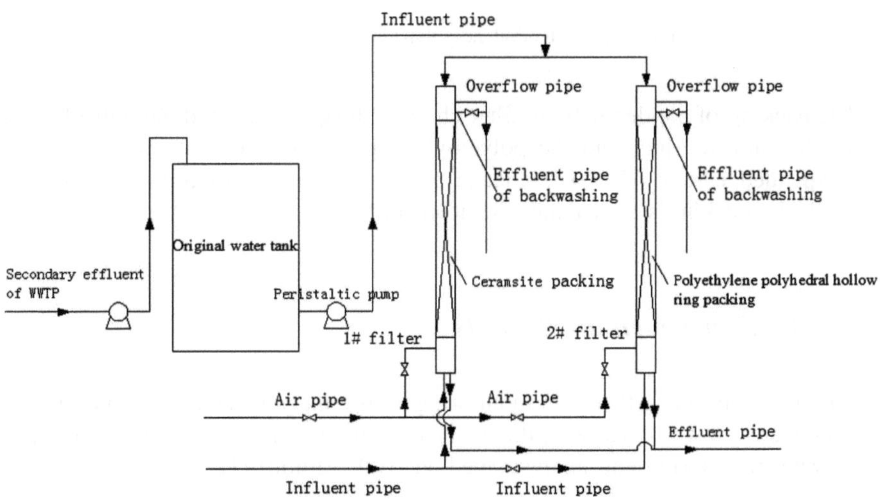

Fig. 1 Experimental schematic diagram

Table 2 The properties of filter media

Properties	Diameter (mm)	Particle bulk density (g/cm^3)	Porosity (%)	Bulk density (g/cm^3)	Accumulation specific surface (m^2/cm^3)	SiO_2 (%)	Al_2O_3 (%)	Fe_2O_3 (%)
Ceramsite	3–5	1.52	55–78	0.75–0.8	>3000	62–68	18–22	6–9
Poly-ethylen polyhedral hollow ring	25	0.25	–	0.1–0.15	620	–	–	–

Fig. 2 Pictures of ceramsite and polyethylene polyhedral hollow ring

The packing of the denitrifying filter chooses the ceramsite and the polyethylene polyhedral hollow ring, and the polyethylene polyhedral hollow ring belongs to the suspended carrier. The performance parameters of the packing are shown in Table 2; the packing's appearance is shown in Fig. 2.

2.3 *The Experimental Methods*

During the experiment, the main monitoring items, analysis methods, experimental apparatus are shown in Table 3; the analysis method consulta with the "water and wastewater monitoring and analysis method" (4th edition) [12].

Table 3 Analysis items and methods

Items	Methods	Instrument
COD_{Cr}	Potassium dichromate method	Backflow device and burette
NH_4^+–N	Nessler's reagent spectrophotometry	Vis spectrophotometer
NO_2^-–N	N-(1-naphthyl) ethylenediamine dihydrochloride spectrophotometric method	Vis spectrophotometer
NO_3^-–N	UV spectrophotometry	UV spectrophotometer
TN	Potassium persulfate oxidation—UV spectrophotometry	UV spectrophotometer
Turbidity	Spectrophotography	Turbidimeter
DO	Rapid determination method	DO meter
Temperature	Thermometer method	Thermometer
pH	Glass electrode method	pH meter
Biomass	Phosphatide method	Vis spectrophotometer

3 Results and Discussion

This experiment chooses artificial inoculation formatting biofilm and it discusses the different laws of different packing in denitrifying biological filter formats biofilm.

3.1 Startup of Formatting Biofilm in Denitrifying Filter at Low Temperature

The low temperature of the experiment water is about 10–20 °C. The secondary effluent of WWTP has low COD and a large number of the refractory organics; the refractory organics go against microbes surviving and reproducing. Therefore, this experiment chooses artificial inoculation formatting biofilm. Sludge is from the oxidation ditch of WWTP, and the methanol is placed into two filter columns in the biofilm culturing process.

The biofilm culturing process has two phases:aeration and continuous water to select advantage bacterium. In the first phase, the influent mixed with sludge in accordance with the proportion of 3:1 inject filter column. The water was aerated for 24 h with 20 L/h continuously, changing water ratio is 1/2, working 10 days. This is for formatting biofilm and letting microbes adhere to the surface of the filler [13]. The second phase is to empty the water of the filter column and stop aerating and take the methanol to the original water. According to the number IAWQ1 model's formula $COD/NO_3^-{-}N = 2.86/(1 - Y_H)$ [14, 15] and the best COD/NO_3–N ratio (COD/NO_3–N = 5) which comes from the test, determined that the additive amount of methanol is 120 mL, COD's concentration is about 75 mg/L, the control of the two column's flow speed is 1 m/h. After 24 days the ceramsite filter's removal rate of COD is steadily around 50% and its removal rate of nitrate nitrogen is around 60%. It indicates that the biofilm surface of the filter has grown to maturity, and the formatting of the biofilm is successful. The polyethylene polyhedral hollow ring filter gets the same treatment effect at the 30th day.

3.2 The Removal Characteristics of Denitrifying Biological Filter During the Biofilm Formatting

The Removal Effect of COD_{Cr}. In the phase of biofilm formatting, 1# and 2# filter column's removal effect of COD_{Cr} is shown in Fig. 3.

Figure 3 shows that during the experiment, the mean concentration of COD_{Cr} of the secondary effluent is 26 mg/L. In the first phase of biofilm formatting, the two-filter column does not add a carbon source. The figure shows that the removal efficiency of COD_{Cr} is less than 10%. The reason is the following two points: firstly, the main component of the secondary effluent of the treatment plant is refractory organics; it is not easy to be used and degraded by microorganisms; secondly, in the start-up phase, organic matter can only be removed by physical removal. From the figure it is also known that the effluent COD_{Cr} of 2# filter column slightly increases in the early stage of the start-up period, and the removal efficiency is less than 5%. It is because the interspace of polyethylene polyhedral hollow ring is bigger than the one of ceramsite, the effect of interception is not strong, inoculation sludge is easy to flow out with the effluent and it will increase the concentration of COD_{Cr}; then it leads to a low removal efficiency. In second

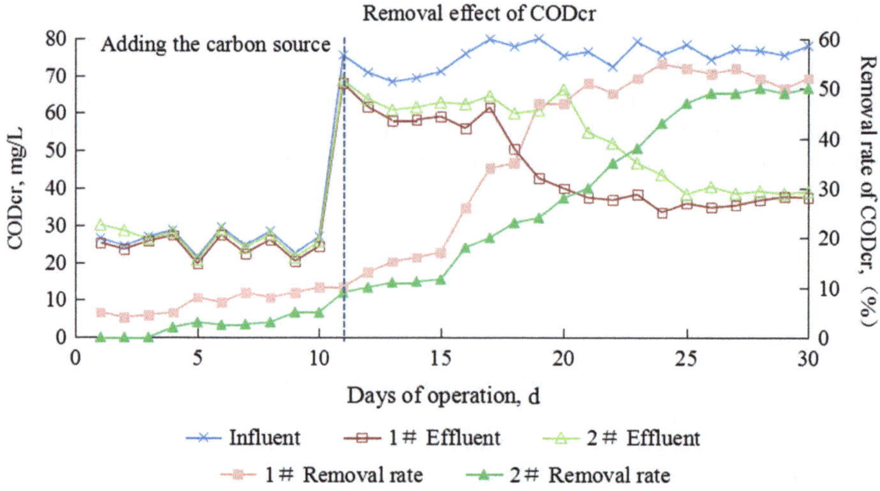

Fig. 3 The removal of COD_{Cr} during the start-up period

phase of biofilm formatting, a carbon source is added in day 14, after stopping aeration, the COD_{Cr} removal rate of #1 filter column is about 15%; the COD_{Cr} removal rate of #2 filter column is about 10%. It is because when using methanol as the carbon source, denitrification bacteria need a period of adaptation time [16]. Five days after adding the carbon source, the COD_{Cr} removal efficiency of the two filter columns is rapidly increased. When it comes to day 25, the COD_{Cr} removal rate of #1 filter column reaches 50%; and when it comes to day 30, the COD_{Cr} removal efficiency of 2# filter column also reaches 50%. All of them operate with stability after that.

From Fig. 3 it is known that during biofilm formatting, the COD_{Cr} removal efficiency of 2# filter column is lower than the one of 1#. It is because the polyethylene polyhedral hollow ring packing is suspended in the column; the interspace of it is bigger than ceramsite; while the specific area of it is 1/5 of ceramsite. It is not easy for denitrifying bacteria to attach on the surface of the packing as its growth rate is low, and the effluent turbidity of the polyethylene polyhedral hollow ring is higher than the ceramsite filter column.

The Removal Effect of NO_3^-–N. In the phase of biofilm formatting, 1# and 2# filter column's removal effect of NO_3^-–N is shown in Fig. 4.

From Fig. 4, the NO_3^-–N of influent is 15 mg/L. In the first phase, two filter columns have not the removal effect of NO_3^-–N, and the concentration of NO_3^-–N is increased. The reason is that the aeration environment is more suitable for the growth of nitrifying bacteria; nitrifying bacteria can transform NH_4^+–N to NO_3^-–N and oxidate NO_2^-–N. In the second phase, aeration is stopped; the aerobic environment of the filter column is destroyed; the growth of nitrifying bacteria is restrained; denitrifying bacteria's growth and reproduction cause themselves

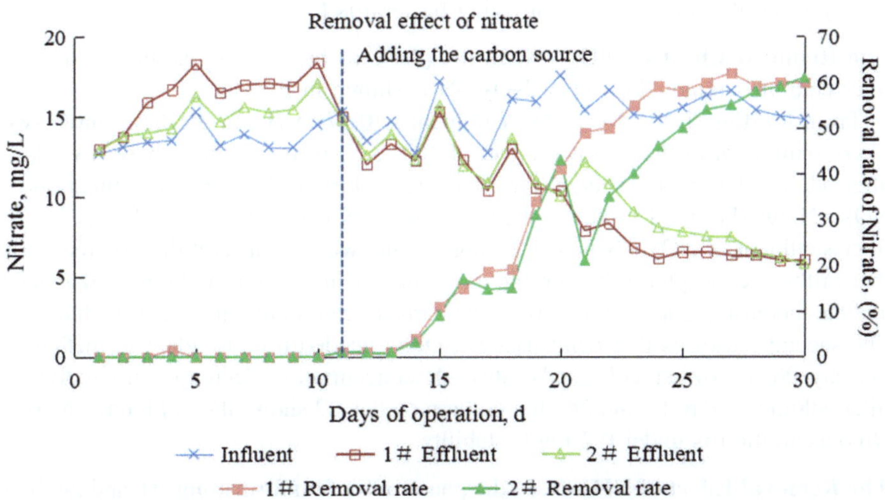

Fig. 4 The removal of NO_3^-–N during the start-up period

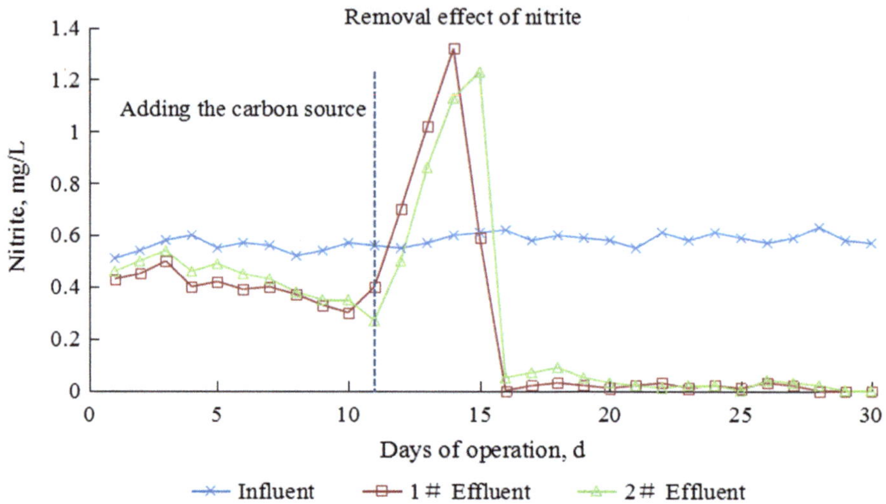

Fig. 5 The removal of NO_2^-–N during the start-up period

to become dominant bacteria; denitrification is gradually strengthened. On day 15, 1# filter's removal rate of NO_3^-–N is to 11%; 2# filter's removal ratio of NO_3^-–N is to 9%. On day 25, 1# filter's removal ratio of NO_3^-–N is to 60%; and on day 30, 2# filter's removal ratio of NO_3^-–N is also to 60%, the formatting biofilm is successful.

In the biofilm culturing process, #1 filter column's removal rate of NO_3^-–N is higher than 2# filter column. The reason is that the polyethylene polyhedral hollow ring is in suspended state; its surface is not easy to grow and adherent biofilms. The small number of denitrifying bacteria, denitrification reaction cannot react completely when the number of denitrifying bacteria is less.

The Removal Effect of NO_2^-–N. In the phase of biofilm formatting, #1 and #2 filter column's removal effect of NO_2^-–N is shown in Fig. 5.

From the Fig. 5, the NO_2^-–N of influent is 0.57 mg/L. In the first phase, two filter columns have the removal effect of NO_2^-–N. In the second phase, five days after adding the carbon source, the two filter columns show the accumulation of NO_2^-–N. On day 14, 1# filter's effluent of NO_2^-–N is to 1.32 mg/L. On day 15, 2# filter's effluent of NO_2^-–N is to 1.23 mg/L. The two reasons are: the first reason is that in the second phase the aerobic environment in the filter column is damaged and the increasing concentration of organic matter can inhibit the nitrifying bacteria. The second reason is that denitrifying bacteria needed time to adapt to methanol. On day 06, two filter columns's NO_2^-–N concentration decreases gradually; 1# filter column on day 19 and 2# filter column on day 22 shows their effluent's NO_2^-–N concentration is under 0.2 mg/L stability.

The Removal Effect of NH_4^+–N. In the phase of biofilm formatting, 1# and 2# filter column's removal effect of NH_4^+–N is shown in Fig. 6.

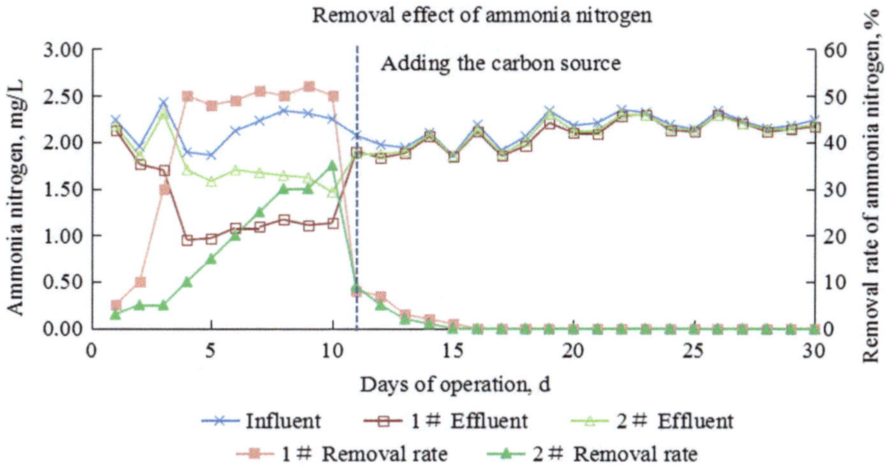

Fig. 6 The removal of NH_4^+–N during the start-up period

From Fig. 6, the NH_4^+–N of influent is 2.14 mg/L. In the first phase, 1# filters the removal rate of NH_4^+–N is to 50%; the effluent of NH_4^+–N is about 1.00 mg/L stability; 2# filters the removal rate of NH_4^+–N is to 35%; the effluent of NH_4^+–N is about 1.50 mg/L stability. Environment of the filter column is suitable for the growth and reproduction of nitrifying bacteria. In the second phase, adding a carbon source and stopping aeration, two filter columns do not have the removal effect of NH_4^+–N. The reason is that adding a carbon source and stopping aeration causes DO to decrease; the low concentration of DO and the high concentration of organics restrain the nitrifying bacteria growth and reproduction. Then the nitrifying bacteria is eliminated, and denitrifying bacteria becomes dominant bacteria. So, there is only the assimilation of microorganism to remove NH_4^+–N.

The Removal Effect of TN. Studies have shown that temperature has obvious effects on denitrification, thus affecting denitrification effect [17]. The suitable temperature of denitrification is 20–40 °C; the temperature of this experiment is between 10 and 20 °C. In this temperature, denitrifying bacteria proliferation and metabolic rate is reduced and denitrification is affected; 1# and 2# filter column's removal effect of NH_4^+–N is shown in Fig. 7.

Figure 7 shows that two filter columns do not have the removal effect of TN; the effluent of TN and the influent of TN do not have an obvious difference. The reason is that in this phase the filter column is aerated; high dissolved oxygen environment makes denitrifying bacteria use oxygen to breathe and restrain denitrifying bacteria synthesizing a reduction in nitrate. It makes nitrogen not to be thoroughly transformed and influence the effect of nitrogen. In the second phase, two filter columns' removal effect of TN improve obviously. Within the first five days, the removal effect of TN does not work well; 1# filter's removal rate is under 10% and 2# filter's removal rate is under 5%. The reason is that the denitrifying bacteria needs a period of time to adapt methanol. However, when it comes to day 15, two filter columns'

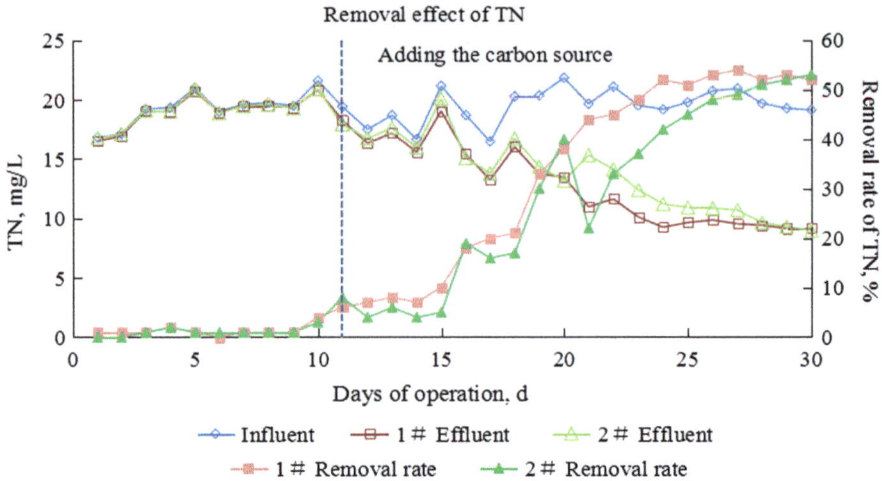

Fig. 7 The removal of TN during the start-up period

removal rate of TN rapidly increases. As formatting biofilm is successful, two filter columns' effluent of TN is under 15 mg/L stability, the mean removal rate is to 50%.

4 Conclusions

This experiment discusses the startup of formatting biofilm in denitrifying filter by two different packing processes. The following conclusions are drawn from the experiment: in this experiment, it uses the aeration phase and selecting advantage bacterium phase to startup formatting biofilm by artificial inoculation sludge. The temperature of influent is 16 °C; the mean concentration of COD_{Cr}, NO_3^-–N, NO_2^-–N, NH_4^+–N, TN is 26 mg/L, 15.03 mg/L, 0.57 mg/L, 2.14 mg/L and 19.16 mg/L. The removal rate of COD and NO_3^-–N 50, 60% is a sign; 1# ceramsite filter needs 25 days and 2# polyethylene polyhedral hollow ring filter needs 30 days to make biofilm to attach on the surface of packing. A comparative study of the two filter columns' decontamination effect during formatting biofilm indicates that the time of startup of formatting biofilm is short in 1# ceramsite filter; the filter's biofilm adheres to ceramsite easily. It has good decontamination effect and the time of startup of formatting biofilm in1# filter is longer than 2# filter; the filter's biofilm is easy to fall off and the decontamination effect is poorer.

References

1. Mann, A.T., Stephenson, T.: Modeling biological aerated filters for wastewater treatment. J. Water Res. **31**, 2443–2448 (1997). https://doi.org/10.1016/S0043-1354(97)00095-X

2. Zhao, B., Yang, W.: Procreative practices of trickling filter's hanging biomembrance. J. Environ. Prot. Transp. **20**, 25–27 (1999)
3. Liu, Q.: Study on biological characteristics and biological nitrification performance of UBAF in treating urban sewage, Master's thesis of Hebei University of Engineering (2009)
4. Yang, Y., Xu, X., Liu, C.: Engineering and Example Analysis of Wastewater Treatment. Chemical Industry, Beijing (2002)
5. Zhang, Z., Lin, R., Jin, R.: Drainage Engineering, 4th edn. China Architecture & Building, Beijing (1999)
6. Wu, W., Ge, H., Zhang, K.: Biological Treatment Technology of Wastewater. Chemical Industry, Beijing (2003)
7. Dahab, M.F.: Nitrate Treatment Methods: An Overview, Nitrate Contamination Exposure, Consequence and Control, pp. 349–368. Springer, Berlin, Germany (1991). https://doi.org/10.1007/978-3-642-76040-2_26
8. Soares, M.I.M.: Biological denitrification of groundwater. Water Air Soil Pollut. **123**, 183–193 (2000). https://doi.org/10.1023/A:1005242600186
9. Huang, X., Li, W., Zhang, D., et al.: Ammonium removal by a novel oligotrophic Acinetobacter sp. Y16 capable of heterotrophic nitrification-aerobic denitrification at low temperature. Bioresour. Technol. **146**, 44–50 (2013). https://doi.org/10.1016/j.biortech.2013.07.046
10. Sahinkaya, E., Kilic, A.: Heterotrophic and elemental-sulfur-based autotrophic denitrification processes for simultaneous nitrate and Cr (VI) reduction. J. Water Res. **50**, 278–286 (2014). https://doi.org/10.1016/j.watres.2013.12.005
11. Tang, S., Liu, N., Cheng, B. et al.: Study on the denitrification effect of the post-denitrification filter. Biotech. World **7**, 91–94 (2012)
12. State Environmental Protection Administration of China: Water and Wastewater Monitoring and Analysis Method, 4th edn. China Environmental Science, Beijing (2002)
13. Wang, Z., Yu, B., Gan, Y., et al.: Study on startup of pre-denitrification biofilter. J. China Water Wastewater **28**, 11–14 (2012)
14. Cai, B., Li, X., Yang, D., et al.: A review on carbon source supplement and optimization in denitrification process. J. Water Purif. Technol. **26**, 37–40 (2007)
15. Ji, F.: Experimental study on BAF with post-denitrification to treat wastewater, Master's thesis of Wuhan University of Science and Technology (2012)
16. Lin, Q., Li, D., Si, H., et al.: Pilot study on start-up of pre-denitrification biological aerated filter. J. China Water Wastewater **28**, 125–128 (2012). https://doi.org/10.3969/j.issn.1000-4602.2012.13.031
17. Liu, X., Gan, Y., Yang, Q., et al.: Effects of carbon source types on the operation and nitrifying microbial community of denitrifying biofilter for advanced nitrogen removal. J. Technol. Water Treat. **39**, 36–40 (2013). https://doi.org/10.3969/j.issn.1000-3770.2013.11.008

Study on Geochemical Evolution and Isotopic Characteristics of Groundwater in Arid and Semi-arid Areas

Lishu Wang, Qingjiao Cao, Yunxin Zhang, Fei Liu, Yingying Shen, Qiang Ma and Yunpu Zheng

Abstract The utilization ratio of water resources is remarkably high in the Shiyang River Basin, which is a typical zone susceptible to water resources crisis in the arid area in China. Owing to the continuous deterioration of the ecological environment in the Shiyang River Basin, it is of great significance to investigate the groundwater circulation and geochemical evolution law in this area. Samples were collected from the top of the Qilian Mountains to Hongyashan Reservoir along the flow cross section. Isotopic survey and geochemical technology were employed to analyze hydrochemistry and environmental isotopes of precipitation, surface water and groundwater. Results indicated that precipitation was the provenance of water resources in the Shiyang River Basin. The groundwater runoff system accepted river leakage and irrigation water infiltration recharge constantly in the process of the runoff to the downstream, meanwhile drained through the river overflow and pumping with a primary vertical movement. The downstream of the Basin was the main groundwater discharge area. The groundwater in the northern Basin

L. Wang · Q. Cao · Y. Zhang · F. Liu · Y. Zheng (✉)
School of Water Conservancy and Hydropower, Hebei University of Engineering, 178 S. Zhonghua Street, Handan 056021, Hebei, China
e-mail: zhengyunpu_000@sina.com

L. Wang
e-mail: fc8232@126.com

Q. Cao
e-mail: 13931061206@163.com

Y. Zhang
e-mail: zyx14315@163.com

F. Liu
e-mail: liufei@hebeu.edu.cn

Y. Shen
China Irrigation and Drainage Development Center, Beijing 100000, China
e-mail: shenyingying_cugb@163.com

Q. Ma
Yahoo! Inc., 701 1st Ave, Sunnyvale, CA 94089, USA
e-mail: qiang@yahoo-Inc.com

W. Dong et al. (eds.), *Sustainable Development of Water Resources and Hydraulic Engineering in China*, Environmental Earth Sciences,
https://doi.org/10.1007/978-3-319-61630-8_25

overflowed into the surface water due to the basement uplift. The TDS (Total Dissolved Solids) values were between 131 and 1750 mg/L, gradually increasing along the flow direction. The hydrochemical types showed an obvious zonation from the upper reaches to the lower, which gradually changed from bicarbonate to sulphates.

Keywords Environmental isotopes · Hydrochemistry · Groundwater circulation
Wuwei Basin

1 Introduction

The Shiyang River Basin, located in the eastern part of the Hexi Corridor in the middle of Gansu province, Northwestern China, is considered a typical arid and semi-arid area. As a result of the long-term water resources development in the Basin, the Shiyang River Basin has become one of the most over-exploited inland basins. Water resources shortage is an increasingly serious issue in China, especially in the arid and semi-arid areas of the northwest. The Shiyang River is an eco-environmentally fragile area, which is characterized by low and irregular rainfall, high temperatures and evaporation, and notable drought periods [1]. In such arid and semi-arid environment, groundwater is a significant part of the total water resource and plays a key role as a water supply for drinking and irrigation [2].

Research interests have been focused on aspects such as utilization survey, evolutionary prediction, reserve estimation and systematic assessment of natural water resources. Some reasonable allocation and utilization plans have been proposed [3]. The effective development and management of the valuable groundwater resources in the area require a better understanding of hydrochemical characteristics of the groundwater and its evolution under natural water circulation processes. During the last thirty years, environmental isotopic techniques have been commonly and largely used in the overall domain of water resources development and management [4].

The application of these relatively new techniques in the case of arid and semi-arid zones has proved to be an attractive tool for the quantitative evaluation of groundwater systems [5]. However, comprehensive approaches to groundwater understanding using chemical and isotopic indicators are relatively few, although there are several investigations on the interaction and mechanism among surface water, groundwater and rocks in recent years [6–9].

The systematic analysis of the hydrochemical types and features of the whole river basin were combined using computer simulation and isotopic geochemical methods in this study. The objective was to elucidate the controls of groundwater quality evolution, recharge, circulation and mixing processes and, furthermore, to explain the mechanism of these processes from the standpoint of environmental geochemistry. The results of the study may provide a theoretical basis for the government to make utilization strategies for water resources and development policies in northwest China in the new century. The purpose of this paper is, thus, to make a new assessment of the availability of groundwater resources in the Wuwei Basin as the typical area of arid northwestern China based on a better understanding of hydrogeological features.

2 Geology and Hydrogeology of the Study Area

The Shiyang River Basin, located in the eastern part of Hexi Corridor in the middle of Gansu province, is a typical arid and semi-arid area (Fig. 1). The Shiyang River originates from Lenglongling Glacier of the Qilian Mountains, and ends gradually in the Tenggeli Desert. The Qilian Mts. form the southern part of the Basin. Many modern glaciers are distributed above 4500 m. The middle and northern part of the Basin is a plain that can be divided into two sub-basins (i.e., Wuwei Basin and Mingqing Basin).

Most of the basin belongs to the arid climate zone. Solar radiation is strong and sunny. Summer is hot and short, while winter is cold and long. The temperature difference is extremely significant. Less rainfall and evaporation intense lead to dry air.

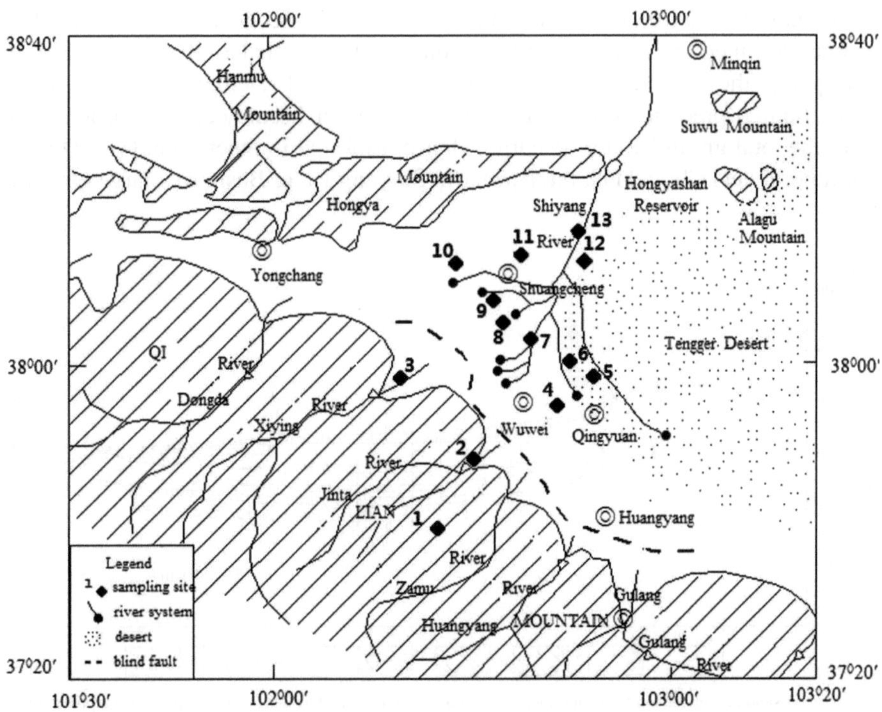

Fig. 1 Geographic location of the study area and water quality sampling point

The annual precipitation is 200–800 mm, while the evaporation capacity is 2000–3000 mm.

Consisting of Quaternary alluvial and pluvial sediments, the Wuwei Basin lies in a depression at the foot of the Qilian Mts. with an elevation ranging from 1400–2500 m. The loose Quaternary sediments filling the basin are of great thickness, which provide an ideal place for groundwater preservation. In terms of the hydrogeological characteristics, two geomorphologic units are recognized in the Wuwei Basin, a faulting terrace zone and its sub-basin. The former is in the vicinity of Qilian Mts. and lies on the head of a pluvial fan. The aquifer there is composed of gravel and sand. Well-connected porosity of the aquifer provides a good conduit for groundwater flow, which is the basin's recharge region. The sub-basin is the enrichment zone of the groundwater. As the grain size of the aquifer gradually decreases from south to north and the flow rate becomes slower, the hydraulic gradient becomes smaller. Along with the uplifting of the Quaternary basement, the burial depth of the aquifer becomes shallower, resulting in the springs of groundwater from the middle part of the Wuwei Basin (Fig. 2).

In the southern part of the Wuwei Basin, the diluvial aquifer is formed from highly permeable cobble and gravel deposits that are between 200 and 300 m thick. The aquifer specific capacity is about 3–30 $l \cdot s^{-1} \cdot m^{-1}$, and aquifer permeability is up to 100–600 $m \cdot d^{-1}$ [10]. This allows a large amount of surface runoff in the piedmont fan to seep down and recharge the aquifer. From the northern edge of this diluvial fan, the aquifer, comprising interbedded cobble gravel, fine sand and clay, becomes confined or semi-confined with piezometric levels less than 5 m depth. In many places the ground-water overflows as springs and re-emerges as streams.

The limited precipitation in the plains of the river basin accounts for less than 7% of the total ground-water recharge. Hence, as found in other inland river basins in arid northwest China [11], the aquifer in Shiyang River Basin to a large extent is

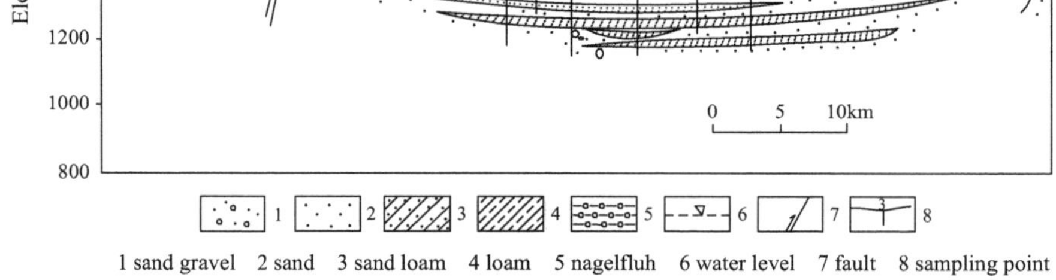

Fig. 2 Geologic section of the Wuwei Basin

recharged by surface water via river infiltration, canal system seepage and farmland irrigation water seepage. In other words, surface water and groundwater are integrated in the middle and lower basins, and the upper, middle and lower reaches are united in using water resources. The water resources in the different parts of the basin are inseparably linked: the ground-water resources in the plain area are recharged from precipitation in the Qilian Mountains. From the Qilian Mountain reach to the Hongyashan section, groundwater and surface water are closely linked due to the gradient and the uniquely permeable nature of the sediments near to the mountains. To a large extent, the groundwater is transferred back to surface water before flowing out of the section. Surface water in the Hongyashan reservoir, therefore, mainly comes from surface water and spring water originally present as short residence time ground-water within the Wuwei Basin.

3 Materials and Methods

According to the topography, geomorphology and hydrogeology, 13 groups of sampling locations from Qilian Mountain to Hongyashan Reservoir were chosen (Fig. 1). The sample points were positioned using GPS locator. The well depth, latitude, longitude, groundwater depth, water temperature as well as PH value (pH) were measured, respectively, on the scene. Sample bottles were washed three times with water sample before collecting water samples. All water samples were packed in plastic bottles, which were then sealed with wax. No air bubbles were allowed in the bottles to avoid exchanging with carbon dioxide in the air.

Precipitation materials were collected from the global precipitation isotope which was established by the International Atomic Energy Agency (IAEA) and the World Meteorological Organization (WMO). Ca^{2+}, Na^{+}, K^{+}, Mg^{2+}, SO_4^{2-}, Cl^{-} and NO_3^{-} were measured using Ion Chromatograph. Deuterium (D) and oxygen-18 (^{18}O) were measured by MAT-252 mass spectrometry. Hydrogen samples were prepared by the zinc reduction method under temperatures of up to 700 °C, and ^{18}O samples by a H_2O–CO_2 equilibrium method. Here, SMOW (standard mean ocean water) was adopted as the criterion for measuring H and O isotopic compositions. Duplicate analyses of δD and $\delta^{18}O$ were within ±1‰ and ±0.1‰, respectively.

4 Results and Discussion

4.1 Hydrochemical Characteristics of Groundwater

The hydrochemical properties of groundwater samples collected from the Quaternary aquifer system are shown in Table 1.

Table 1 TDS and isotope analysis results of water samples in the study area

Number	Site	Type	PH	TDS (mg/L)	$\delta^{18}O$ (‰)	δD (‰)
I	Wuwei1	Rain	7.1	59.3	−1.662	−22.95
II	Wuwei2	Rain	7.2	56.7	−2.45	−25.16
III	Gulang	Snow water	6.9	69.9	−8.6	−56.74
IV	Hongyashan	Surface water	–	–	−3.1	−27
1	Qilianshan	Snow water	5.5	131	−11.99	−76.7
2	Nanyingshuiku	Surface water	6.0	171	−10.35	−70.1
3	Xiyingshuiku	Surface water	6.5	444	–	–
4	Shiyanzhan	Groundwater	7.0	385	–	–
5	Dawancun	Groundwater	6.0	1063	−9.6	−64.5
6	Gaogoucun	Groundwater	6.0	1480	−9.21	−61.1
7	Yujiawancun	Groundwater	6.0	313	−9.71	−62.8
8	Hongshuihecun	Groundwater	6.5	467	−9	−60.5
9	Baiyuncun	Groundwater	5.5	906	−8.61	−56
10	Qijiahucun	Groundwater	6.0	1750	−8.4	−58.3
11	Xigoucun	Groundwater	6.5	455	−9.4	−63.2
12	Liuquancun	Groundwater	6.5	987	−9.51	−62.3
13	Caiqi	Surface water	6.0	1297	−9.69	−61.4

Note The date of I-IV from IAEA

The pH value in groundwater ranged from 5.5 to 7.0 with an average value of 6.2. The total dissolved solids (TDS) of groundwater varied from 313 to 1750 mg/l with an average of 769 mg/l. The plot of chemical analyses on a trilinear diagram of groundwater and surface water is shown in Fig. 3.

Based on the long-term observation of groundwater hydrochemistry, the hydrochemical types showed an obvious zonation from the upper to lower reaches. It changed from HCO_3^- through to SO_4^{2-} to Cl^- types (Fig. 3). As shown in Fig. 4, HCO_3^-–Ca^{2+} or HCO_3^-–Ca^{2+}–Mg^{2+} water was mainly to the south of the basin where the recharge zone was located. Meteoric water can easily flow through the gravel and pebble aquifer, and groundwater was actively circulated. As a result, the water-rock reaction time was short, which led to the formation of weakly mineralized hydrocarbonate water. From the south to the middle, the hydrochemical zone gradually transformed into a HCO_3^-–SO_4^{2-}–Ca^{2+}–Mg^{2+} or HCO_3^-–SO_4^{2-}–Ca^{2+}–Mg^{2+}–Na^+ type. Discontinuous SO_4^{2-}–HCO_3^-–Ca^{2+}–Mg^{2+} zone or even SO_4^{2-}–Ca^{2+}–Mg^{2+} water existed in the north of the basin. As a result of runoff leakage, a narrow HCO_3^-–Ca^{2+} or HCO_3^-–Ca^{2+}–Mg^{2+} zone occurred along the bank of Shiyang River. This kind of regional zonation was consistent with the flow direction of the groundwater. A hydrocarbonate (fresh water) zone was located in the south, hydrocarbonate–sulfate (fresh water) zone in the middle, sulfate–hydrocarbonate (sub-fresh water) zone in the north and a sulfate (saline water) zone in the northern margin of the basin.

The variation of TDS concentration in groundwater illustrated that the extent of groundwater-surface water interaction governed groundwater quality in the

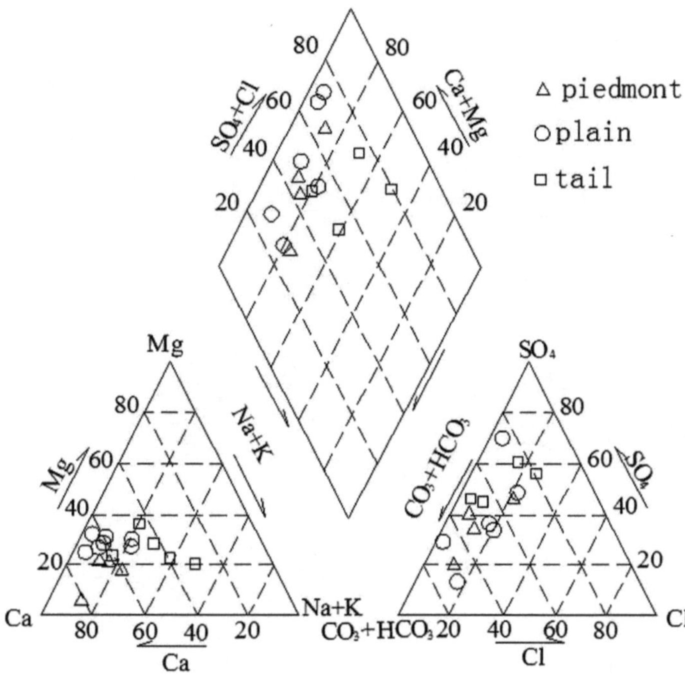

Fig. 3 Piper plot of chemical analysis of groundwater and surface water samples

Wuwei Basin (Fig. 5). The groundwater samples were collected near the river bank, allowing groundwater to be easily recharged from the river, and assuring a high quality of groundwater. Comparatively, for those collected farther from the river, the TDS and concentrations of the major ions generally rose along the flow path of the groundwater, which may be due to its greater residence time in the rock formations bearing it and the absence of the influence of surface water recharge.

4.2 *Isotopic Composition of Groundwater*

Hydrogen (H) and oxygen (O) are two widespread elements in nature. They have lower atomic orders and a much larger difference in isotopic ratio; that is, ΔA/A of H: D up to 100% and ^{16}O:^{18}O to 12.5%, where A is the mass number of the isotope and ΔA is the difference between the light and heavy isotope. The utmost isotopic fractionation can take place in natural status between their isotopes. In a natural water body, the fractionation processes of H and O isotopes were also affected by physical and chemical factors. Therefore, they were sensitive to isotopic tracers, and were widely applied in studying natural water circulation and groundwater movement [12].

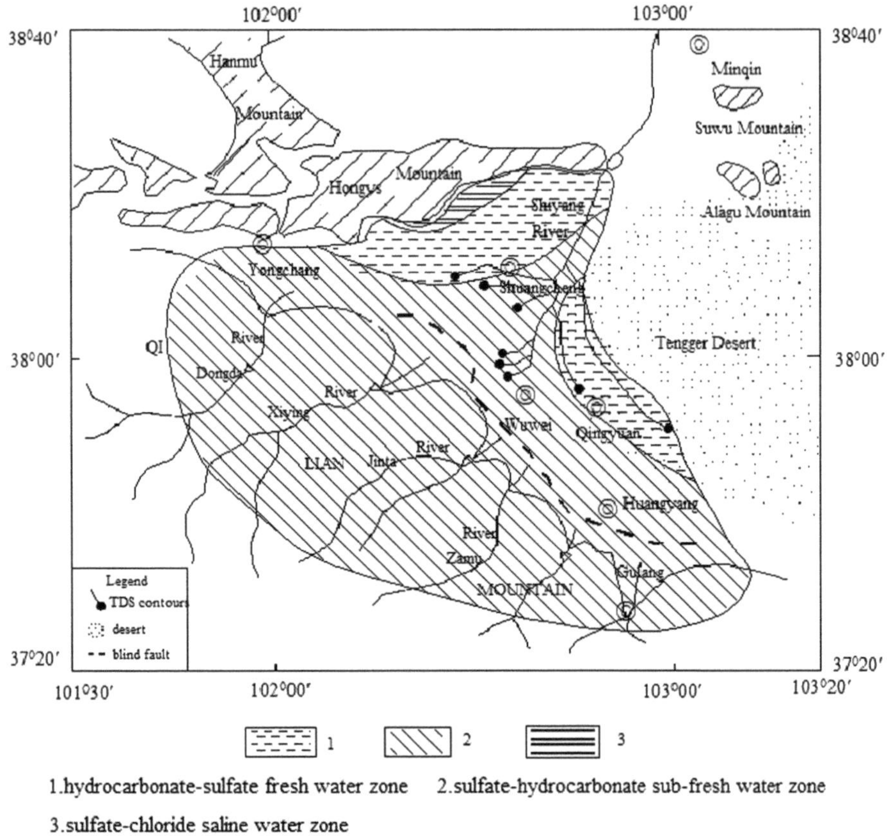

Fig. 4 Hydrochemical zones of the groundwater system in the Shiyang River

The results of stable isotope analysis both for water samples in the basin and precipitation collected in the IAEA were plotted in Fig. 6. Values of $\delta^{18}O$ and $\delta^{2}H$ in precipitation varied from −20.6 to 3.8‰ and from −158 to 33.1‰, respectively, but they are linearly similar to the global meteoric water line (GMWL) with an equation of $\delta^{2}H = 7.8\delta^{18}O + 5$ ($r^2 = 0.92$), which is defined as local meteoric water line (LMWL). These values were slightly enriched and indicated modification due to evaporation in the monsoon rains producing departure from the GMWL. Weighted mean rainfall values lied around $\delta^{18}O$ −7.3‰ and δD −46.9‰. These values were most likely to be the representative of the present local waters.

It showed some deviation of H and O isotopic compositions from the meteoric water line. This was mainly because of the effect of local variations in rainfall. In addition, the poor coverage of vegetation in the Shiyang River Basin was another influencing factor. when the meteoric water recharged into groundwater, drastic evaporation can produce a much larger isotopic fractionation and cause the deviation from the meteoric water line.

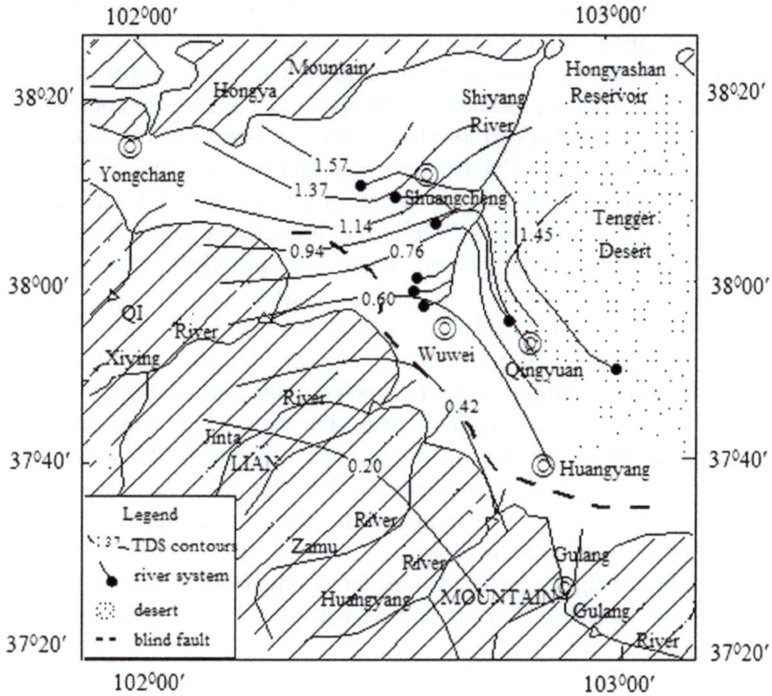

Fig. 5 TDS contours of groundwater in the study area

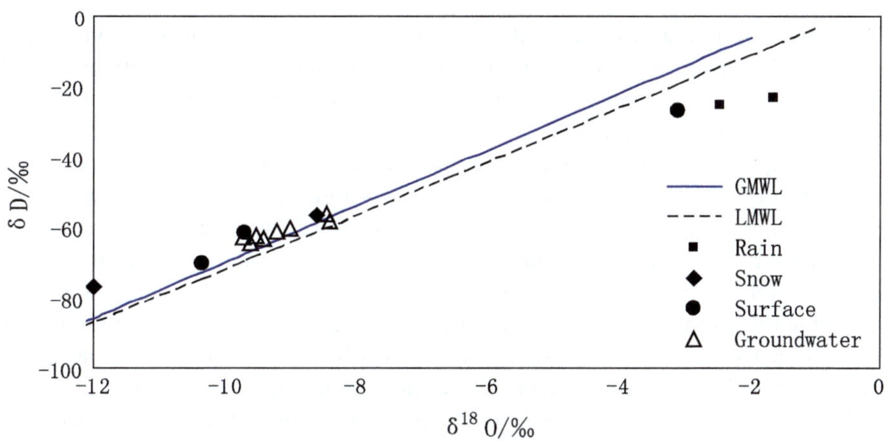

Fig. 6 Plot of $\delta^{18}O$ and δD for groundwater from the Quaternary aquifer system of the Shiyang River with plots of the GMWL and LWML

In the vicinity of Qijiahu, the ^{18}O value was lighter than that of other areas. It may be caused by the distribution of granite and metamorphic rocks. When groundwater contacts with those rocks, water can exchange H and O isotopes with minerals in rocks. Such a water-rock equilibrium fractionation can reduce the content of ^{18}O in water. The isotopic fractionation intensity and the water-rock system depended on the temperature, quantity ratio of water and rock, and reaction time. The water-rock equilibrium fractionation was very small during groundwater leakage. The slow groundwater circulation rate enabled a full isotopic exchange between water and rock.

For the stable isotope composition of the Shiyang River running course, the mean values of $\delta^{18}O$ increased along the river course. The $\delta^{18}O$ and δD values for these samples ranged from −11.99 to −8.4‰ and from −76.7 to −56‰, respectively (Table 1). Figure 5 illustrated that the samples from the Shiyang River running course had the similar stable isotope compositions to those from the wells drilling into the unconfined aquifers near the river bank. This indicated that evaporation probably occurred in channels, reservoirs and fields in the study area.

The majority of the isotope compositions of the groundwater samples plot were close to the GMWL (Fig. 6). This indicated that the groundwater recharge resource mainly originated from precipitation and was weakly affected by evaporation. These more depleted signatures were considered to be from the late Pleistocene and Holocene periods, during which the climate was wetter and colder than that of the present day.

4.3 Analysis on the Evolution of Groundwater Circulation

The surface runoff was converged as the atmospheric precipitation down to the surface. It infiltrated into the underground along the mountain rock of fracture and fault fracture zone to supply the groundwater. The mountainous rivers flowed into Wuwei Basin and formed pluvial and alluvial fan belt in the southern piedmont of the basin. The quaternary unconsolidated sediments varied significantly in thickness, and it was mainly composed of sandy gravel and sand gravel with great permeability. Piedmont plain rivers and irrigation water leakage, as well as the canal system were a primary source to supply the groundwater. Because the basin downstream was the main location of the groundwater evaluation, the block water boundary was constituted in the north of the basin where the bedrock exposed. Thus, the groundwater cannot directly flow into the basin downstream, whereas overflow as the basement raise, and finally transformed surface water flowing into the lower basin.

The evolution of groundwater circulation determined the characteristics of groundwater chemical and isotopic distribution in the Shiyang River Basin. TDS and hydrogen and oxygen stable isotope values increased in the flow directions. The groundwater salinity was higher, and water quality was worse with gradually closing to the lower of the downstream. Hydrochemical and isotopic distribution can clearly reflect the relationship among the groundwater (Figs. 7 and 8). TDS and oxygen isotope value gradually decreased with the increasing of depth, which obviously

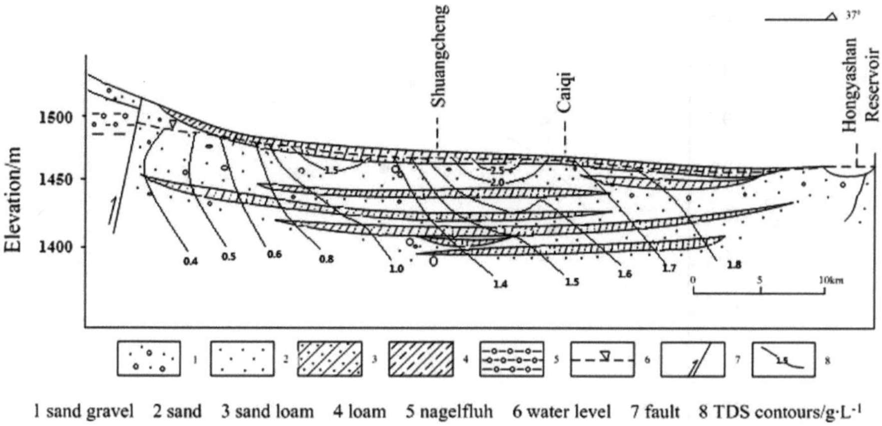

Fig. 7 TDS contours of groundwater in the study area

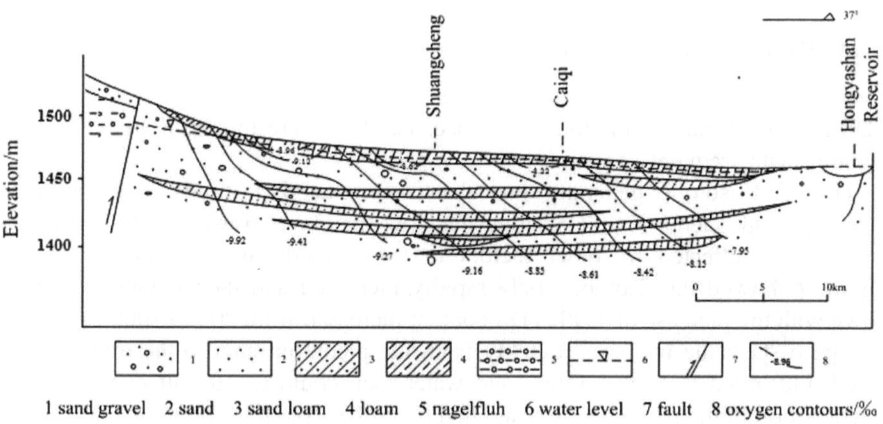

Fig. 8 $\delta^{18}O$ contours of groundwater in the study area

reflected the characteristics of groundwater laterally flowing. Groundwater of the basin flowed to the Hongyashan Reservoir and the desert area that located in the north boundary, which had the same flow law with groundwater flow field.

In the piedmont area, due to the replenishment of rainfall and runoff, groundwater was leached ceaselessly; therefore, the groundwater quality was relatively good and most of the underground water were fresh water resources, with small TDS and $\delta^{18}O$ values, TDS in 400–800 mg/L, and $\delta^{18}O$ in −9.92 to −9.27‰. In the plain area, due to the influence of evaporation and irrigation, TDS and $\delta^{18}O$ values increased continually which resulted in the relatively high contents of TDS and $\delta^{18}O$ in some areas. For example, the content of TDS and $\delta^{18}O$ began to increase in the regions of Qingyuan, Shuangcheng and Caiqi. These regions had influence on

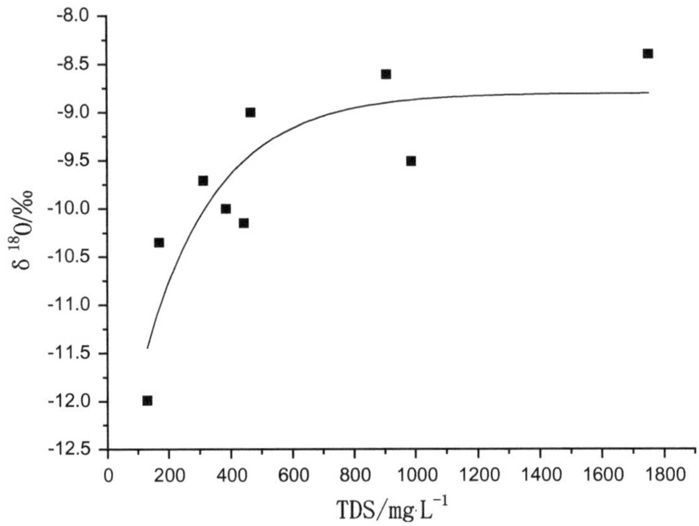

Fig. 9 Major ions versus $\delta^{18}O$ in the study area

TDS and $\delta^{18}O$ value variation trends because they were the main areas for crop planting with frequent irrigations.

TDS and isotope generally had the same changing trends. Isotope value increased with increasing TDS along the flow direction. However, this trend can probably be affected by other factors in the groundwater circulation process. Figure 9 showed that isotope firstly rapidly increased and then gradually slowed down with the increase of TDS. TDS content increased from 200 to 1000 mg/L and isotope $\delta^{18}O$ value increased quickly from the piedmont area to the middle area, which illustrated that evaporation and water-rock could greatly affect the isotope fractionation in the piedmont area and had little effect on the content of TDS. When the TDS > 1000 mg/L in the middle area, the isotope value grew slowly, and TDS value increased rapidly, which showed that the exploiting of groundwater and recharge had significantly effect on TDS but had negligible effect on isotope fractionation. This is probably due to the supply of the deep confined water with lower isotope value and irrigation and permeability of the surface water.

5 Conclusions

The hydrochemical and isotope investigation was conducted in the Wuwei Basin to understand the geochemical evolution and the residence time of the groundwater. The main conclusions are as follows:

(1) Hydrochemical characteristics of the groundwater in Shiyang River Basin showed apparent zonation from the south to north. Hydrocarbonate-type water occurred in the gravel-pebble zone of the mount front, and sulfate-type water was mainly distributed in the alluvial, pluvial and lacustrine plain in the middle reach. Sulfate-chloride-type water appeared in the marginal area of Tenggeli Desert in the lower reach, and even chloride-type brine existed in some parts of the lower reach area.
(2) Local meteoric water line (LMWL) equation of the Shiyang River Basin was $\delta^2H = 7.8\delta^{18}O + 5$ ($r^2 = 0.92$), which was similar to the global meteoric water line (GMWL). In the basin plain, $\delta^{18}O$ in −8.61 to −9.71‰, δD in −56 to −62.8‰, isotope values increased along the flow direction. This is mainly because of evaporation caused by isotope enrichment.
(3) Precipitation was the main factor that controlled the groundwater hydrochemistry of the Wuwei Basin, whereas drastic evaporation influenced the geochemistry of the Wuwei Basin. Most of the groundwater isotopic compositions of O and H were meteoric water features, indicating that rainwater was the main recharge source to the groundwater system.
(4) Coarser aquifers of distributing in the south of the Wuwei Basin enabled rapid recharge into groundwater. In the middle-lower reaches, the recharge rate was much slower, and it had a long circulation period. The formation of some of the groundwater was closely related to the mixing of the crevice water and brine.

Acknowledgements This study was sponsored by Natural Science Foundation of China (91125017) and Natural Science Foundation of Hebei (E2015402128).

References

1. Edmundes, W.M., Ma, J.Z., Aeschbach, H.W., et al.: Groundwater recharge history and hydrogeochemical evolution in the Minqin Basin, North West China. Appl. Geochem. **21**, 2148–2170 (2006). https://doi.org/10.1016/j.apgeochem.2006.07.016
2. Ding, Z.Y., Ma, J.Z., Zhang, B.J., et al.: Analysis on the climate change in the Shiyang River Basin since resent 50 years. Arid Zone Res. **24**(6), 779–784 (2007)
3. Liu, F.J., Bai, S.F., Cheng, G.D.: Utilization and exploitation of water resources and impact of human activities in Shiyang River catchment, Hexi Corridor, Gansu. In: Li, J.J., Cheng, G.D. (eds.) Papers on the Western Resources and Environment and Sustainable Development (in Chinese with English summary), pp. 74–82. Lanzhou University Press, Lanzhou (1998)
4. Fritz, P., Fontes, J.C.: Handbook of Environmental Isotope Geochemistry. Elsevier Scientific Publishing Company, Amsterdam-Oxford-New York (1980)
5. IAEA: Arid zone hydrology: investigations with isotope techniques. In: Proceeding of an Advisory Group Meeting, International Atomic Energy Agency, Vienna, p. 265 (1980)
6. Shi, J.A., Wang, Q., Chen, G.J., et al.: Isotopic geochemistry of the groundwater system in arid and semiarid areas and its significance: a case study in Shiyang River Basin, Gansu Province, Northwest China. Environ. Geol. **40**(4–5), 557–565 (2001)
7. Feng, Q., Wei, L., Su, Y.H., et al.: Distribution and evolution of water chemistry in Heihe River Basin. Environ. Geol. **45**, 947–956 (2004). https://doi.org/10.1007/s00254-003-0950-7

8. Zhang, Y.H., Wu, Y., Su, J., et al.: Groundwater replenishment analysis by using natural isotopes in Ejina Basin, northwestern China. Environ. Geol. **48**, 6–14 (2005). https://doi.org/10.1007/s00254-004-1214-x
9. Zhu, G.F., Li, Z.Z., Su, Y.H., et al.: Hydrogeochemical and isotope evidence of groundwater evolution and recharge in Minqin Basin, Northwest China. J. Hydrol. **333**, 239–251 (2007). https://doi.org/10.1016/j.jhydrol.2006.08.013
10. Ma, J.Z., Wang, X.S., Edmunds, W.M.: The characteristics of ground-water resources and their changes under the impacts of human activity in the arid Northwest China–a case study of the Shiyang River Basin. J. Arid Environ. **61**, 277–295
11. Ma, J.Z., Gao, Q.Z.: Groundwater vulnerability and its assessing method in the arid land of NW China. Arid Land Geogr. **26**(1), 44–49 (2003)
12. Gu, W-z., Pang, Z-h., Wang, Q-j. et al.: Isotope Hydrology, pp. 105–107. Science Press, Beijing (2011)

A Bibliometric Analysis of Publications on Solar Pumping Irrigation

Yunxin Zhang, Zhanyi Gao and Yan lei Jia

Abstract Solar energy is considered as one of the cleanest forms of energy sources. Meanwhile, many studies have used solar power for water pumping, which is an economically attractive solar energy application since the year of 1615. To better understand the history of the development of solar energy irrigation, a bibliometric analysis of publications on solar energy pumping research from 1956 to 2015 in the sci-direct database is presented here. The analysis informs of the growing trends and indicates that "*water*", "*solar*", and "*soil*" are hot topics of research on solar energy irrigation during this period. The language of the publications, publication output, journal distribution, countries and territories of these publications, hot topics and highly cited papers have been assessed. The top 10 countries/territories were ranked according to their total number of articles (TA), single country articles (SCA), internationally collaborative articles (ICA) and first author articles (FAA). Meanwhile, the processes of solar energy installations for pumping irrigation water can be divided into three stages, including photo-thermal conversion stage, solar energy-thermal-electric power conversion stage and photovoltaic conversion stage.

Y. Zhang (✉) · Z. Gao
Department of Irrigation and Drainage, China Institute of Water Resources and Hydropower Research, Beijing 100038, China
e-mail: zyx14315@163.com

Z. Gao
e-mail: gaozhy@iwhr.com

Y. Zhang
College of Conservancy and Hydropower, Hebei University of Engineering, Handan 056021, China

Y. Zhang
School of Water Conservancy and Hydropower, Hebei University of Engineering, Handan 056038, China

Y. l. Jia
Linxi Experimental Irrigation Station, Xingtai Water Bureau, Xingtai 054000, China
e-mail: jyl1212@126.com

W. Dong et al. (eds.), *Sustainable Development of Water Resources and Hydraulic Engineering in China*, Environmental Earth Sciences,
https://doi.org/10.1007/978-3-319-61630-8_26

Keywords Solar energy · Bibliometric analysis · Photovoltaic pumping
Irrigation

1 Introduction

Solar energy is one of the cleanest forms of energy sources and considered as a green source of energy. From of old, humans have learned that solar energy can be used to dry the cereal, and so on. With the rapid industrialization worldwide, the energy crisis is presented in the foreseeable future, especially conventional energy, such as oil, coal, etc. To improve the inappropriate energy structure, it is important to find an exchangeable energy. Therefore, the renewable sources are regarded, such as solar energy, wind energy, tide energy, bio-energy, etc. Solar energy has especially been regarded as it is renewable and friendly to installation. An abundant solar energy resource gives a chance to improve the inappropriate energy structure for the future. However, because the intensity of solar radiation is low, it needs to use some installation to collect the energy; therefore, the collector to gather the solar energy must be studied. The aim of this study is to systematically analyze all the peer-reviewed articles on solar irrigation published in the SCI-direct database from 1956 to 2015 to: (1) characterize the spatial and temporal dynamics of these publications by country/territory, institution, journal and research field; (2) identify popular topics and the most-cited articles; and (3) recommend guidelines for future potential research directions on solar irrigation and related fields, especially under future global climate change and increasing environmental pressures. Based on the publications that have been collected from sci-direct database and from solar study initial publication to the end of 2015, the total of 63,748 publications have been identified (Fig. 1).

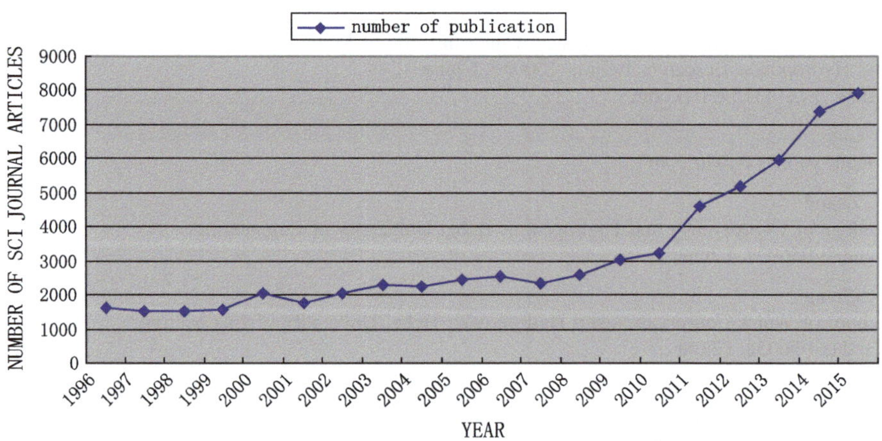

Fig. 1 World SCI journal publications per year with “solar” from 1996 to 2015

Applying solar energy to irrigation also has a long history. The pumping of irrigation water is an operation where normally an intermittent operation may be acceptable. This fact, plus the potential economical benefits to be obtained, has instigated the research in this area. Based on the publications collected from sci-direct database, the process of research about using solar energy to pump water is mainly divided into three stages.

1.1 Photo-Thermal Conversion Stage

Pytlinski [1] reviewed the past efforts to develop solar powered systems from 1615 to 1978. In the history, the earliest reported effort to convert solar energy into mechanical power for raising water was that of Solomon de Caux, a French engineer in 1615 [2]. Then, more and more researchers began to input more time into the technology regarding the conversion to the solar energy. Mouchot [3–6], Pifre [7], Langley [8] also experimented with concentrating reflectors and boilers that drive a rotary pump raising water. Hence, the sun energy conversion efficiency was obviously enhanced. For the remainder of the 19th century, there were still many researchers who input much energy and financial resources into the improvement in the efficiency of sun energy converting systems [1].

In the twentieth century, a number of important solar experiments were undertaken. The most famous of the solar installations built by Eneas was the one erected at the Ostrich Farm in Pasadena, California in 1901 [9]. The reflector's inner surface was composed of 1788 small, flat mirrors arranged to approximate the conical surface. The system could produce a peak power of about 10 h.p. Abbot [10, 11], Tabor and Zeimer [12], and Francia [13] also devoted themselves to the study of solar energy conversion. The National Physical Laboratory of Israel commercially developed Bronicki's [14] concept and produced what are known as ORMAT Rankine Power Units. In April 1977, a large installation for pumping irrigation water was put into operation on Gila River Ranch near Phoenix, Arizona [15]. The system is equipped with parabolic tracking collectors having a total area of 564 m^2. The collector design efficiency was 55% while the actual field efficiency is 44–50%.

Because of the shortcoming of poor efficiency and pollution of the system, the application of this system in irrigation was not expanded on a large scale.

1.2 Solar Energy-Thermal-Electric Power Conversion Stage

The largest advantage of this system in that time was that the unused current could be employed to charge batteries; then these, in turn, could maintain the normal current supply of cloudy days and at night. Masson and Girardier [16] operated a small solar engine which vaporized the water to a secondary fluid which expanded through a turbine to generate electricity which drove the pump via an electric motor. In 1976, a solar water pumping station built by the French company SOFRETES began operation in Guanajuato, Mexico. In this system, water was used as the primary heat transfer fluid. A vapor turbine and electric generator drives two pumps. Fenton et al. [17] analyzed and evaluated the operation of the Willard solar power system in 1984. He pointed out some critical segments of efficiency losses. However, due to the expensive cost of the batteries, this system was not rapidly extended.

1.3 Photovoltaic Conversion Stage

With the increasing concerns about environmental awareness, and due to the high price of conventional energy, interest in the use of solar energy for irrigation pumping was renewed in the 1970s. In 1977, the first-scale use of photovoltaic cells to generate electricity for pumping irrigation water was found near Mead, Nebraska. The electric system consists of 120,000 individual silicon-type photovoltaic cells mounted on 28 flat panels arranged in two rows. Inverters are used to convert direct electrical current produced by cells and batteries into alternating current to power the irrigation pump motor.

The cost of the system must be a significant fact for the consumer [18–20]. Lots of researches had compared solar energy with the conventional energy [21–24]. Results indicated that solar energy was more economical, especially in off-grid area [25, 26]. In addition, due to the expensive price of battery, system without electricity storage also was studied [27, 28], and it was used with various irrigation pattern [29, 30]. Therefore, analysis of system must be essential [24, 31–33] which would decide the system whether was suitable [34–39].

With the increased use of water pumping systems, more attention has been paid to their design and optimum utilization to achieve the most reliable and economical operation [40, 41]. Lots of models were applied to evaluate the reliability and economical efficiency [42–47], and researchers introduced diverse algorithm [48, 49] to study the integrated system. The study analyzed the pay-back period [50] and investigated the economic feasibility of stand-alone photovoltaic systems [51–53]. Meanwhile, some new concept was proposed, such as value proposition [54], Photovoltaics Opportunity Irrigation (PVOI) [55].

2 Methodology

In this study, documents used were derived from Science Direct (SCI-Direct) database which is published by Elsevier Science Company. SCI-Direct is a website which provides subscription-based access to a large database of scientific and medical research which delivers over 14 million publications from over 3,800 journals and more than 35,000 books from Elsevier [56]. The documents were downloaded from the database searched with the words "solar" and "irrigation" in titles, abstracts, and keywords plus from 1956 to 2015 and were analyzed with the bibliometrical method in the current study. Moreover, articles coming from England, Scotland, Northern Ireland, and Wales were classified as being from the United Kingdom (UK). The collaboration type was categorized and determined by the addresses of the authors as: (1) single country articles (SCA) with address(es) from the same country, (2) internationally collaborative articles (ICA) with author addresses from more than one country or territory. The impact factor of a journal was determined for each document as reported in the Thomson Reuters Journal Citation Reports of 2016.

3 Results and Discussion

3.1 Document Types and Language of Publications

A total of 378 publications were identified during the 60-year study period. Journal articles (333) accounted for 88.1% of the total publications and books (45) for 11.9%. Journal articles (306) accounted for 91.9%, with the remainder of 8.1% including review articles (13), short communications (7), discussions (2), and others (5). Because a majority of the publications were journal articles, the focus was on the journal articles in the current study. All journal articles used were published in English during the study period.

3.2 Characteristics of Publication Output

A long-term trend analysis from 1956 to 2015 was conducted by using article titles, abstracts and keywords. In the database, 378 related documents were found in the total published from 1956 to 2009 and 333 of which were articles, accounting for 88% of the overall documents. The annual publication of articles increased slightly before 1978 and jumped nonlinearly from 2006 to 2015 (Fig. 2). The scarcity of articles published before 1978 might be attributed to the low price of conventional energy and the lack of environmental awareness. A first breakthrough in the field was marked by two articles entitled "Solar energy installations for pumping irrigation water" [1] and "solar water pumping" [57] which were all published on solar energy, with a total citation of 28 and 26 times from its initial publication to the end of 2015, respectively. With the hot debating about climate change, the second breakthrough in the field was in 2006. The annual publication of articles increased significantly from 2006 to 2015. More attention had been paid to the solar pump irrigation system design and optimum utilization to achieve the most economical operation. Ghoneim [42] developed a

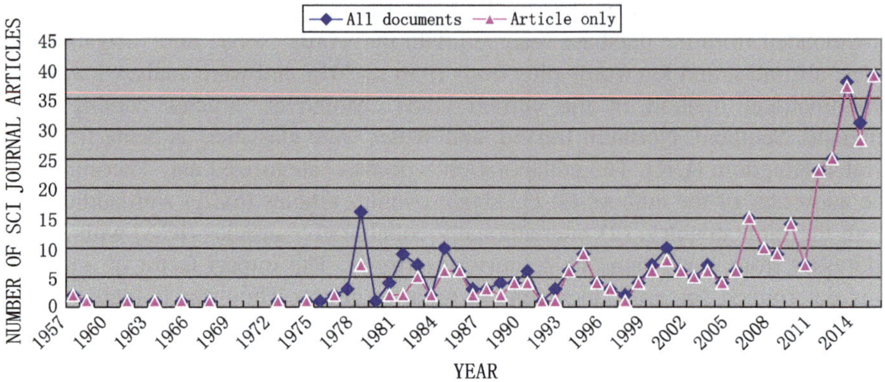

Fig. 2 World SCI-Direct journal publications per year with "solar irrigation" in titles, abstracts and key words during 1956–2009

computer simulation program that consists of a component model for the PV array with maximum power point tracker and component models for both the DC motor and the centrifugal pump to determine the performance of the proposed system in the Kuwait climate. Hamidat and Benyoucef [49] used a method of the load losses probability (LLP) to optimize sizing of the photovoltaic pumping systems with a similarity between the storage energy in batteries and water in tanks in Algeria. Bakelli et al. [43] recommended an optimal sizing model to optimize the capacity sizes of different components of photovoltaic water pumping system (PWPS) using water tank storage. Campana et al. [46] proposed a novel optimization procedure which took into consideration not only the availability of groundwater resources and the effect of water supply on crop yield, but also the investment cost of the photovoltaic water pumping system and the revenue from crop sale. Aman et al. [58] presented the overview of solar energy technologies and addressed the Health and Environmental (SHE) impact of solar energy technologies to the sustainability of human activities and recommended the possible ways to reduce the effect of potential hazards of widespread use of solar energy technologies.

3.3 Output in Journals

There were 333 articles published in 74 different SCI-Direct journals. *Agricultural Water Management,* with an impact factor of 2.063, was ranked first with 51 (15.3%) articles on "solar irrigation" published in the journal. *Solar Energy* (IF = 3.685) was ranked second with 24 (7.2%) articles published on the subject, followed by *Agricultural and Forest Meteorology* (IF = 4.461) with 22 articles and *Field Crops Research* (IF = 2.927) with 20 articles. Moreover, *Renewable and Sustainable Energy Reviews* with 18 published article had the highest impact factor

(6.798). Other journals with a high impact factor, such as *Water Research* (IF = 5.991), *Applied Energy* (IF = 5.746), *Energy Conversion and Management* (IF = 4.801) and *Energy* (IF = 4.292) published 6, 12, 6, and 6 articles, respectively. These statistical results will help researchers select journals when publishing their articles on solar irrigation-related research.

3.4 Publication Performance of Countries/Territories

The contribution with countries/territories index was estimated by the location of the affiliation of at least one author of the published papers. However, in all 333 articles, there were 7 (2.1%) articles without author address information. The other 326 articles published from 1956 to 2015 originated from 63 countries. Of all the articles, 252 (77.3%), from 44 countries had authors of the same country (i.e., were SCA) and 74 (22.6%) involved international collaboration with 50 other countries. Eighteen countries had no SCA, while fourteen countries had no ICA. Twenty-two countries contributed only one or two SCA and 36 countries contributed only one or two ICA. The top 10 countries/territories were ranked according to their number of total articles (TA). This included four indicators: TA, SCA, ICA and first author articles (FAA) [59]. The rank and percentage of internationally collaborative articles among the total articles for each country were also considered (Table 1).

Table 1 Top 10 most productive countries based on the total number of articles published

Country/territory	TA	TAR (%)	SCAR (%)	ICAR (%)	FAAR (%)	CR (%)
USA	65(19.5)	1(19.5)	1(12)	1(7.5)	1(14.7)	8(38.5)
Spain	45(13.5)	2(13.5)	2(10.2)	3(3.3)	2(12.9)	9(24.4)
China	45(13.5)	2(13.5)	3(8.1)	2(5.4)	3(11.4)	7(40.0)
India	28(8.4)	4(8.4)	4(7.2)	10(1.2)	4(7.2)	10(14.3)
Australia	21(6.3)	5(6.3)	5(3.3)	5(2.7)	5(4.2)	3(47.6)
UK	17(5.1)	6(4.8)	6(3.0)	6(2.1)	6(3.6)	6(41.2)
Germany	14(4.2)	7(4.2)	8(1.5)	7(1.8)	9(1.8)	1(64.3)
Sweden	13(3.9)	8(3.9)	7(2.1)	7(1.8)	7(2.7)	4(46.2)
France	12(3.6)	9(3.6)	7(2.1)	9(1.5)	8(2.1)	5(41.7)
Mexico	11(3.3)	10(3.3)	8(1.5)	10(1.2)	9(1.8)	2(54.5)

Note TA the number of total articles; *TAR* the rank of total articles and percentage; *SCAR* the rank of single country articles and percentage; *ICAR* the rank of internationally collaborative articles and percentage; *FAAR* the rank of first author articles and percentage; *CR* the percentage of internationally collaborative articles in its total articles for each country and rank

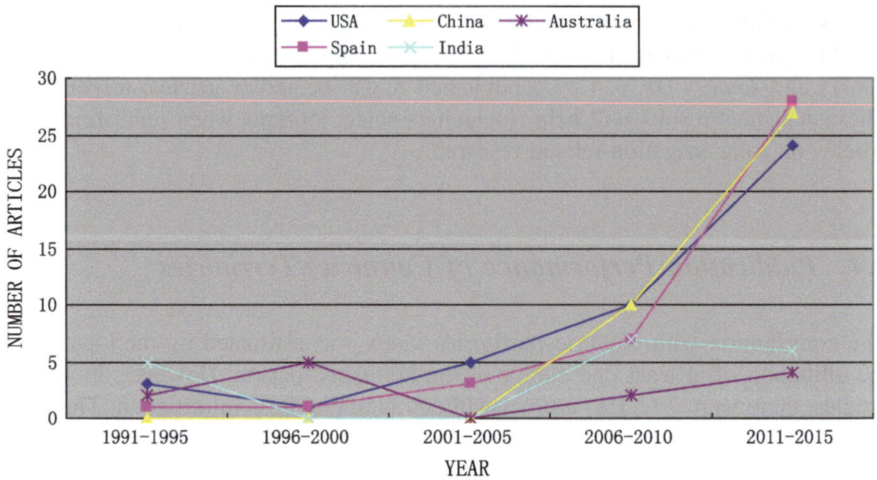

Fig. 3 Comparison of total articles (TA) of the top five most productive countries in the world

In addition, the increasing number of articles from universities and collaboration between universities and institutions [60] were found.

Figure 3 shows the top five productive countries which have published on "solar irrigation". China, the second on the list ranked by TA, has published 45 (13.5%) articles with the fastest growing trend. Zero articles from China were published before 2008 but the number increased sharply by 2015. Meanwhile, for the other countries' number of published articles there also appears to be a sharp growth, especially. This indicates that with the increasing price of conventional energy and the awareness of climate exchange, the study about solar irrigation becomes a hot research topic.

3.5 The Future Trend of Hot Topics

Keywords supply the main information in articles and thus research trends can be obtained by analyzing the keywords [61]. "Water", "solar" and "soil" were the most popular keywords among all the 333 keywords examined (Fig. 4). Articles referring to water, such as water cycle and water use, are the most popular ones with 66 counts (20%) and a continuously rising trend is expected in the coming years; "solar" is the second most popular keyword which appeared in 44 (13%) articles. Based on Fig. 4, TA on "water", "solar", and "soil" have grown remarkably. This suggests that study about solar energy applying on agricultural irrigation will be expected to be one of the hottest trends in the future.

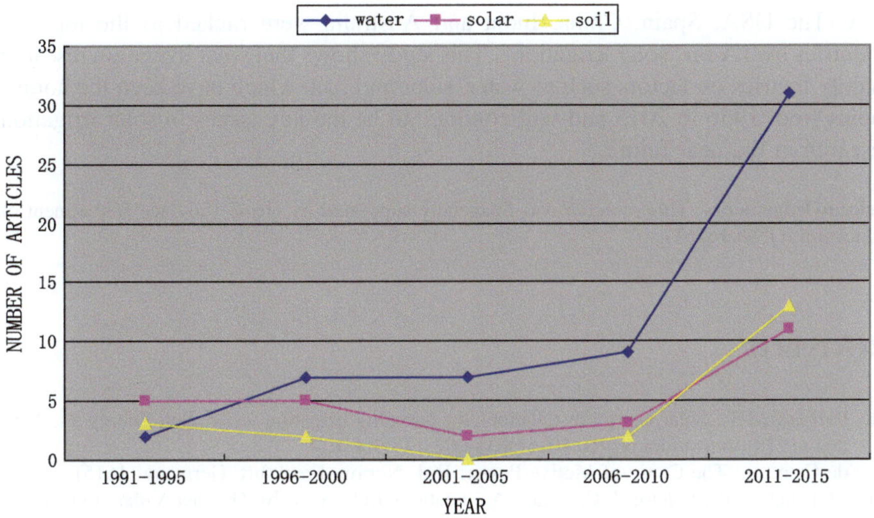

Fig. 4 Trends of hot topics of solar irrigation

3.6 *The Most Frequently Cited Articles*

During 1956–2015, five most frequently cited articles were cited, and their cited times are 140, 127, 107, 103 and 85, respectively. The most frequently cited article was published in *Computers and Electronics in Agriculture* and was originated in the USA. Other four articles were published in *Journal of Hydrology, Field Crops Research and Agricultural Water Management* and were originated in France, Australia, Mexico, India and Spain, respectively. Among all the articles in the SCI-Direct database from 1956 to 2009, the most frequently cited article was "Imaging from an unmanned aerial vehicle: agricultural surveillance and decision support" by Herwitz et al. in 2004 [62] with 140 times of TC2015. However, this article does not actually concentrate on "solar irrigation" but simply mentions the subject in its abstract. This was a major source of bias in this work but could not be avoided. Although several flaws produced by Keywords Plus by separating words related to "solar irrigation" in article title, abstract and keywords have been successfully eliminated, limitations are still inevitable due to some inherent problems of citation analysis.

4 Summary and Conclusions

Based on the SCI-Direct database, 378 publications were found during the period from 1956 to 2015. Articles were published in 70 journals from 60 countries/territories. Seventy-seven percent of the articles were SCA and 23% were

ICA. The USA, Spain, China, India and Australia were ranked as the top five countries by TA on "solar irrigation". This study shows that solar irrigation research mainly focuses on factors such as water, solar and soil, which have been the hottest topics from 1956 to 2015 and will continue to be the key issues in solar irrigation research in the near future.

Acknowledgements This research was financially supported by Hebei Province Department of Education (QN2016261).

References

1. Pytlinski, J.T.: Solar energy installations for pumping irrigation water. Sol. Energy **21**, 255–262 (1978). https://doi.org/10.1016/0038-092x(78)90001-4
2. de Caux, S.: The Cause of Motive Power. E. I. Norton, Frankfurt, Germany (1615)
3. Mouchot, A.B., Solaire, L.C., et al.: Applications Industries. In: Gauther-Villars (ed.) Paris, France (1869)
4. Mouchot, A.B.: Comptes Rendus de l'Acad ~ mie des Sciences, Paris, France **81**, 74 (1875)
5. Mouchot, A.B.: Comptes Rendus de I'Acad ~ mie des Sciences, Paris, France **86**, 132 (1878)
6. Mouchot, A.B.: Comptes Rendus de l'Acad ~ mie des Sciences, Paris, France **87**, 481 (1878)
7. Pifre, A.: A solar printing press. Nature **26**, 503–504 (1882)
8. Langley, S.P.: The New Astronomy, Houghton, Mi-lin, New York, pp. 91–116 (1896)
9. Anon: The utilization of solar heat for the elevation of water. Sci Am. **53**(14), 214 (1927)
10. Abbot, C.G.: Smithsonian Miscellaneous Collections **98**(5), 1 (1939)
11. Abbot, C.G.: Ann. Rep. Smithsonian Inst. 99 (1943)
12. Tabor, H., Zeimer, H.: Low-cost focusing collector for solar power units. Sol. Energy **6**, 55–59 (1962). https://doi.org/10.1016/0038-092x(62)90004-x
13. Francia, G.: Pilot plants of solar steam generating stations. Sol. Energy **12**, 51 (1968). https://doi.org/10.1016/0038-092x(68)90024-8
14. Bronicki, L.Y.: In: Proceedings of the 7th Intersociety Energy Conversion Conference, San Diego, California, Paper 729057, pp. 327 (1972)
15. McClure, G.M.: In: Proceedings of the Solar Irrigation Workshop, Albuquerque, New Mexico, pp. 28 (1977)
16. Masson, H., Girardier, J.P.: Solar motors with flat-plate collectors. Sol. Energy **10**, 165–169 (1966). https://doi.org/10.1016/0038-092x(66)90003-x
17. Fenton, D.L., Abernatnv, G.H., Karvokaplch, G.A.: Operation and evaluation of the Willard solar thermal power irrigation system. Sol. Energy **32**, 735–751 (1984). https://doi.org/10.1016/0038-092x(84)90248-2
18. Moore, R.M.: Cost predictions for photovoltaic energy sources. Sol. Energy **18**, 225–234 (1976). https://doi.org/10.1016/0038-092x(76)90021-9
19. Sudhakar, K., Muralikrishna, M., Rao, D.P.: Analysis and simulation of a solar water pump for lift irrigation. Sol. Energy **24**, 71–82 (1980). https://doi.org/10.1016/0038-092x(80)90022-5
20. Karmeli, D., Atkinson, J.F., Todes, M.: Economic feasibility of solar pumping. Sol. Energy **27**, 251–260 (1981). https://doi.org/10.1016/0038-092x(81)90126-2
21. Howes, M.: The potential for groundwater exploitation by solar-powered pumps in Pakistan. Agric. Adm. **16**, 229–248 (1984). https://doi.org/10.1016/0309-586x(84)90095-5
22. Ramos, J.S., Ramos, H.M.: Solar powered pumps to supply water for rural or isolated zones: a case study. Energy Sustain. Dev. **13**, 151–158 (2009). https://doi.org/10.1016/j.esd.2009.06.006

23. Khan, M.T.-U.-I., Pathik, B.B.: A comprehensive study on photovoltaic irrigation system for different crop cultivation: financial evaluation perspective. Electr. Eng. **2**, 228–237 (2014). https://doi.org/10.17265/2328-2223/2014.05.005
24. Hossaina, M.A., Hassan, M.S., Mottalib, M.A., Ahmmed, S.: Technical and economic feasibility of solar pump irrigations for eco-friendly environment. Procedia Eng. **105**, 670–678 (2015). https://doi.org/10.1016/j.proeng.2015.05.047
25. Chandratilleke, T.T., Ho, J.C.: A study of a photovoltaic array for water pumping. Sol. Wind Technol. **3**, 59–71 (1986)
26. Alajlan, S.A., Smiai, M.S.: Performance and development of PV-plant for water pumping and desalination for remote area in Saudi Arabia. WREC 441–446 (1996). https://doi.org/10.1016/0960-1481(96)88895-1
27. Bennouna, A., Ijdiyaou, Y.: Water pumping using a photovoltaic D.C. solar pump without energy storage. Renew. Energy **4**, 847–854 (1994). https://doi.org/10.1016/0960-1481(94)90237-2
28. Amer, E.H., Younes, M.A.: Estimating the monthly discharge of a photovoltaic water pumping system: model verification. Energy Convers. Manag. **47**, 2092–2102 (2006). https://doi.org/10.1016/j.enconman.2005.12.001
29. Al-Ali, A.R., Rehman, S., Al-Agili, S., et al.: Usage of photovoltaics in an automated irrigation system. Renew. Energy **23**, 17–26 (2001). https://doi.org/10.1016/s0960-1481(00)00110-5
30. Cuadros, F., López-Rodríguez, F., Marcosb, A., et al.: A procedure to size solar-powered irrigation (photo-irrigation) schemes. Sol. Energy **76**, 465–473 (2004). https://doi.org/10.1016/j.solener.2003.08.040
31. Fedrizz, M.C., Sauer, I.L., Zilles, R.: Economic analysis of photovoltaic and gasoline pumping systems. WREC 424–427 (1996). https://doi.org/10.1016/0960-1481(96)88891-4
32. Odeh, I., Yohanis, Y.G., Norton, B.: Economic viability of photovoltaic water pumping systems. Sol. Energy **80**, 850–860 (2006). https://doi.org/10.1016/j.solener.2005.05.008
33. Campana, P.E., Leduc, S., Kim, M., et al.: Optimal grassland locations for sustainable photovoltaic water pumping systems in China. Energy Procedia **75**, 301–307 (2015). https://doi.org/10.1016/j.egypro.2015.07.355
34. Pulfrey, D.L., Ward, P.R.B., Dunford, W.G.: A photovoltaic-powered system for medium head pumping. Sol. Energy **38**, 255–265 (1987). https://doi.org/10.1016/0038-092x(87)90047-8
35. Mustafizul Karim, A.N., Rahman, M.M.: Cost-effective analysis on the suitability of photovoltaic pumping systems in Bangladesh. Sol. Energy Mater. Sol. Cells **30**, 177–188 (1993). https://doi.org/10.1016/0927-0248(93)90019-y
36. Hamidat, A., Benyoucef, B., Hartani, T.: Small-scale irrigation with photovoltaic water pumping system in Sahara regions. Renew. Energy **28**, 1081–1096 (2003). https://doi.org/10.1016/S0960-1481(02),00058-7
37. Mahmoud, E., el Nather, H.: Renewable energy and sustainable developments in Egypt: photovoltaic water pumping in remote areas. Appl. Energy **74**, 141–147 (2003). https://doi.org/10.1016/S0306-2619(02)00140-X
38. Roy, A., Kabir, M.A.: Relative life cycle economic analysis of stand-alone solar PV and fossil fuel powered systems in Bangladesh with regard to load demand and market controlling factors. Renew. Sustain. Energy Rev. **16**, 4629–4637 (2012). https://doi.org/10.1016/j.rser.2012.03.068
39. Benghanem, M., Daffallah, K.O., Alamri, S.N., Joraid, A.A.: Effect of pumping head on solar water pumping system. Energy Convers. Manag. **77**, 334–339 (2014). https://doi.org/10.1016/j.enconman.2013.09.043
40. Fraidenraich, N., Costa, H.S.: Procedure for the determination of the maximum surface which can be irrigated by a photovoltaic pumping system. Sol. Wind Technol. **5**, 121–126 (1988). https://doi.org/10.1016/0741-983X(88),90069-0

41. Al-sagir, B.E., Hasson, A.M.: A procedure for minimizing the photovoltaic pumping system in irrigation and leaching applications in the Baghdad area. Sol. Wind Technol. **7**, 13–145 (1990). https://doi.org/10.1016/0741-983X(90)90081-C
42. Ghoneim, A.A.: Design optimization of photovoltaic powered water pumping systems. Energy Convers. Manag. **47**, 1449–1463 (2006). https://doi.org/10.1016/j.enconman.2005.08.015
43. Bakelli, Y., Arab, A.H., Azoui, B.: Optimal sizing of photovoltaic pumping system with water tank storage using LPSP concept. Sol. Energy. **85**, 288–294 (2011). https://doi.org/10.1016/j.solener.2010.11.023
44. Stoppato, A., Cavazzini, G., Ardizzon, G., Rossetti, A.: A PSO (particle swarm optimization)-based model for the optimal management of a small PV (Photovoltaic)-pump hydro energy storage in a rural dry area. Energy **76**, 168–174 (2014). https://doi.org/10.1016/j.energy.2014.06.004
45. Haddad, S., Benghanem, M., Mellit, A., Daffallah, K.O.: ANNs-based modeling and prediction of hourly flow rate of a photovoltaic water pumping system: experimental validation. Renew. Sustain. Energy Rev. **43**, 635–643 (2015). https://doi.org/10.1016/j.rser.2014.11.083
46. Campana, P.E., Li, H., Zhang, J., Zhang, R., Liu, J., Yan, J.: Economic optimization of photovoltaic water pumping systems for irrigation. Energy Convers. Manag. **95**, 32–41 (2015). https://doi.org/10.1016/j.enconman.2015.01.066
47. Olcan, C.: Multi-objective analytical model for optimal sizing of stand-alone photovoltaic water pumping systems. Energy Convers. Manag. **100**, 358–369 (2015). https://doi.org/10.1016/j.enconman.2015.05.018
48. Glasnovic, Z., Margeta, J.: A model for optimal sizing of photovoltaic irrigation water pumping systems. Sol. Energy **81**, 904–916 (2007). https://doi.org/10.1016/j.solener.2006.11.003
49. Hamidat, A., Benyoucef, B.: Systematic procedures for sizing photovoltaic pumping system, using water tank storage. Energy Policy **37**, 1489–1501 (2009). https://doi.org/10.1016/j.enpol.2008.12.014
50. Kaldellis, J.K., Zafirakis, D., Kondili, E.: Energy pay-back period analysis of stand-alone photovoltaic systems. Renew. Energy **35**, 1444–1454 (2010). https://doi.org/10.1016/j.renene.2009.12.016
51. Qoaider, L., Steinbrecht, D.: Photovoltaic systems: a cost competitive option to supply energy to off-grid agricultural communities in arid regions. Appl. Energy **87**, 427–435 (2010). https://doi.org/10.1016/j.apenergy.2009.06.012
52. Kelley, L.C., Gilbertson, E., Sheikh, A., et al.: On the feasibility of solar-powered irrigation. Renew. Sustain. Energy Rev. **14**, 2669–2682 (2010). https://doi.org/10.1016/j.rser.2010.07.061
53. Xu, H., Liu, J., Qin, D., Gao, X., Yan, J.: Feasibility analysis of solar irrigation system for pastures conservation in a demonstration area in inner Mongolia. Appl. Energy **112**, 697–702 (2013). https://doi.org/10.1016/j.apenergy.2013.01.011
54. Zhang, C., Yan, J.: Business model innovation on the photovoltaic water pumping systems for grassland and farmland conservation in China. ICAE2014, 1483–1486 (2014). https://doi.org/10.1016/j.egypro.2014.12.152
55. López-Luque, R., Reca, J., Martínez, J.: Optimal design of a standalone direct pumping photovoltaic system for deficit irrigation of olive orchards. Appl. Energy **149**, 13–23 (2015). https://doi.org/10.1016/j.apenergy.2015.03.107
56. Information on https://www.elsevier.com/solutions/sciencedirect
57. Bahador, M.: Solar water pumping. Sol. Energy **21**, 307–316 (1978)
58. Aman, M.M., Solangi, K.H., Hossain, M.S., et al.: A review of safety, health and environmental (SHE) issues of solar energy system. Renew. Sustain. Energy Rev. **41**, 1190–1204 (2015). https://doi.org/10.1016/j.rser.2014.08.086
59. Malarvizhi, R., Wang, M.H., Ho, Y.S.: Research trends in adsorption technologies for dye containing wastewaters. World Appl. Sci. J. **8**, 930–942 (2010)

60. Yu, J.J., Wang, M.H., Xu, M., et al.: A bibliometric analysis of research papers published on photosynthesis: 1992–2009. Photosynthetica **50**, 5–14 (2012). https://doi.org/10.1007/s10099-012-0010-1
61. Zhang, G.F., Xie, S.D., Ho, Y.S.: A bibliometric analysis of world volatile organic compounds research trends. Scientometrics **83**, 477–492 (2010). https://doi.org/10.1007/s11192-009-0065-3
62. Herwitz, S.R., Johnson, L.F., Dunagan, S.E., et al.: Imaging from an unmanned aerial vehicle: agricultural surveillance and decision support. Comput. Electron. Agric. **44**, 49–61 (2004). https://doi.org/10.1016/j.compag.2004.02.006

Applicability of Runoff Simulation in the Zhanghe Upstream Based on SWAT Model

Wu Haixia, Li Qingxue and Sun Yuzhuang

Abstract Sequential Uncertainty Fitting-2 (SUFI-2) with ArcSWAT2009 was used to test the performance of SWAT model for predicting runoff in the Zhanghe upstream. Parameter calibration and distributed hydrologic model building for the Zhanghe upstream were performed by coupling manual and auto-calibration methods. Monthly simulation values of R^2 and NSE were 0.83 and 0.79 during the calibration period, and 0.83 and 0.76 during the validation period, respectively. The results showed that SWAT model could be successfully used to model long-term continuous runoff in the study area. The calibrated model can be used for further analysis of the effects of the climate and land use change, water quality analysis and sediment yield analysis.

Keywords Runoff simulation · SUFI-2 algorithm · SWAT-CUP
SWAT model

1 Introduction

With the improvement of living standards, climate changes, land and water use policies, water shortages have become the major crises of sustainable development of communities all over the world. It is momentous for sustainable development of human society to reasonably develop and utilize the water resources. In recent

W. Haixia
Hebei University of Engineering, Handan 056038, Hebei, China
e-mail: whx2307@126.com

L. Qingxue (✉) · S. Yuzhuang
Hebei Collaborative Innovation Center of Coal Exploitation,
Key Laboratory of Resource Exploration Research of Hebei,
Hebei University of Engineering, Handan 056038, Hebei, China
e-mail: 745259266@qq.com

S. Yuzhuang
e-mail: syz@hebeu.edu.cn

W. Dong et al. (eds.), *Sustainable Development of Water Resources and Hydraulic Engineering in China*, Environmental Earth Sciences,
https://doi.org/10.1007/978-3-319-61630-8_27

years, the rapid social and economic development as well as human activities produced great change within the catchment storage water diversion; thereby, more and more complex water resources problems have developed in the Zhanghe upstream, which is located in the southwest of Haihe River Basin (HRB), China and belongs to the sub-basin of HRB. It is necessary to conduct hydrological research in the Zhanghe upstream that can support the improved catchment management programs and safeguard the alarming degradation of soil and water resources.

Researches on water resources in HRB can be found in many previous studies [1–5]. Weng et al. [1] developed an integrated scenario-based multi-criteria decision support system (SMC-DSS) applied to planning water resources management in HRB. Guo and Shen [2] developed an operational model to simulate water and energy fluxes in the HRB over the past 28 years using multiple data sources of meteorological data, remotely sensed NDVI data, soil parameters, and land use data. A model was developed in the companion paper to analyze the changes of water and energy budgets using multi-source data. Based on the analysis, it is found that climate change and human activities have resulted in massive changes in energy balance and hydrological cycle in HRB [3]. Historical simulations from 1965 to 2000 with different water demands, and future simulations were conducted to investigate the effects of water overexploitation and the possible responses to different exploitation scenarios in the Basin [4]. According to the Chinese National Standards of GB 3838-2002 and GB 18918-2002, the comprehensive water quality in HRB has been described [5]. However, relevant hydrological simulation that is quite important to forecast the flood events in the future in the Zhanghe upstream is relatively rare.

Hydrological models are important tools for planning sustainable use of water resources to meet various demands. The Soil and Water Assessment Tool (SWAT) is a physics-based, long-term, and distributed hydrological model and has been applied worldwide as an excellent assessment model for hydrological modeling and water resource management [6]. This model is applied to runoff and soil loss prediction [7, 8], water quality modeling [9, 10], land use change effect assessment [11, 12] and climate change affects water quality modeling [13, 14]. These models, due to a large number of immeasurable parameters (owing to measurement limitations and scaling issues) and various sources of uncertainties, require careful calibration and uncertainty analysis before they can be used to understand and investigate the underlying system. In the previous studies, several sensitivity analyses, calibration and uncertainty analysis techniques can be found [15, 16]. The current modeling requires that models are transparently described, and that calibration, validation, sensitivity and uncertainty analyses are routinely performed as part of the modeling work. As calibration is affected by many factors (i.e. the model structure, model inputs, analyst's assumptions, calibration data, etc.) and not uniquely determined, sensitivity analysis and uncertainty analysis are essential to evaluate the strength of a calibrated model. SWAT-CUP (Calibration and Uncertainty Procedures) is a standalone program developed for calibration of SWAT. The program contains three different calibration procedures and functionalities for validation and sensitivity analysis as well as visualization of the study area. These procedures include: GLUE, ParaSol and Sequential Uncertainty Fitting (SUFI-2). The program SUFI-2 is more methods used for model calibration and uncertainty analysis [17]. The SUFI-2 algorithm is an effective method, but it requires additional iterations as well as the need for adjustment of the parameter ranges [18].

In this study, SWAT model and the uncertainty analyses were carried out in the Zhanghe upstream. The focus was particularly on calibration, evaluation and application of SWAT2009 model for hydrological simulation in the study area. The main objective of this study is to test the performance of the SWAT model for predicting streamflow in the mountainous basin, which will contribute to the water resources management in the Zhanghe upstream and thereby is useful for the sustainable development of the country.

2 Study Area

The upper reaches of Zhanghe basin are seen in Fig. 1.

The area covers a total area of 18,284 km^2 and equals to 6% out of 318,200 km^2 of the HRB area. The area includes 22 counties (cities, districts) in Shanxi, Hebei, Henan provinces; the basin has many tributaries with large tributaries of the Qing Zhanghe and Zhuo Zhanghe. The elevation of the basin ranges from 133 to 2149 m above the mean sea level.

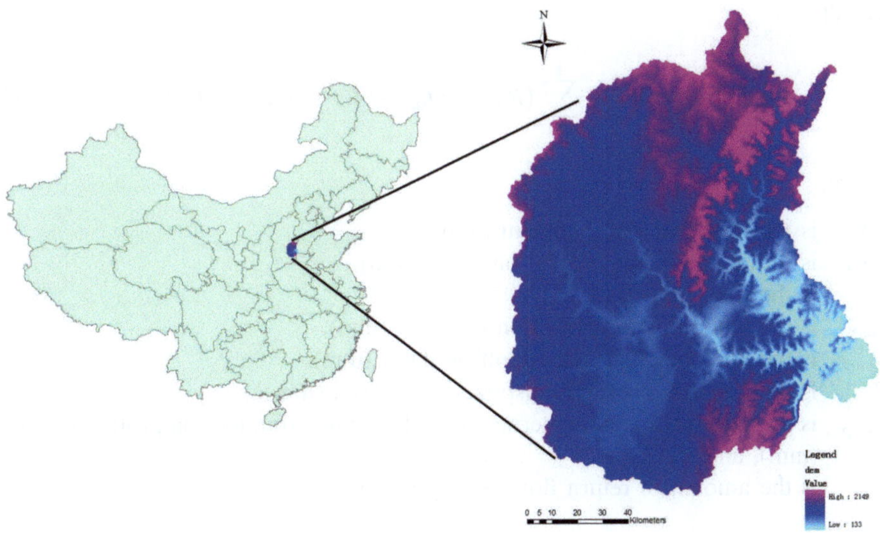

Fig. 1 Location of the study area

Within the basin most of the runoff yield area is mountainous; and the geological composition for limestone, pad formed under complex surface area. The area involves the provinces in the region with a large number of water storage and water diversion projects.

3 Methods and Data

3.1 SWAT Model

SWAT is a distributed hydrological and water quality model. It has played important roles in various fields, such as modeling water availability and quality, plant growth, evaluating effects of watershed management options, and land use and climate change impacts [19].

It uses hydrological response units (HRUs) that consist of specific land use, soil and slope characteristics. The HRUs are used to describe the spatial heterogeneity in terms of land cover, soil type and slope class within a watershed. The model estimates relevant hydrological components such as evapotranspiration, surface runoff and peak rate of runoff, groundwater flow and sediment yield for each HRU. ArcGIS extension is a graphical user interface for the SWAT model. The SWAT model is developed and refined by the U.S. Department of Agricultural Research Service (ARS) and scientists at universities and research agencies around the world. The water balance equation is the base of the hydrologic cycle simulation in SWAT:

$$SW_t = SW_0 + \sum_{i=1}^{t} \left(R_{day} - Q_{surf} - E_a - w_{seep} - Q_{gw}\right)_i \quad (1)$$

where,

SW_t is the final soil water content (mm),
SW_0 is initial soil water content on day i (mm),
t is the time (days),
R_{day} is the amount of precipitation on day i (mm),
Q_{surf} is the amount of surface runoff on day i (mm),
E_a is the amount of evapotranspiration on day i (mm),
w_{seep} is the amount of water entering the vadose zone from the soil profile on day i (mm), and
Q_{gw} is the amount of return flow on day i (mm).

3.2 SWAT-CUP Model

SWAT-CUP (Calibration and Uncertainty Procedures) is a standalone program developed for calibration of SWAT [20]. The program contains five different calibration algorithms, such as Sequential Uncertainty Fitting SUFI-2, Particle Swarm Optimization (POS), Generalized Likelihood Uncertainty Estimation (GLUE), Parameter Solution (ParaSol), and Mark chain Monte Carlo (MCMC) procedures to SWAT. It includes functionalities for validation, uncertainty and sensitivity analysis. The sub-basins, simulated rivers, and outlet, rainfall, and temperature stations can be visualized. For time-consuming large-scale models, SUFI-2 was found to be quite efficient. Therefore, the SUFI-2 methods are applied in this study.

3.3 Data Input

The spatially distributed data (GIS input) needed for the ArcSWAT interface include the Digital Elevation Model (DEM), soil data, land use, stream network layers and data of the weather and river discharge. Descriptions of these data are given in Table 1.

The land use map year 2005 and soil type maps were shown in Fig. 2a, b. The observed monthly discharge data period 1980–2010 used in this model was divided

Table 1 Basic data for model development

Data type	Resolution/ proportion/scale	Format	Description of these data
Figure DEM	30 m × 30 m	ESRI grid	Elevation and slope
Water System Images	30 m × 30 m	ESRI grid	Channel distribution
Land use map	30 m × 30 m	ESRI grid	Space distribution of Land use/ vegetational cover
Vegetation data	——	dBase table	Vegetational cover;growth data
Soil type maps	1:100000	ESRI grid	Space distribution of soil type
Soil properties	——	dBase table	Soil grouping and physical and chemical properties
Weather	Observed daily data	dBase table	Rainfall, air temperature, wind speed, humidity, solar radiation
Reservoir	Observed monthly data	dBase table	Observed monthly discharge data
Hydrology	Observed monthly data	dBase table	Runoff

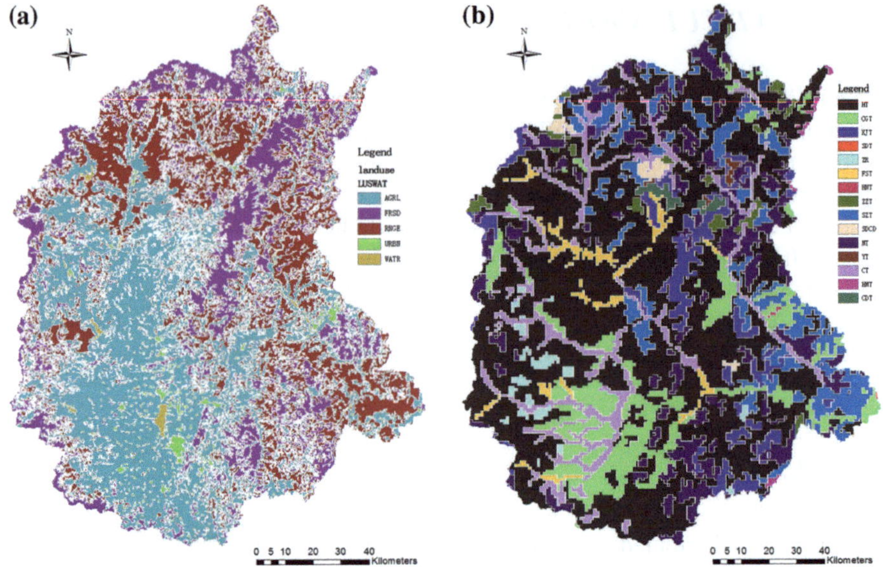

Fig. 2 Landuse (**a**) and Soil types (**b**) in the Zhanghe upstream

into three periods, that is, model preheating period (1980–1984), calibration (1985–2004) and validation (2005–2010). As the input data for the weather generator of SWAT, the daily temperature, humidity, wind speed and precipitation data are used. These hydrological and meteorological data were provided by the Ministry of Hydrology of upper reaches of Zhanghe. The original polygon land use data were combined and transferred to the land use raster map by using ArcGIS 9.3.

3.4 *Model Development*

Basin Delineation. Basin delineation depends on DEM to delineate the watershed and to analyze the drainage patterns of the land surface terrain. The ArcSWAT interface uses the mask area for stream delineation, and the stream networks are delineated from the DEM by using an automatic delineation to SWAT model. The model fills all of the non-draining zones to create a flow direction and superimposes the digitized stream network into the DEM to define the location of stream networks.

The ArcSWAT proposes the minimum, maximum and suggested size of the sub-watershed area in hectare to define the minimum drainage area. Generally, the smaller the threshold area is, the more detailed the drainage networks and the number of sub-basins and HRUs are. In addition, more processing times and spaces are needed. In this study, the 12,000-ha area is provided to define all sub-basins of upper reaches of Zhanghe and outlet in which it is later taken as a point of calibration of the simulated flows. As a result, there are 74 sub-basins of upper reaches of Zhanghe.

HRU Definition. The analysis of HRU definition indicates that the dominant type of HRU definition results in a single HRU for each sub-basin where the dominant land use, soil and slope within the basin are considered to be the land use, soil and slope of each sub-basin. This single HRU within each sub-basin is not able to properly represent the characteristics of the sub-basins. Accordingly, the simulated stream flow shows an unsatisfactory result as compared to the measured stream flows in the observed stations of upper reaches of Zhanghe. The multiple scenarios that account for 5% land use and 10% soil threshold combination give a better estimation of streamflow.

The upper reaches of Zhanghe results in 504 HRUs in the whole basin. This scenario results in the detailed land use, slope and soil database, containing many HRUs, which in turn represent the heterogeneity of the study area. The comparison between the default model predictions and measured discharge produces the highest Nash-Sutcliffe efficiency (NSE). The distribution of land use, soil and slope characteristics within each HRU has the greatest impact on the predicted stream flow. As the percentage of land use and soil threshold increases, the actual evapotranspiration decreases due to eliminated land use classes. Hence, the characteristics of HRUs are the key factors affecting the streamflow.

4 Results and Discussion

The calibration is the modification or adjustment of model parameters, within the recommended ranges, to optimize the model output so that it matches with the observed set of the data. The calibration provides several different parameters for adjustment through user intervention. These parameters can be adjusted manually or automatically until the model output best matches with the observed data. The validation is the process of determining the degree in which a model or simulation is an accurate representation of the observed set of data from the perspective of the intended uses of the model. In this study, SWAT-CUP is applied for calibrating and validating outlet streamflow. The discharge data were recorded during the years 1980–2010 at Guantai station, and the monthly discharges from 1985–2004 were used for calibration; 2005–2010 were used for validation, but for the years 1980–1984 it was skipped for model warm-up.

A general performance rating of statistics for streamflow monthly time-step was recommended: $0.75 < NSE < 1.00$ considered as very good. Values of R^2 greater than 0.5 were considered acceptable [21]. The comparison between the simulated monthly stream flow and the observed data is a good result for the calibration and validation periods, respectively. The simulated monthly flow matches the observed values for the calibration and validation periods with R^2 = 0.83, 0.79 and NSE = 0.83, 0.76, respectively (Table 2). They were acceptable between the observations and the final "best" simulation. The results showed that SWAT was able to

Table 2 Statistical results of the calibration and validation periods

	P-factor	R-factor	R^2	NSE
Calibration period	0.47	0.46	0.83	0.83
Validation period	–	–	0.79	0.76

simulate the hydrological characteristics of the Zhanghe upstream very well. Hence, the model can be used for further hydrological studies in the basin.

To offer a pictorial comparison of measured and simulated flows, calibration and validation results were shown graphically in addition to tabulating quantitative statistical values. Comparison of the observed monthly streamflow time series hydrographs were in good agreement with predicted one for most parts of calibration and validation period (Figs. 3 and 4). The strength of calibration analysis as explained by R-factor (95PPU band) was also illustrated on the same figure by the shaded region.

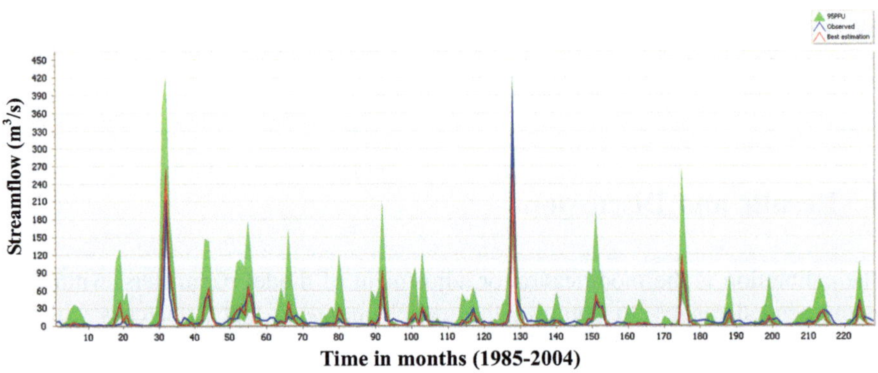

Fig. 3 Observed, best estimation and 95PPU of monthly streamflow for model calibration

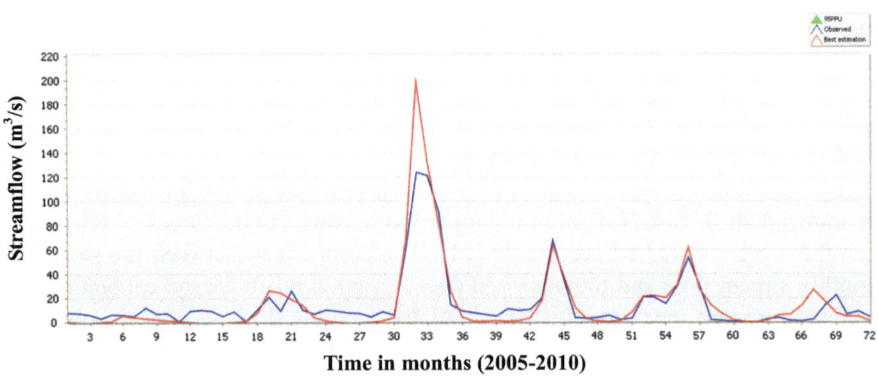

Fig. 4 Observed and best estimation of monthly stream flow for model validation

Often hydrologic conditions of a validation period were different from those of a calibration period, which may lead to differences in performance of the parameter solutions for the respective periods [22]. Likewise, in this study, SUFI-2 achieved a bit higher R^2 and NSE values (0.83 and 0.83, respectively) for the calibration period than did in validation period (0.79 and 0.76, respectively) (Table 2). Effective techniques were used for calibration and prediction uncertainty analysis of SWAT model in the study watershed which fine-tuned the parameters to provide reasonable verification in the validation stage.

5 Conclusions

Efficient and effective algorithms for optimization of computationally intensive hydrologic models like SWAT are becoming increasingly more important because of limited time and computational resources. Application of SUFI-2 algorithms to simultaneously conduct calibration and validation of SWAT in the Zhanghe upstream was perfectly done. The good result is shown with the likelihood measure of the model calibration and validation for two periods: 1986–2004 and 2005–2010. Monthly simulation values of R2 and NSE were 0.83 and 0.79 during the calibration period, and 0.83 and 0.76 during the validation period, respectively. The calibrated model can be used for further analysis of the effects of the climate and land use change, water quality analysis and sediment yield analysis.

Acknowledgements This study was supported by Ministry of Water Resources' special funds for scientific research on public causes of China (No. 201401030), Key Educational Commission of Hebei Province, China (No. ZH2012044), Youth Foundation Project of Hebei Provincial Department of Education China (No. 2017039).

References

1. Weng, S.Q., Huang, G.H., Li, Y.P.: An integrated scenario-based multi-criteria decision support system for water resources management and planning—a case study in the Haihe River Basin. Expert Syst. Appl. **37**, 8242–8254 (2010). https://doi.org/10.1016/j.eswa.2010.05.061
2. Guo, Y., Shen, Y.: Quantifying water and energy budgets and the impacts of climatic and human factors in the Haihe River Basin, China: 1. Model and validation. J. Hydrol. **528**, 206–216 (2015). https://doi.org/10.1016/j.jhydrol.2015.06.039
3. Guo, Y., Shen, Y.: Quantifying water and energy budgets and the impacts of climatic and human factors in the Haihe River Basin, China: 2 Trends and implications to water resources. J. Hydrol. **527**, 251–261 (2015)
4. Zou, J., Xie, Z., Zhan, C. et al.: Effects of anthropogenic groundwater exploitation on land surface processes: A case study of the Haihe River Basin, northern China. J. Hydrol. **524**, 625–641 (2015). doi:https://doi.org/10.1016/j.jhydrol.2015.03.026

5. Ke, X., Wang, C., Jing, D., et al.: Assessing water quality by ratio of the number of dominant bacterium species between surface/subsurface sediments in Haihe River Basin. Mar. Pollut. Bull. **98**, 267–273 (2015). https://doi.org/10.1016/j.marpolbul.2015.06.003
6. Arnold, J.G., Srinivasan, R., Muttiah, R.S., Williams, J.R.: Large area hydrologic modeling and assessment, part I: model development. J. Am. Water Resour. Assoc. **34**, 73–89 (1998). https://doi.org/10.1111/j.1752-1688.1998.tb05961
7. Morgan, R.P.C.: A simple approach to soil loss prediction: a revised Morgan–Morgan–Finney model. CATENA **44**, 305–320 (2001). https://doi.org/10.1016/S0341-8162(00)00171-5
8. Grønsten, H.A., Lundekvam, H.: Prediction of surface runoff and soil loss in south eastern Norway using the WEPP Hillslope model. Soil Tillage Res. **85**, 186–199 (2006). https://doi.org/10.1016/j.still.2005.01.008
9. Debele, B., Srinivasan, R., Yves Parlange, J.: Coupling upland watershed and downstream water hydrodynamic and water quality models (SWAT and CE-QUAL-W2) for better water resources management in complex River Basins. Environ. Model. Assess. (2006). https://doi.org/10.1007/s10666-006-9075 -1
10. Zhang, Y., Xia, J., Shao, Q., Zhai, X.: Water quantity and quality simulation by improved SWAT in highly regulated Huai River Basin of China. Stoch. Environ. Res. Risk Assess. **27**, 11–27 (2011). https://doi.org/10.1007/s00477-011-0546-9
11. Sheng, X.B., Sun, J.Z., Liu, Y.X.: Effect of land-use and land-cover change on nutrients in soil in Bashang area China. J. Environ. Sci. **15**, 548–553 (2003). https://doi.org/10.1007/s40333-013-0143-5
12. Wu, W., Hall, C.A.S., Scatena, F.N.: Modelling the impact of recent land-cover changes on the stream flows in Northeastern Puerto Rico. Hydrol. Process. **21**, 2944–2956 (2007). https://doi.org/10.1002/hyp.6515
13. Shrestha, B., Babel, M.S., Maskey, S., van Griensven, A., Uhlenbrook, S., Green, A., Akkharath, I.: Impact of climate change on sediment yield in the Mekong River Basin: a case study of the Nam ou Basin, Lao PDR. Hydrol. Earth Syst. Sci. **17**, 1–20 (2013). https://doi.org/10.5194/hess-17-1-2013
14. Abbaspour, K.C., Yang, J., Maximov, I., Siber, R., Bogner, K., Mieleitne, J., Zobrist, J., Srinivasan, R.: Modelling hydrology and water quality in the pre-alpine/alpine Thur watershed using SWAT. J. Hydrol. **333**, 413–430 (2007). https://doi.org/10.1016/j.jhydrol.2006.09.014
15. Yesuf, H.M., Assefa, M.M., Zeleke, G., Alamirew, T.: Streamflow prediction uncertainty analysis and verification of SWAT model in a tropical watershed. Environ. Earth Sci. **75**, 806–820 (2016). https://doi.org/10.1007/s12665-016-5636-z
16. Zhang, D., Chen, X., Yao, H., et al.: Moving SWAT model calibration and uncertainty analysis to an enterprise Hadoop-based cloud. Environ. Model Softw. **84**, 140–148 (2016). https://doi.org/10.1016/j.envsoft.2016.06.024
17. Abbaspour, K.C., Johnson, C., van Genuchten, M.T.: Estimating uncertain flow and transport parameters using a sequential uncertainty fitting procedure. Vadose Zone J. **3**, 1340–1352 (2004). https://doi.org/10.2113/3.4.1340
18. Vilaysane, B., Takara, K., Luo, P., et al.: Hydrological stream flow modelling for calibration and uncertainty analysis using SWAT model in the Xedone River Basin Lao PDR. Procedia Environ. Sci. **28**, 380–390 (2015). https://doi.org/10.1007/s10666-006-9075-1
19. Gassman, P.W., Reyes, M.R., Green, C.H., Arnold, J.G.: The soil and water assessment tool: historical development, applications, and future research directions. Trans. Asabe. **50**, 1211–1250 (2007). https://doi.org/10.13031/2013.23637
20. Abbaspour, K.C.: SWAT-CUP2. SWAT Calibration and Uncertainty Programs, Version 2. 2009
21. Beven, K., Binley, A.: The future of distributed models—model calibration and uncertainty prediction. Hydrol. Process. **6**, 279–298 (1992). https://doi.org/10.1002/hyp.3360060305
22. Alamirew, C.D.: Modeling of hydrology and soil erosion in upper Awash River Basin. Univ. Bonn, Institut für Städtebau, Bodenordnung und Kulturtechnik. **223**, 235–241 (2006)

Research on Reservoir Water Temperature Simulation and Fish Response

Wei Li, Jia-Hong Liu, Yu-fei Zhang and Wei-Hua Xiao

Abstract Temperature is an important parameter for a body of water, affecting chemical and biological reactions as well as the material exchange. It is also important with respect to the water environment and the growth and propagation of microbes. Water temperature research addresses two important aspects: the impact on the environment due to water temperature change caused by human activities, and better analysis of reactions related to water temperature. Deep reservoirs have both vertical and horizontal temperature stratification. This paper establishes a 2D water temperature model for the Panjiakou Reservoir and simulates the stratification phenomena. Temperature was simulated from May to October 2010, and the diffusion parameters were confirmed. Results showed distinct stratification from May to August with vertical equality from September to October. From May to June, water temperature downstream declined because of discharging. Discharged water temperature significantly affects the spawning time of fishes downstream and can delay spawning by up to a month.

Keywords Temperature stratified · Water temperature model · Fish response Panjiakou Reservoir

W. Li (✉) · J.-H. Liu · W.-H. Xiao
China Institute of Water Resources and Hydropower Research, Beijing, China
e-mail: yixin22769@sina.com

J.-H. Liu
e-mail: liujh@iwhr.com

W.-H. Xiao
e-mail: xwsen998@163.com

Y. Zhang
College of Environment, Hohai University, Nanjing, China
e-mail: zyfhhu1996@163.com

W. Dong et al. (eds.), *Sustainable Development of Water Resources and Hydraulic Engineering in China*, Environmental Earth Sciences,
https://doi.org/10.1007/978-3-319-61630-8_28

1 Introduction

Temperature, which is stratified vertically in lakes, reservoirs and rivers, is one of the most important characteristics of water. Water temperature can affect the material exchange ability, chemical reactions, and the entire ecosystem, especially those species sensitive to temperature. Research into a water temperature distribution law could help the integrated management and regulation of lakes or reservoirs, which is important to keep fish healthy.

Water temperature ecosystem research comprises two main aspects: firstly, the effect of water temperature variation leading to climate change at ecosystem level [1, 2], such as a rise in temperature from global warming influencing fish habitats [3]; and secondly, the influence of human activities [4, 5], such as reservoir regulation, influencing the ecosystem [6]. Research to date has focused mainly on qualitative analysis, while ignoring quantitative analysis. In China, most of the relevant research areas are in the south [7, 8], although deep reservoirs in the north also have temperature delamination. At the same time, storage of water in and discharge from the reservoirs has had an effect on downstream ecologies.

Experimental methods, investigative and statistical methods, model simulations, etc., form the basis of the research conducted in this study. The model that this paper proposes can simulate water temperature changes under different weather conditions, leading to advantageous applications. Based on the theory of heat conduction and aquatic animals' temperature response research, this paper simulated the water temperature of the Panjiakou Reservoir in North China using a vertical 2D model. The spatial and temporal distribution of water temperature in the reservoir was analyzed. By investigating local aquatic animals, the effect of reservoir discharge on fish downstream is presented with a view to minimizing any adverse effects of reservoir discharge by regulation.

2 Water Temperature Simulation Model

2.1 Theory of Heat Conduction

The temperature and rate of flow of incoming water, and the weather are the main parameters affecting a reservoir's water temperature. Depending on their temperature delamination, lakes and reservoirs can be classified as mix type, stable stratified type, or transition type. The temperature of a mix type is well distributed with a temperature gradient of less than one. The stable stratified type has a temperature difference between surface water and the lower layer, especially in summer. The transition type lies in between. The method of $\alpha - \beta$ index could be used to identify the stratification type of lakes or reservoirs, where:

$$\alpha = \frac{\text{average income flow}}{\text{total storage capacity}}. \tag{1}$$

$$\beta = \frac{\text{total valume of flood}}{\text{total storage capacity}}. \tag{2}$$

α indicates the time that water stayed in the reservoir, with $\alpha < 10$ a stable stratified type, $10 < \alpha < 30$ a transition type, and $\alpha > 30$ a mix type. The stable stratified type could be subdivided into two types according to β, with $\beta > 1.0$ a temporary stratified type, and $\beta < 0.5$ a stable stratified type.

The main heat conductors that could cause water temperature change include heat exchange in water surface, heat exchange with the riverbed, heat generated inside the water, and artificial heat exchange. Sources of heat include both solar and atmospheric radiation. Heat dissipation factors include water radiation, surface water reflection and evaporation heat consumption.

2.2 2D Vertical Water Temperature Model

A 2D vertical water temperature model could predict the distribution of water temperature in a body both vertically and horizontally. It is used to calculate water temperatures in deep water, such as channel reservoirs and deep lakes. It is assumed that water temperature and current velocity are constant in the cross-sectional direction, while they change in the vertical and horizontal directions. The water body is divided into a discrete number of small horizontal and vertical units. The model simulates heat exchanges, current velocities and silt of each unit.

The heat flux into the water can be expressed by:

$$\phi = \phi_0 - \phi_e - \phi_c - \phi_{ra} \tag{3}$$

where ϕ_0 is solar radiation flux, ϕ_e evaporative heat flux, ϕ_c heat transfer flux, and ϕ_{ra} effective inverse radiation flux,which equals water radiation flux, ϕ_{rw}, minus atmospheric radiation, ϕ_a.

$\phi_0 = (1 - a_r)\phi_s\phi_s$ is heat generated by solar radiation, which will enter the interior water body. The amount of radiation received at the depth of y can be expressed as $\phi_y(y) = (1 - \beta)\phi_0 e^{-\eta(y_s - y)}$ where $\beta\phi_0$ is surface water absorption, a_r is surface water reflectance, β is surface absorption ratio, and η is water transparency.

$$\phi_e + \phi_c = (0.000308 + 0.000185w)\rho(e_s - \psi e_a)[L_v + C_w T_s + \frac{269.1(T_s - T_a)}{e_s - \psi e_a}] \tag{4}$$

$$\phi_{ra} = 0.97k[T_w^4 - 0.937 \times 10^{-5} T_A^6 (1 + 0.17C_1^2)]. \tag{5}$$

This is total heat flux of the water body. The heat transfer process is carried out only on the surface layer. w is the wind speed, e_a the saturated vapor pressure of air, e_s the saturated vapor pressure of water, ψ relative humidity, L_v the latent heat of vaporization of water, T_s the water temperature, T_a the surface air temperature, k the Stefan-Boltzmann constant, T_w the absolute surface water temperature, T_A the absolute surface air temperature, and C_1 a cloudiness factor (equal to 10 with complete overcast and 0 when cloudless).

The water temperature at a depth of y (meters) can be expressed as:

$$\begin{aligned} \frac{\partial T_y}{\partial t} = & \frac{(\alpha + D)}{A_y} \frac{\partial}{\partial y}\left(A_y \frac{\partial T_y}{\partial y}\right) - \frac{1}{\rho C_p A_y} \frac{\partial}{\partial y}(A_y \phi_y) \\ & - \frac{1}{A_y} \frac{\partial}{\partial}(\upsilon_y A_y T_y) + \frac{1}{A_y}(u_{iy} B_y T_i' - u_{iy} B_y T_y) \end{aligned} \tag{6}$$

where T_y is the water temperature at depth y, u_{iy} flow velocity in the horizontal direction at depth y, υ_y vertical flow velocity at depth y, T_i' the inflow water temperature, A_y the water area at the depth y, ϕ_y the daily radiation dose at the depth y, α the molecular diffusion coefficient, D the eddy diffusion coefficient, ρ the water density, and C_p the specific heat of water.

The 2D vertical water temperature simulation model developed by Gao and Zhang [9] is used in this paper.

3 Case Study

3.1 Panjiakou Reservoir

The Panjiakou Reservoir is located in Taoyuan, in the Qianxi County of Hebei Province. It is a large water conservancy project in the middle reaches of the Luan River. When full it has a surface area of 68 km^2, and extends for 64 km. Panjiakou, a typical river-type reservoir, has a total capacity of 2.93 billion m^3, with a dam wall height of 107.5 m. Flood discharge structures in the middle of the river are fitted with radial gates. The spillway has 18 floodgates with 4 exit holes (4 m × 6 m). The reservoir's catchment area is 33 700 km^2, which is 75% of the whole Luan River basin.

Applying the $\alpha - \beta$ index method to the Panjiakou Reservoir gives α equal to 0.95, far less than 10 and thus indicating a stable stratified type, and β equal to 0.09 which, being much smaller than 5, indicates that flooding has little effect on the distribution of water temperature.

3.2 Data Collection

Topographic Data. The longitudinal section of the generalized reservoir is shown in Fig. 1. When full, the maximum water depth is 69 m; and, as mentioned earlier, the reservoir extends up to 64 km. The reservoir width reduces gradually upstream from the dam wall. At Qinghekou, the surface width is 1150 m, and the corresponding bottom width is 970 m. The width of the reservoir is 200 m at the narrowest part. The simulation model takes 1 km (horizontal) by 1 m (vertical) as a computing unit. There are thus 64 horizontal and 69 vertical cells which gives a total of 2246 computing units. The cross sections were changed into trapeziums, and the bottom cell depth was determined according to the nearest measured section terrain. The width of each unit was determined by linear interpolation of the width at the top and bottom.

Meteorological Data. The meteorological data used in the simulation is mainly from the Tangshan and Leting Stations, which were part of the Global Meteorological Observation Project of the World Meteorological Organization (WMO), as well as from the China Meteorological Administration. Figures 2 and 3 show the 2010 air temperature and the average wind speed at Tangshan respectively. Over this period, the average air temperature was 13 °C, and average wind speed was 5.5 m/s. The daily radiation data, calculated solely from the amount of sunshine due to the lack of measured cloud data, came from the Leting Station. The average relative humidity in 2010 was approximately 50%, with the lowest measured value 15%. The rain was concentrated in July and August.

Hydrological Data. The runoff data and water temperature data of inflow at the end of the reservoir is from the Haihe River Basin Hydrological Yearbook. Inflow from this tributary is the greatest in June, July, and August.

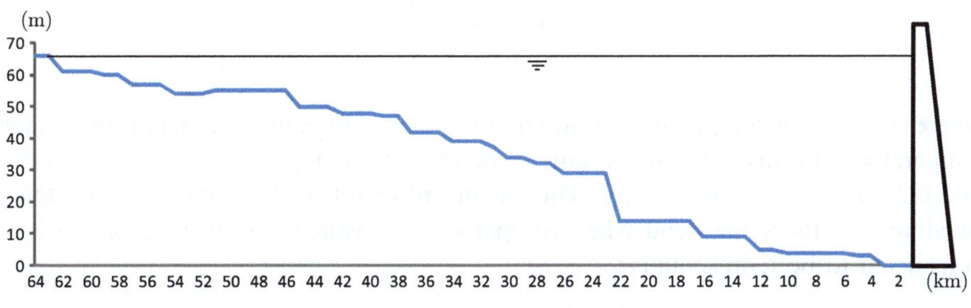

Fig. 1 Longitudinal geometric shape of the Panjiakou Reservoir

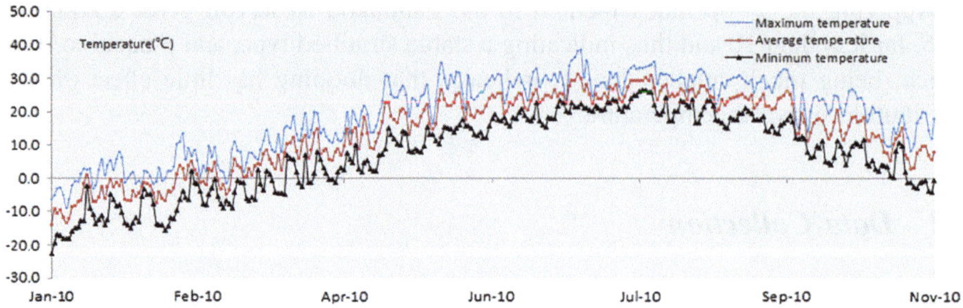

Fig. 2 The temperature at Tangshan in 2010

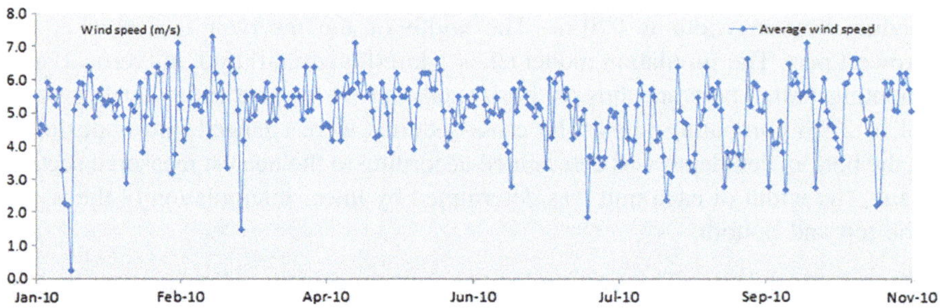

Fig. 3 Wind speed at Tangshan in 2010

During the simulation period the reservoir water level varies between 53.3 and 59.1 m. The reservoir discharges water in spring in order to have a low level in June. The water level rises about 5 m through the summer, peaking in September. Discharges of up to about 60 m^3/s are done mid-August, late-September, and October.

Inflow Data. The inflow water temperature data was approximated using air temperature data due to the lack of measured data in 2010. Mohseni et al. [10] have studied the nonlinear correlation between air temperature and water temperature as follows:

$$T_s = \mu + \frac{\alpha - \mu}{1 + e^{\gamma(\beta - T_a)}} \tag{7}$$

where T_s is water temperature, μ minimum water temperature, α maximum water temperature, β inflection temperature, γ the maximum slope (which is between 0.1 and 0.2), T_a the air temperature. The calculated result of 2010 inflow water temperatures has the same trend when compared to measured data from 2006, so it is considered to be representative.

3.3 Verification

The main sensitivity parameters of the model are surface and bottom temperature, diffusion coefficient (X, Y direction), surface reflectivity, surface absorption ratio and water transmission rate. The diffusion coefficient has a large effect on the temperature distribution structure. The last three parameters affect mainly the water energy absorption from solar radiation (Fig. 4).

The model's simulation results were validated using measured water temperature data from October 29–31, 2010. Figure 4 (left) shows the simulation's water temperature compared with that measured at the Section 1 km before the dam wall. There is a 0.5–1.5 °C difference between the two temperatures. On October 29, the water temperature at 12:00 a.m., was 0–0.5 °C higher than that at 10:00 a.m., which has the same trend as the measured data. Water temperature above the 20 m level rose between 0 and 1.4 °C more on October 31 than on October 29 according to the measured data, while the simulated values are higher. Figure 4 (right) shows the simulation's water temperatures compared to the temperatures measured at the section of South Guojiazhuang on October 29 and Jiajiaan on October 31. The surface temperature results compare well, but the bottom results differed between 0 and 4 °C with the measured data.

The model's simulation results are acceptable at sections near the dam, with the errors in the 6–10% band. At the section of South Goujiazhuang, the simulation error is within 10% above the 5 m level, while at the section of Jiajiaan, the error is between 2 and 10% above the 13 m level, with a larger bottom error.

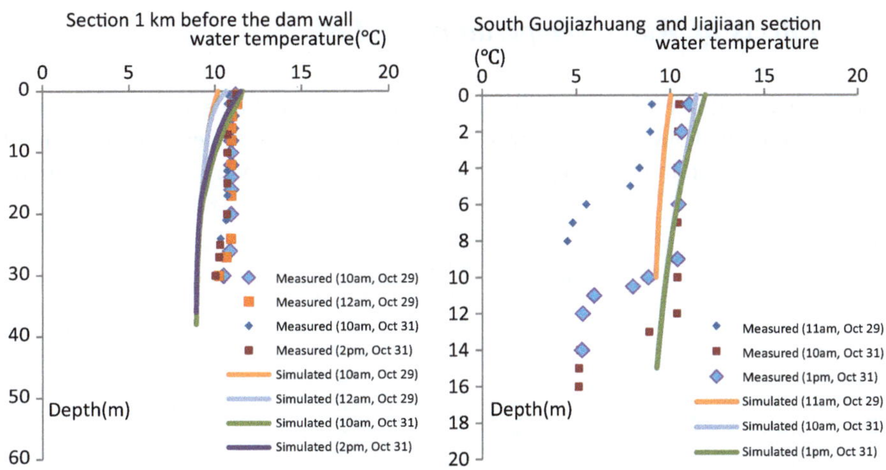

Fig. 4 Model verification

4 Results

4.1 *Water Temperature Distribution*

The vertical water temperature distribution in different sections is displayed in Figs. 5, 6 and 7. The sections are located, respectively, at 1 km, 11 km, 21 km, 31 km and 41 km upstream of the dam wall. The Panjiakou Reservoir has vertically temperature stratification from May to August, especially in July and August. The temperature also differs in the horizontal direction. The water temperature is highest near the dam wall, and decreases upstream where it is affected by the temperature of the inflowing water. From May to October, the surface water temperature increased from 16 to 24 °C, and then decreased to 9 °C, and the bottom water from 11 to 13 °C, and then to 6 °C.

From the beginning of May, surface water temperature increases due to the warming weather, and also transfers heat to the bottom. The surface water temperature peaks in July and August, and decreases from September. The bottom water temperature decreases from October.

Figures 8, 9 and 10 show isotherm diagrams of Panjiakou Reservoir at different times. The isograms in May are inclined which indicated horizontal differences in water temperature. The average temperature in May is about 12 °C. Bottom water temperature reached 15 °C in July. In August, while the surface water had increased by 1–2 °C, the bottom water began decreasing by 3–4 °C. The water temperature to a depth of 10 m was above 20 °C.

4.2 *Effects of Water Temperature on Fish*

The Panjiakou Reservoir discharges water through the hydroelectric power station in both spring and autumn. The spring discharge runs from the end of March until June, while the autumn discharge is in September and October. Assuming a minute

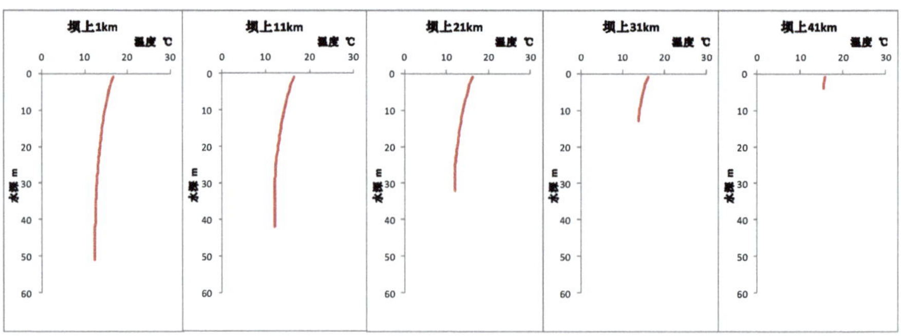

Fig. 5 Water temperature curve on vertical at 08:00 a.m., May 20

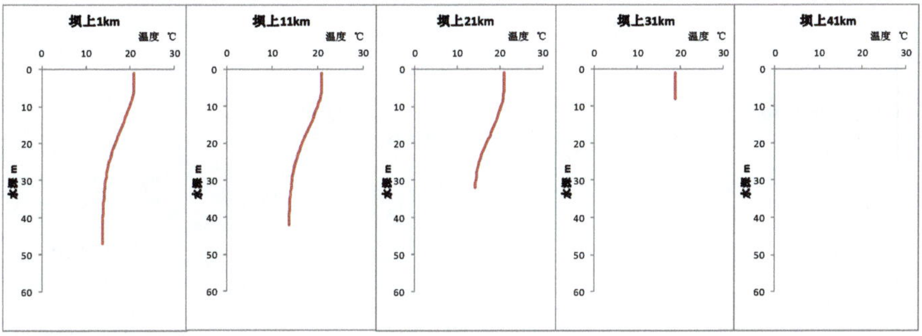

Fig. 6 Water temperature curve on vertical at 08:00 a.m., July 20

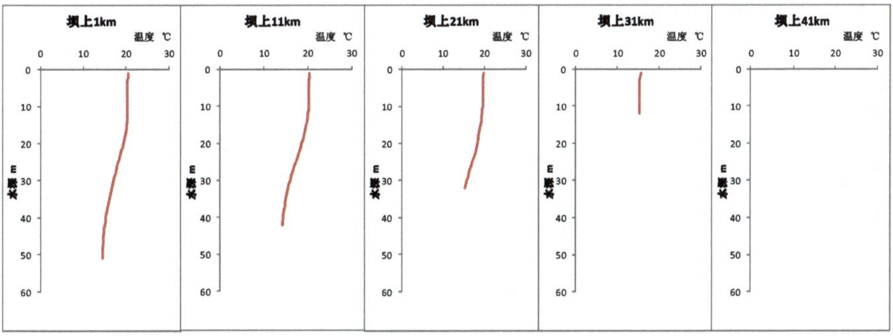

Fig. 7 Water temperature curve on vertical at 08:00 a.m., August 20

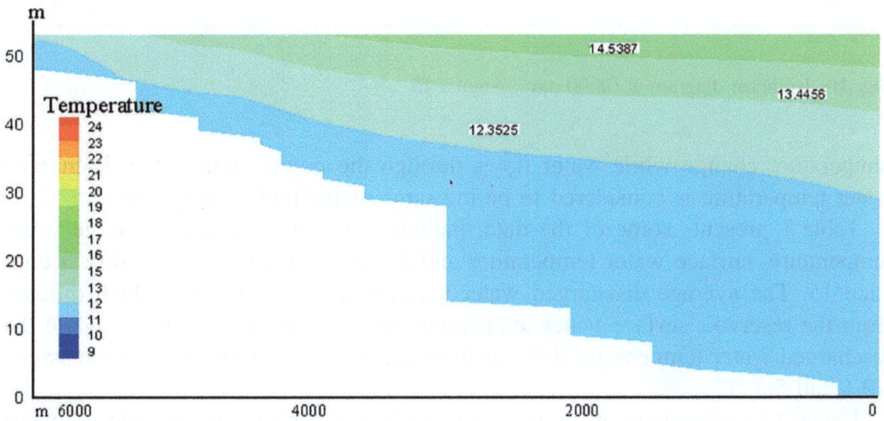

Fig. 8 Isotherm diagram at 08:00 a.m., May 15

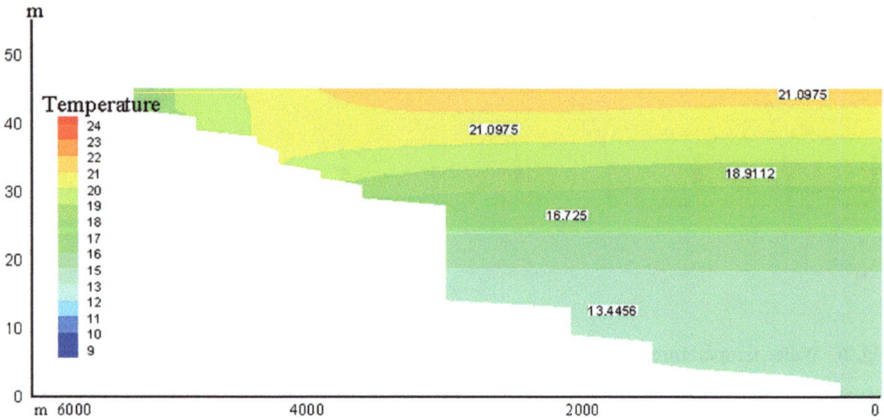

Fig. 9 Isotherm diagram at 08:00 a.m., July 15

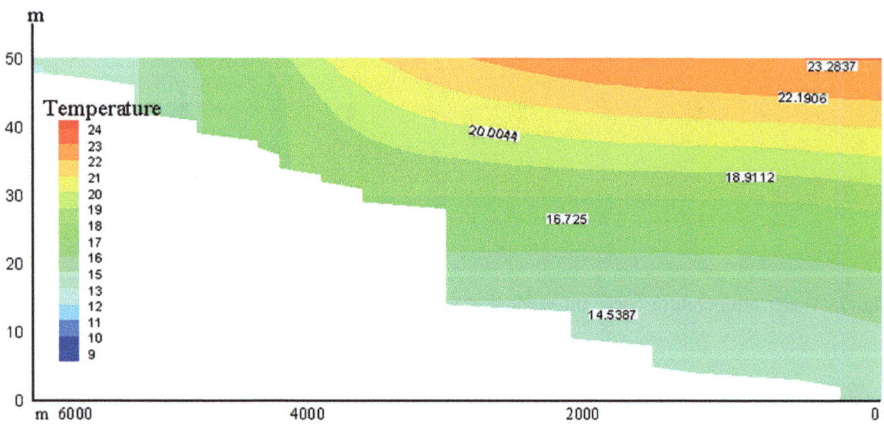

Fig. 10 Isotherm diagram at 08:00 a.m., August 15

temperature change while water flows through the power station, the discharged water temperature is considered to be the same as the intake temperature.

Table 1 presents some of the data, including air temperature, discharge water temperature, surface water temperature and the deltas for the period May 3 until June 15. The average discharged water temperature was 13.7 °C, which differed from the reservoir surface water temperature by between 0 and 7.6 °C, while the discharged water temperature differed from the reservoir inflow water by between −0.1 and 5.3 °C.

Using the comparison of discharged to surface water temperature, the influence on fish downstream is evaluated. The investigation showed that the fishes in Panjiakou Reservoir and downstream are primarily carp, grass carp, crucian carp, silver carp, loach, and Pseudorasbora parva. Fish spawning and embryo

Table 1 Contrast of water temperature (°C)

Time	Air temperature	Discharge water temperature	Surface water temperature	ΔT_1	Inflow water temperature	ΔT_2
May 3	18.7	13.2	17.2	4.0	13.6	0.4
May 7	20.7	14.1	15.7	1.7	14.8	0.8
May 11	16.1	12.1	13.9	1.9	13.8	1.7
May 15	22.1	12.2	15.6	3.4	14.6	2.4
May 19	23.4	12.8	16.2	3.4	17.8	5.1
May 23	21.7	13.1	18.7	5.6	17.2	4.1
May 25	25.0	13.3	19.6	6.3	17.4	4.2
May 27	21.2	13.4	18.6	5.2	14.4	1.0
May 31	18.4	13.7	18.4	4.7	19.0	5.3
June 4	21.9	14.5	18.3	3.8	18.0	3.5
June 7	24.3	15.0	20.0	5.0	19.2	4.2
June 10	19.6	15.6	20.2	4.6	16.4	0.8
June 15	25.7	15.8	21.2	5.5	17.8	2.0

Table 2 Suitable water temperature to fish downstream

Fish	Spawning time	Minimum temperature for spawning (°C)	Optimum temperature for embryonic development (°C)
Grass carp	May–July	18	20
Silver carp	May–June	18	22–26
Pseudorasbora parva	April–June	20	20–25
Carp	April–June	14	25
Crucian carp	March–September	14	22–25
Loach	May–July	18	20

development occur primarily in spring, coinciding with the reservoir discharge period. The spawning time, minimum spawning temperature and optimum development temperature for embryo are listed in Table 2. The minimum spawning temperature for carp and crucian carp is 14 °C. Grass carp, silver carp, loach, and Pseudorasbora parva have a higher temperature requirement (>18 °C) for spawning. The optimum temperature for fish embryonic development is above 20 °C.

The discharge water temperature from Panjiakou is 11–13 °C in May, rises to 15–16 °C in the first half of June, and reaches 18 °C in the second half of June. Compared with the surface water, which would be above 14 °C in May and be stable at 18 °C in late May, the discharge water has a temperature rise delay. For all the fish listed, the reservoir discharge water would delay their spawning time by up to one month.

The inflow (tributary) water temperature will reach 14 °C in the second half of May, and run up to 18 °C in June. To carp and crucian carp, the reservoir discharge water compared with tributary would delay the spawning time up to half a month. To the other fishes, the delay is not significant.

5 Conclusions

(1) The water temperature of the Panjiakou Reservoir between May and October 2010 was simulated using a 2D simulation model. The results indicated that the Panjiakou Reservoir had vertical temperature stratification from May to October. Over this period the surface water temperature varied between 9 and 24 °C, while the bottom water temperature varied between 6 and 13 °C. The vertical stratification was most prominent during July and August.
(2) The water temperature declined because of Panjiakou Reservoir discharge. The temperature of reservoir discharge water is 0–7.6 °C (4 °C average) lower than the surface water. The difference between reservoir discharge and inflow water is −0.1–5.3 °C (2.3 °C average).
(3) The discharge water in spring from the Panjiakou Reservoir could have an impact on spawning of some fish. For crucian carp, carp, and other low-temperature spawning fish, the impact of discharged low-temperature water is obvious; the spawning time can be postponed by 0.5–1 month. The impact on fish spawning in higher temperature water is less, typically the Pseudorasbora parva and loach.

Acknowledgements The researchers would like to extend thanks to the Chinese National Natural Science Foundation (No. 51522907, No. 51279208) and National Social Science Foundation Youth Project (NO.13CGL091). The study was also supported by the Research Fund of the State Key Laboratory of Simulation and Regulation of Water Cycle in River Basin, China Institute of Water Resources and Hydropower Research (No. 2016ZY02).

References

1. Sinokrot, B.A., Stefan, H.G., McCormick, J.H., et al.: Modeling of climate change effects on stream temperatures and fish habitats below Dams and near groundwater inputs. Clim. Change **30**, 181–200 (1995). https://doi.org/10.1007/BF01091841
2. Stefan, H.G., Hondzo, M., Sinokrot, B.A., Fang, X.: A Methodology to Estimate Global Climate Change Impacts on Lake and Stream Environmental Conditions and Fishery Resources with Application to Minnesota. University of Minnesota, St. Anthony Falls Hydraulic Laboratory 1992, R No. 323
3. Mohseni, O., Stefan, H.G., Eaton, J.G.: Global Warming and Potential Changes in Fish Habitat in U.S. Streams 389–398. https://doi.org/10.1023/A:1024847723344
4. Congqing Academy of Environment Science: Environmental Impact Report of Chongqing Key Water Source Project in Recent Planning, vol. 12 (2010) (In Chinese)

5. Yunnan Institute of Water and Hydropower Engineering Investigation Design and Research.: Environmental Impact Report of the Southwest Key Water Source Project in Recent Planning, vol. 11 (2010) (In Chinese)
6. Deng, Y., Li, J., Li, R.: Impact of Reservoir operation on water temperature downstream of Xiluodu power station. J. Sichuan Univ. (Engineering Science Edition) **38**(5), 65–69 (2006) (In Chinese). https://doi.org/10.3969/j.issn.1009-3087.2006.05.010
7. Chang, L., Zhong, X., Zhang L.: Analysis and prediction of Guangzhao power station water temperature and stratified water taking measures. Des. Hydroelectr. Power Stat. **23**(3), 31–32 (2007) (In Chinese). https://doi.org/10.3969/j.issn.1003-9805.2007.03.007
8. Chen, X.H., Liu, M.N., Lin, Y.S.: Study on two-dimensional water quality distribution in reservoirs. J. Hydraul. Eng. **4**, 9–16 (1997) (In Chinese). https://doi.org/10.3321/j.issn:0559-9350.1997.04.002
9. Gao, Z.X., Zhang, D.: Numerical Simulation of Reservoir Water Environment, vol. 12. Seismological Publishing House, Beijing (2005)
10. Mohseni, O., Stefan, H.G., Erickson, T.R.: A nonlinear regression model for weekly stream temperatures. Water Resour. Res. **34**, 2685–2692 (1998). https://doi.org/10.1029/98WR01877

Calculation of Early Warning Index of Mountain Torrent Disaster in Yushe County

Xuyang Yang, Dandan Yan and Lingling Sun

Abstract According to the investigation of the disaster of mountain torrents in Yushe County, the disaster water level of the flood disaster prevention objective is determined with regard to the disaster situation of the village and the historical flood investigation of the small watershed. According to the "Shanxi Hydrological Calculation Manual" analysis of a design storm, combined with the actual situation in Shanxi, analysis determined the warning period and soil moisture content. Based on the hypothesis of rainstorm and flood of the same frequency, the same frequency anti-push method is used to reverse the disaster flow from the designed storm flood. The designed rainfall is the early warning rainfall in the early warning period, that is, the small rainfall warning index (immediate transfer of indicators). Through the analysis and calculation, the rainfall warning index of 0.5, 1 and 1.5 h of Yuhong County flood disaster prevention objective is determined, and the rationality of an early warning index is analyzed. The results show that the early warning index of the mountain flood disaster in Yuqi County is reasonable and reliable.

Keywords Mountain torrent disaster · Critical rainfall · Rainfall warning index Yushe County

1 Introduction

A mountain torrents disaster is caused by floods and flash floods induced debris flow and landslides in the hilly area, which cause loss of the national economy and people's lives and property [1]. The formation of mountain torrents is the result of

X. Yang (✉) · D. Yan · L. Sun
College of Water Conservancy and Hydroelectric,
Hebei University of Engineering, Handan 056038, China
e-mail: 799686644@qq.com

D. Yan
e-mail: 806635967@qq.com

L. Sun
e-mail: 455766897@qq.com

W. Dong et al. (eds.), *Sustainable Development of Water Resources and Hydraulic Engineering in China*, Environmental Earth Sciences,
https://doi.org/10.1007/978-3-319-61630-8_29

the combination of various factors such as climate, topography, rainfall and underlying surface conditions [2]. A flash flood warning is an important part of the prevention of system flood disasters, which is an effective means of reducing casualties and property losses. The mountain flood disaster warning is based on weather, hydrological and other forecast information to predict the occurrence of flash floods that occur and to release the emergency instructions or signals. In the method of judging the critical flow or water level method and critical rainfall method, the critical rainfall method is the most widely used one at home and abroad [3, 4]. The concept of critical rainfall is widely found in the study of generalized mountain flood disaster prediction [4–7], including landslide and debris flow, which is the key index of the forecast of mountain flood disaster and directly affects the rate of omission and empty reporting rate, which is the center of the work of mountain flood disaster prevention and control work.

The warning index of mountain flood disaster plays a key role in the operability of the flood protection plan and determines whether sufficient time is required when flash floods or debris flows occur. The general situation is based on historical rainfall and flash flood disaster data, terrain, topography, vegetation, and soil type to determine the revised and improved practical application. For the more abundant areas, it is usually measured by the actual rainfall statistics, storm critical curve method [8], water level or flow reversal method [7, 9] and comprehensive analysis and other calculations. For less data or no data areas, interpolation [5, 6], comparison [5, 6], disaster and rainfall frequency analyses were used [7, 10, 11]. Due to the scarcity of rain stations in Yushe County, many areas are monitoring the blank area, and the construction time is relatively late. The observation series is short and the observation means is backward, which is because of the severe shortage of rainstorms. In this paper, the critical rainfall is calculated by using the watershed model method.

2 Overview of Mountain Torrent Disaster in Yushe County

A mountain flood disaster is the highest frequency, the greatest harm to natural disasters in Yushe County, which are affected by the climate and the underlying surface. The main source of floods in Yushe County is the formation of rainfall, the convergence of the mountain slope steep, more rivers, ferocious flash floods convergence. In a few hours or less it can form a torrential flood, resulting in flash floods. Yushe County flood season is in early May to the end of September; the main flood season is concentrated in June–August. This time is also the Yushe County torrential disasters of the multiple period, the regional heavy rain intensity, steep mountains, quick convergence, flood steep rise and fall, with obvious mountain river characteristics. In the county since the founding of the disaster occurred in the year 1963, 1976, 1989, 1991, 1992, 1996, 1998, 2004, 2008, 2009,

2012, 2013, torrential disasters directly caused major casualties and property damage.

According to the investigation, there are 49 mountain flood disaster prevention objects that need to be heavily guarded by the threat of mountain torrent disaster in Yushe County to carry out analysis of early warning indicators. This paper lists the calculation of analyzed results as a typical example of early warning index analysis for Luo Xiu Village, Yushe County and Simma township.

3 Research Methods

3.1 Calculating Designed Flood

Calculation of the Design of Rainstorm. The statistical parameters of villages threatened by mountain torrential floods are read on the isoline map of statistical parameters of point rainstorm in Shanxi Hydrology Manual. The design point rainfall of various diachronic periods is calculated according to the parameters, and the design area rainfall is calculated according to the point surface reduction coefficient. Time history distribution is according to the design rain pattern. It mainly includes three steps: design point rainfall, design area rainfall and design rainstorm time distribution. According to the design rain pattern and the design rainfall results of the time period, the time history distribution of the time period design rainfall of each frequency is carried out by using the method of precipitation sequence of the time period.

Calculating Designed Flood. Design floods were the design of heavy rain by the production flow and the convergence calculation derived. In the calculation of the runoff yield of the watershed model, the design of net rainfall depth is calculated by using the hyperbolic tangent model. The net rain process is calculated by the variable loss rate deduction method. The integrated instantaneous unit hydrograph is used to calculate the confluence of the watershed model. In the design flood analysis, the river cross section along the river village is taken as the control section, and the calculation and analysis of the different frequency design floods are carried out. Each frequency designed flood result of Luo Xiu village control section is shown in Table 1.

Table 1 Luo Xiu village control section of the frequency design of the flood results

Flood elements	Reproduced flood element values				
	100 years	50 years	20 years	10 years	5 years
Peak flow (m^3/s)	52.07	43.6	32.34	23.54	15.64
Flood volume (million m^3)	28	23	16	12	8
Flood hydrograph (H)	1.5	1.5	1.5	1.5	1
Flood duration (h)	9	8.5	8.5	8.5	8
Peak water level (m)	–	–	–	–	–

3.2 Calculation of Rainfall Early Warning Index

Determination of Disaster Water Level and Control Section. According to the actual situation of the village section along the river, the hydraulic method is used to calculate the water surface line. The river section of the village along the river in Yushe County did not investigate the highest flood level in history, so the scope of design flood inundation once in 100 years is now defined as the dangerous area. It is determined that there are 49 dangerous river villages in Yushe county from the range of dangerous areas. The design flood water surface line is designed for 50 years, 20 years, 10 years and 5 years, respectively. According to the results of design flood surface line of each frequency, combined with the topography of villages along the river and the elevation of residential households, the flood inundation range of each frequency is drawn.

A comparison of the elevation of the residents on the side of the river and the water surface level of the river reach were used to determine flood level, and the specific method is:

(1) According to the frequency design flood, the flooding range was used to determine the minimum design flooding period that can threaten households.
(2) The residential households on the side of the river submerged by the design flood during the recurrence period are projected onto the vertical section, and the comparison between the elevation of the residential household and the design flood surface line during the recurrence period is drawn. The comparison between the elevation of residents and the line of water surface in Luo Xiu Village is shown in Fig. 1. Residential households below the water line represent the submerged.
(3) The residential household height farthest from the water surface line determined in the last step is the flood level, and the nearest cross section from the household is the control section.

The water level-flow relationship of the control section is derived from the analysis of the frequency of the water surface line. The results are shown in Fig. 2.

According to the above water level-flow curve (Fig. 2), it can be seen that Luo Xiu village control sections of the water level flow relations are gentle. There are no bifurcations and tributaries of the river along the river control section, and they are relatively straightforward, with good representation; therefore, there is no need to correct the relationship diagram.

According to the above-mentioned water level-flow curve, the peak flow rate corresponding to the flood water level is deduced; at the same time, it refers to the flow frequency curve, and the disaster water level of the flood flow corresponds to the peak flow, which is determined by the interpolation method. The flood return period is then obtained as the river flood control capacity. The table of the flood level and its corresponding flood frequency in Yushi County is given in Table 2.

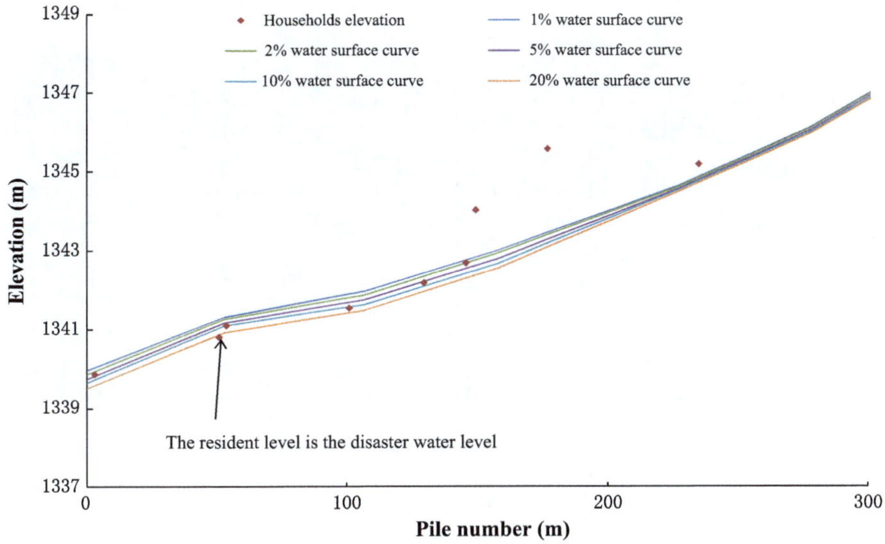

Fig. 1 Luo Xiu village residents' elevation and water line comparison diagram

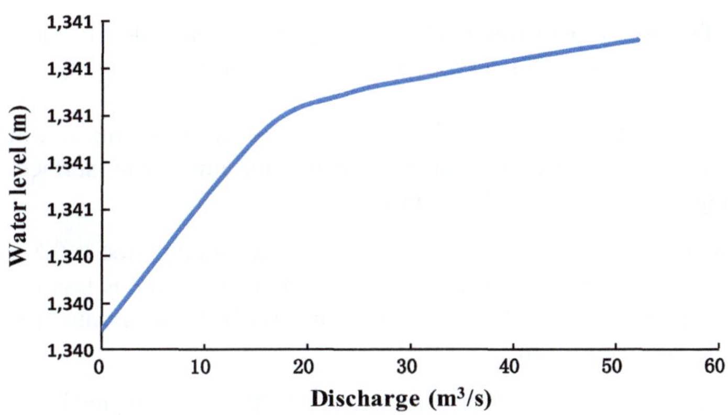

Fig. 2 Luo Xiu village control section water level-discharge curve

Table 2 Achievements of flood water level and corresponding flood frequency in Yushe County

Name of administrative division	Administrative division code	Disaster water level (m)	Peak flow (m3/s)	Frequency (%)
Luo Xiu village	140721202226100	1340.36	5.3	>20.0

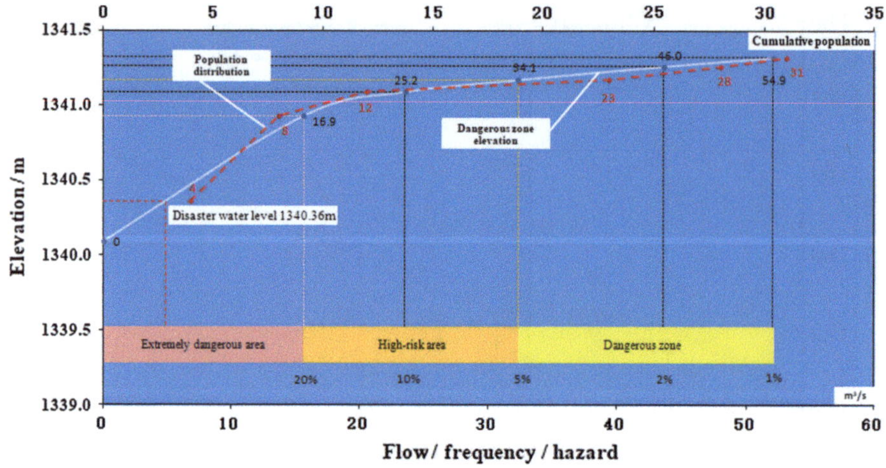

Fig. 3 Luo Xiu village flood control status evaluation chart

According to the water level flow curve, as well as the disaster water level, the frequency of the design of the flood level of population and household statistics for Luo Xiu village, the flood control status evaluation is shown in Fig. 3.

Warning Period to Determine. The warning period is related to the confluence time of the river basin and is determined according to the following principles:

(1) According to Yushe County characteristics of a rainstorm, the size of the basin area, the average ratio, the shape factor and the underlying surface, the basic warning period is 0.5, 1, 2, 3 and 6 h;

(2) If the confluence time is not less than 6 h, the warning period is 0.5, 1, 2, 3 and 6 h as well as the confluence time. If the confluence time is less than 6 h, the warning period is set as the confluence time and the basic warning period is less than the confluence time.

Watershed soil water content: the use of "hydrological manual" in the basin before the water holding capacity B_0 is an indirect indicator of soil moisture or soil moisture. B_0 values of 0, 0.3 and 0.6, respectively, on behalf of the soil moisture is more dry, generally and more wet in three cases.

Calculation of Critical Rainfall. After determining the disaster water level, the warning period and the production and confluence analysis method, it is possible to calculate the water-holding capacity of different watersheds (B_0) for each typical period of the critical area of the critical rainfall. The specific calculation steps are as follows:

(1) A maximum value is assumed in the second hours to maximum value in the sixth hours of the total rainfall initial value H. According to the design

Table 3 Range of rainstorm parameters

Rainstorm parameters	Ranges	Precision
S_p	P_2–100	0.1
N_s	0.01–1	0.01
λ	0.001–0.12	0.001

hyetograph, respectively, the maximum value in the second hours to maximum value in sixth hours the corresponding rainfall is P2′–P6′.

(2) Calculation of the rainstorm parameters: The total rainfall value H1–H6 and rainfall value P2–P6 of the maximum value in the second hours to maximum value in sixth hours of different rainstorm parameters are calculated from the formulas (1) and (2). According to the range of rainstorm parameters in Table 3, it is possible to obtain multiple sets of P2–P6, and each group P2–P6 is compared with P2′–P6′. The parameter of P2–P6 with the smallest sum of squares is the required. The rainstorm parameters are as follows:

$$H_p(t) = \begin{cases} S_p \cdot t^{1-n}, \lambda \neq 0 & 0 \leq \lambda < 0.12 \\ S_p \cdot t^{1-n_s}, \lambda = 0 \end{cases} \tag{1}$$

$$n = n_S \frac{t^{\lambda} - 1}{\lambda \ln t} \tag{2}$$

In the formula, n, n_S, respectively, the slope of the curve between rainstorm duration and rainstorm intensity and the slope at t = 1 h in double logarithmic coordinate system; s_p for the design of rain force, that is, 1 h design rainfall, mm/h; t for the storm lasted, h; λ for empirical parameters.

(3) For the value of the rainstorm parameter calculated from (2), the rainfall value of the maximum value in the second hours to maximum value in the sixth hours can be calculated by the Eqs. (1) and (2). According to the design hyetograph, a typical period of time per hour of rainfall Hp1, Hp2, …, Hp6 must be obtained.

(4) The hyperbolic tangent flow model and the unit line basin convergence model were used to calculate the peak flow rate Qm formed by the rainfall in the typical period.

(5) If $|Q_m - Q| > 1$ m^3/s, then with dichotomy to re-assume H, where Q is the flood level corresponding to the peak flow.

(6) Repeat steps (2)–(5) until the $|Q_m - Q|$ is less than 1 m/s. the total amount of rainfall in each hour in a typical period is the critical rainfall.

According to the above calculation steps, the dynamic critical rainfall of 49 dangerous villages along the river village is obtained, and the dynamic critical rainfall results of Luo Xiu village are shown in Table 4. The warning of critical

Table 4 Results of dynamic critical rainfall in Yushe County

Name of administrative division	Administrative division code	B_0	Time period (h)	Rainfall (mm)
Luo Xiu village	140721202226100	0	0.5	19
			1 (τ)	27
		0.3	0.5	17
			1 (τ)	22
		0.6	0.5	14
			1 (τ)	18

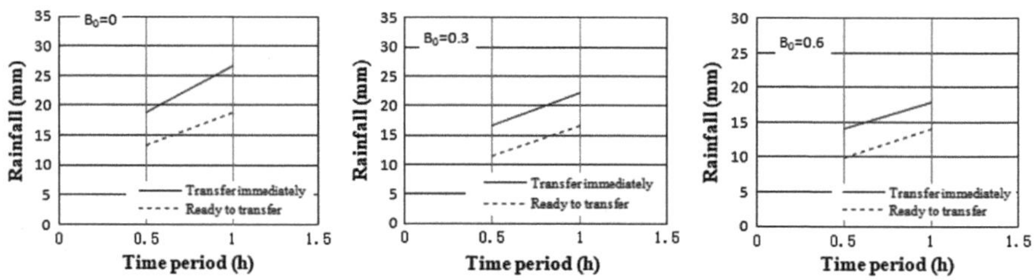

Fig. 4 Luo Xiu village warning rainfall critical curve

rainfall is drawn from the dynamic critical rainfall. The warning of critical rainfall curve is shown in Fig. 4.

Estimation of Rainfall Early Warning Indicators.

(1) Immediate transfer of indicators

Since the critical rainfall is calculated from the flood of the corresponding disaster water level, in the numerical form the critical rainfall is considered to be an immediate transfer of indicators.

(2) Prepare transfer indicators

When the warning period is 0.5 h, prepare the transfer index = immediate transfer index × 0.7.

When the warning period is 1, 2, 3 and 6 h and the confluence time, the immediate transfer index of the first 0.5 h is the preparation index of the warning period.

Calculation results can be seen from Table 5.

Analysis of Rationality of Rainfall Early Warning Index. Because of the limitations of the data conditions with only three years of data of Yushe County village along the river floods, the data series is shorter, and should not be used as a basis for comparison of design flood results. However, the field investigation of two historical floods in 2006, Shecheng town of Xiyadi river section of the catchment area

Table 5 Yushe County early warning indicators results table

Name of administrative division	Categories	B_0	Time period (h)	Early warning indicator (mm)		Rainfall (mm)
				Prepare for transport	Immediate transfer	Water level (m)
Luo Xiu cun	Rainfall amount	0	0.5	13	19	18.9
			1 (τ)	19	27	26.7
		0.3	0.5	12	17	16.5
			1 (τ)	17	22	22.3

is 51.7 km^2 with peak flow of 35.6 m^3/s, the watershed model method was used to calculate the design of floods in 100 years. The catching area of Shawang village in the town is 23.16 km^2 and the peak flow rate is 44.8 m^3/s in 5 years. Through the area ratio method, it can be seen that with the historical flood marks close; and the recurrence period is 3 years. The calculation result is reasonable. In 1970, Qicheng town of Shizhandao river section of the catchment area is 702 km^2, the peak flow is 1190 m^3/s, the catchment area of Wangjing Village in Qicheng town is 55.25 km^2, and the peak flow rate is 137.15 m^3/s in 5 years. Through the area ratio method, it can be seen that with the historical flood marks close and the recurrence period of 4 years, the calculation results are reasonable.

Based on the analysis, the calculation results of Shawang and Wang Jing Village are reasonable. The data and methods used in the calculation of the torrential floods in the villages threatened by the torrential floods, which are derived from the Handbook of Hydrology in Shanxi Province and Detailed verification, the design of the storm flood calculation results is reasonable.

4 Conclusion

In view of the complexity of the mountain channel, such as more wading buildings and serious river siltation, part of the river is occupied by man-made buildings, channel irregularity, bank slope low easy floodplain. In addition, there are factors such as measurement error, soil moisture content, roughness, specific drop error, design error, etc. The early warning index of a mountain flood disaster in a small watershed of Yushe County is not completely accurate, but through the analysis the results are more reasonable. According to the actual situation of Yushe County, the following suggestions are put forward: (1) The early warning index of a mountain flood disaster in Yushuan County can be placed on the early warning platform after the accumulation of hydrological data and rainfall data over a certain period of time. In the application process, there must be step-by-step optimization, in order to more accurately guide the mountain flood disaster prevention. (2) For the objects of defense in the same river basin, in order to facilitate management and application, if the current flood control capacity is basically the same, the same early warning index may be used; for a river basin, the villages along the upper and lower rivers must be linked up and down, and the information should be shared which is common defense.

References

1. Tan, C., Xu, C., Liu, H.: Mountain torrent disaster prevention benefit evaluation index and its calculation method. J. Water Conservancy Econ. **2007**(1), 124 (2007). https://doi.org/10.3969/j.issn.1003-9511.2007.01.001. (in Chinese)
2. Zhao, R., Wang, M., Lu, X.: Research on determination method for rainfall warning indexes of torrential flood disaster. J. Water Resour. Power **29**(9), 49253 (2011). https://doi.org/10.3969/j.issn.1000-7709.2011.09.014. (in Chinese)
3. Hapuarachchi, H., Wang, Q.J., Paganto, T.C.: A review of advances in flash flood forecasting. J. Hydrol. Process. **25**(18), 2771–2784 (2011). https://doi.org/10.1002/hyp.8040
4. Li, C., Sun, D.: Determination of flood warning index for mountain flood. J. China Water Resour. **9**, 54–56 (2012). https://doi.org/10.3969/j.issn.1000-1123.2012.09.015. (in Chinese)
5. Cheng, W.: A review of rainfall thresholds for triggering flash floods. J. Adv. Water Sci. **24** (6), 128 (2013). DOI: 10.14042j.cnki.32.1309.2013.06.012. (in Chinese)
6. Chen, G., Yuan, Y.: Research on critical precipitation amount computation method of mountain torrential flood disaster. J. Yangtze River **36**(12), 40243 (2006). https://doi.org/10.3969/j.issn.1001-4179.2005.12.016. (in Chinese)
7. Ye, Y., Wang, Z., Fan, B.: An analysis method for ascertain critical rainfall of mountain flood disaster of small water shed in ZheJiang Province. J. China Hydrol. **28**(1), 56258 (2008). https://doi.org/10.3969/j.issn.1000-0852.2008.01.013. (in Chinese)
8. Jiang, J., Shao, L.: Standard of mountain flood warning based on the precipitation observation data. J. Hydraul. Eng. **04**, 458–463 (2010). (in Chinese)
9. Chen, Z., Huang, G., Cheng, G.: Research on the calculation methods for critical rainfall of mountain torrent disasters of small watershed. J. China Rural Water Hydropower **06**, 82–85 (2014). https://doi.org/10.3969/j.issn.1007-2284.2014.06.020. (in Chinese)
10. Norbiato, D., Borga, M., Dinale, R.: Flash flood warning in ungauged basins by use of the flash flood guidance and model based runoff thresholds. J. Meteorol. Appl. **16**(1), 65–67 (2009). https://doi.org/10.1002/met.126
11. Mogil, H.M., Monro, J.C., Groper, H.S.: NWS's flash flood warning and disaster preparedness programs. J. Bull. Am. Meteorol. Soc. **59**(6), 690–696 (1978). https://doi.org/10.1175/1520-0477(1978)059<0690:nffwad>2.0.co;2

Impacts of Climate Variability and Human Activities on Runoff: A Case Study in the Jinghe River Basin

Qiu-Bo Long

Abstract Assessing the impacts of climate variability and human activities on streamflow is crucial to regional water resource management. In recent years, significant decreases in streamflow have emerged in the Yellow River basin. The Jinghe River Basin (JRB), a tributary in the middle reaches of the Yellow River, has experienced dramatic land use changes over the past decades. At the same time, the climate has also shown a significant change. The annual streamflow exhibits an apparent decreasing trend in the JRB. In this study, the non-parametric Mann-Kendall test is employed to detect the trend and breakpoint. The effects of climate variability and human activities are assessed quantitatively using the climate elasticity method and the hydrological model. The results showed that annual streamflow in the JRB decreased significantly ($P<0.05$) from 1956 to 2012 and that an abrupt change occurred in 1997. The annual precipitation presents a slightly decreasing trend, whereas the annual mean temperature and potential evapotranspiration show an upward trend. The results of quantitative assessment indicated that climate variability was responsible for 44.0% of the decrease in streamflow, whereas the hydrological modeling method indicated a value of 48.8%. The annual amount of water diversion from the Jinghe River increased, accounting for 20.5% of the decrease in streamflow from 1997 to 2010. Moreover, the land use and land cover change contributed to 30.7–35.5% of the decrease in streamflow. All human activities contributed to 51.2–56.0% of the decrease in streamflow. Therefore, the effect of human activities played a dominant role in the decrease in streamflow in the JRB.

Keywords Streamflow · Climate change · Human activity · Climate elasticity Hydrological model

---

Q.-B. Long (✉)
Institute of Geographic Sciences and Natural Resources Research,
Chinese Academy of Sciences, Beijing 100101, China
e-mail: longqiubo17@126.com

Q.-B. Long
Hunan Hydro and Power Design Institute, Changsha 416000, China

W. Dong et al. (eds.), *Sustainable Development of Water Resources and Hydraulic Engineering in China*, Environmental Earth Sciences,
https://doi.org/10.1007/978-3-319-61630-8_30

1 Introduction

The variance of runoff can be attributed to the joint effects of climate change and human activities in a river basin. Climate variation is expected to extensively alter hydrological processes, primarily by changing the precipitation and actual evapotranspiration levels at the watershed scale [1]. Human activities, such as land use practices, soil and water conservation, dam construction and city expansion, change water infiltration, evapotranspiration and storage at the watershed scale result in significant hydrological alteration [2–4]. Quantitatively assessing the impacts of climate change and human activities on streamflow is vital for water resource management. Numerous studies have been conducted to separate the effects that climate change and human activities have on streamflow; however, the results present high regional differences [5–8]. For instance, Tang et al. (2008) [9] found that climate change played a dominate role in the decrease in streamflow in the upper reach of the Yellow River. Chen et al. (2012) [10] used a hydrological model to study streamflow variations in the Pearl River Basin and noted that the change points of monthly streamflow series are roughly in line with those of annual, winter and summer precipitation, indicating that changes in precipitation have obvious influences on runoff change. Bao et al. (2012) [11] analyzed the reasons for runoff changes in Haihe River Basin and the influence of human activities using hydrological model, and then proved that human activities were the main force driving the reduction of water resources. Zhang et al. (2013) [12] estimated the influence of human activities on runoff change in places where water conservation projects have been constructed. The effect of the Three Gorges Dam (TGD) on streamflow of the Yangtze River and Poyang Lake was investigated, and the results showed that the operation of the TGD changed the seasonal variations of Poyang Lake and Yangtze River; the strongest influence of TGD on the lower reaches of the Yangtze River occurred in October.

Many methods can be used to evaluate the response of streamflow to climate change and human activities. Of these methods, the climate elasticity method and the hydrological model are the two major methods widely used to separate the effects of climate change and human activities on runoff variation [1]. The climate elasticity method, which was first applied by Dunne and Milly in 2002 [13], is primarily applied to evaluate the sensitivity of annual runoff to precipitation and potential evapotranspiration. This method has been proven to be a simple but efficient approach for separating the impact of climate change from that of human activities on streamflow on a yearly time scale, particularly in mountain basins [14]. Compared with the climate elasticity method, the hydrological method generally requires higher input but can provide a more detailed description on the hydrological process. To achieve more reasonable assessment results, the two methods noted above are thus employed in this study to calculate the impacts of climatic and anthropogenic factors on runoff variation. Liu et al. (2013) [15] used the two methods in a sub-basin of the Yellow River and found that human activities contributed 58.4–61.1% to the decrease in annual runoff. Wang et al. (2012) [16] applied the hydrological model method, hydrological sensitivity analysis method and climate elasticity method to assess the response of streamflow to climate change and human activities in the Haihe River and obtained similar results. Zhang et al. (2004) [17] analyzed the reasons for the decrease in streamflow in the Fenhe River using a two-parameter monthly water balance model and climate elasticity method and concluded that human activities were the major driving factor for runoff decrease.

The Jinghe River is the largest tributary of the Wei River, which is a part of the Yellow River. Recently, a series of soil conservation practices have been implemented to reduce soil erosion [18–20]. Water consumption in the river basin has increased dramatically with the expanding population [21]. The climate also showed a significant change [16, 22, 23]. The annual streamflow significantly decreased in the JRB [2]. However, a systematic quantification of the effects of climate change and human activities on changes in runoff in the JRB has not yet been reported. Therefore, the main objectives of this study are to identify trends in such changes and the change points in annual runoff and to separate the effects of climatic variability from those of human activities. The paper is organized as follows: first, a brief description of the study area and data sources is provided in section two. The methods used are described in section three, followed by a presentation of the results, which include trend analysis, change-point detection and inter-comparison of the two models. Finally, discussions and conclusions are given.

2 Study Area and Data

2.1 Study Area

The JRB is located in the middle reach of the Yellow River and hasa basin of 45,421 km^2 (Fig. 1). The basin belongs to a typical continental climate between the semi-humid and semi-arid temperate zones. The average annual air temperature is nearly 8 °C, and the lowest temperature can reach −10 °C. The frost-free span is approximately 150 days long. The average annual precipitation is approximately 500 mm, of which 73% occurs during the wet season. Recently, human activities have become extensive due to rapid population growth and economic development. According to statistical data from the Ministry of Water Resources, the irrigation area in the basin in 2008 is approximately 9500 km^2, accounting for 20.9% of the total area of the basin. In addition, more than 130 reservoirs, with a total storage capacity of 1.67 billion m^3, have been built since 1949 [24]. These reservoirs lead to increasing evaporation due to the increase of water surface area and affect the inter-annual distribution of streamflow.

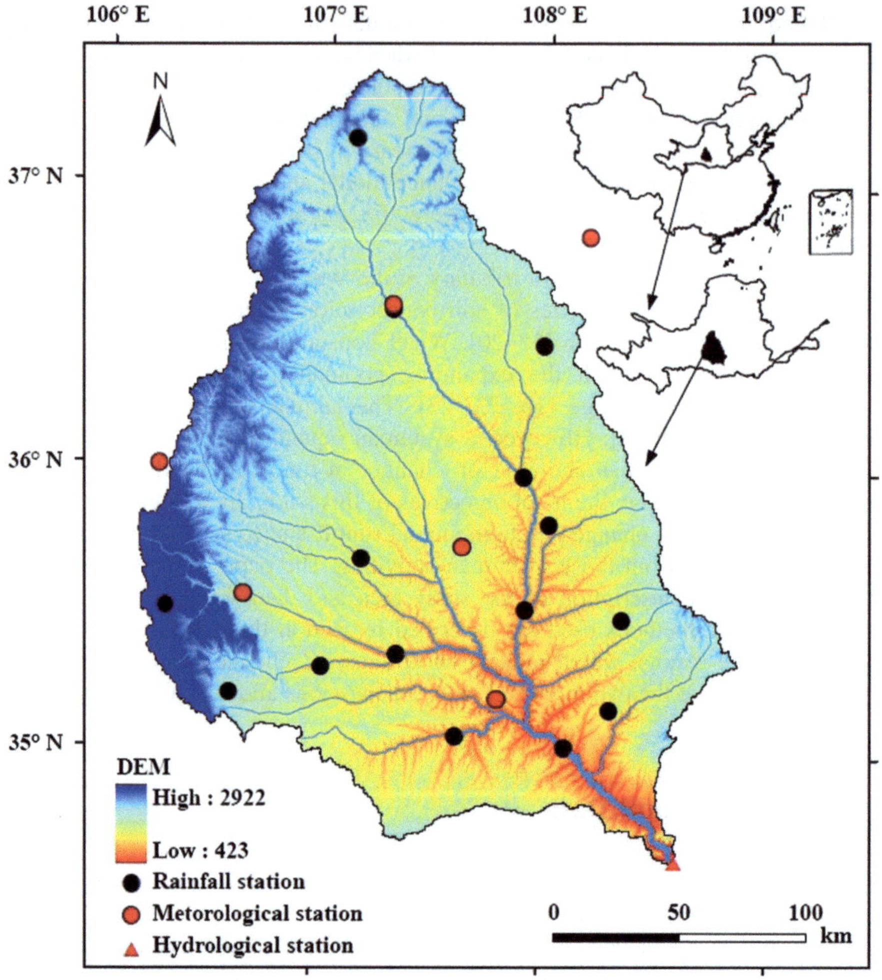

Fig. 1 Study area—Jinghe River basin

2.2 *Data*

The meteorological data and streamflow data were collected from the Yellow River Hydrological Bureau. The meteorological data contain precipitation, temperature, wind speed, vapor pressure and sunshine duration data. Potential evapotranspiration was calculated using the Penman-Monteith method. The spatial averages of the hydro-climatic variables were estimated using the Tiesen polygon. The Zhangjiashan station is situated at the outlet of the basin, where daily streamflow data from 1956 to 2012 were collected.

3 Methodology

3.1 Statistical Analysis Method

Penman-Montieth Method. Penman-Montieth formula is introduced to obtain the potential evapotranspiration in this study:

$$ET_0 = \frac{0.408\Delta(R_n - G) + \gamma \frac{900}{T+273} u_2(e_s - e_a)}{\Delta + \gamma(1 + 0.34u_2)} \quad (1)$$

where ET_0 is evapotranspiration (mm/d); T is average temperature (°C); Δis saturated vapor pressure curve slope (kPa/°C); γ is psychomotor constant (kPa/°C), which can be obtained by elevation calculation; R_n is net solar radiation (MJ m^{-2} d^{-1}); G is the soil heat flux (MJ m^{-2} d^{-1}); u_2 is wind speed around two meters above ground; e_s and e_a is saturated vapor pressure and water vapor pressure (kPa), respectively.

Trend Test. The rank-based Mann-Kendall test [25, 26] was used to identify trends in hydro-climatic series. The method, which is recommended by the World Meteorological Organization (WMO), is broadly used and has generally been adopted to estimate monotonic trends in hydrological and meteorological time series [27–29].

For the sequential time series x_1, x_2, …, x_n, m_i is the number of later terms in the series whose values exceed x_i. The MK rank statistic d_k is calculated as:

$$\mathrm{d}_k = \sum_{i=1}^{k} m_i \quad (2 \le k \le n) \quad (2)$$

The expected value $E(d_k)$ and variance $Var(d_k)$ could be estimated as follows:

$$\begin{cases} E(d_k) = \frac{k(k-1)}{4} \\ Var(d_k) = \frac{k(k-1)(2k+5)}{72} \end{cases} \quad (3)$$

The test statistic (UF_k) is calculated by the following formula:

$$UF_k = \frac{d_k - E(d_k)}{\sqrt{Var(S_k)}} \quad (1 \le k \le n) \quad (4)$$

The positive value of UF indicates an increasing trend, and negative value denotes a decreasing trend. Then according to the inverse time series: x_n, x_{n-1}, …, x_1, the same method is applied to get a series UB. The intersection point of UF and UB curves would be regarded as the breakpoint of the series [30].

The Climate Elasticity Method. Precipitation and evaporation are the dominant factors that determine natural runoff variability [31]. The streamflow variation [32] of a river watershed can be described as follows:

$$\Delta \mathrm{Q} = \Delta Q_H + \Delta Q_C \tag{5}$$

where $\Delta \mathrm{Q}$ is the streamflow change between the natural period and the impacted period and ΔQ_C and ΔQ_H are changes in streamflow due to climatic variation and human disturbances, respectively. The naturalized streamflow is the sum of the observed streamflow and the water direction:

$$Q_{Natural} = Q_{Obs} + Q_{Direct} \tag{6}$$

where $Q_{Natural}$ is naturalized runoff, Q_{Obs} is observed runoff and Q_{Direct} is water diversion.

The impact of human activities on streamflow in the Jinghe River (ΔQ_H) comprises the direct impact (water diversion, ΔQ_{Direct}) and indirect impact (change in land use and vegetation, $\Delta Q_{Indirect}$) and can be written as:

$$\Delta Q_H = \Delta Q_{Direct} + \Delta Q_{Indirect} \tag{7}$$

The impact of climate change on the streamflow (ΔQ_C) can be approximately calculated as follows [31, 32]:

$$\Delta Q_C = \Delta Q_P + \Delta Q_{ET_0} = \left(\frac{\varepsilon p \Delta P}{P} + \varepsilon ET_0 \Delta ET_0 / ET_0\right) Q_{Natural} \tag{8}$$

where ΔQ_P and ΔQ_{ET_0} are the contributions of the changes in precipitation and ET_0 to the change in streamflow, respectively. ΔP and ΔET_0 are the change in precipitation and ET_0 between the periods, respectively. εp and εET_0 are climate elasticity of streamflow due to precipitation and ET_0, respectively, which can be estimated as follows based on the Budyko hypothesis:

$$\varepsilon \mathrm{P} = 1 + \varphi F'(\varphi)/[1 - F(\varphi)] \text{ and } \varepsilon \mathrm{P} + \varepsilon \mathrm{E} \mathrm{T}_0 = 1 \tag{9}$$

where φ is the aridity index, which is the ratio of ET_0 and precipitation and calculated at a yearly timescale in this study. $F(\varphi)$ is the function of φ, and $F'(\varphi)$ is the derivative of $F(\varphi)$ with respect to φ. Based on the Budyko hypothesis, many forms of $F(\varphi)$ were proposed [33–35]. The calculations of $F(\varphi)$ and $F'(\varphi)$ are expressed as follows:

$$\begin{cases} F(\varphi) = (1 + \omega\varphi)/(1 + \omega\varphi + 1/\varphi) \\ F'(\varphi) = (\varphi^{-2} + 2\omega\varphi^{-1} + \omega^{-1})/(1 + \omega\varphi + 1/\varphi)^2 \end{cases} \tag{10}$$

where ω is the plant-available coefficient relating to vegetation type.

Hydrologic Model and Assessment Criteria. Model-based projections of the impacts of hydro-climate variability and non-stationarity on water resources underpin policy-relevant decision making of major national and international significance [36]. This study employed "abcd", a monthly water balance model, as the assessment model because it is comparable with other balance models and because each of its parameters has a physical interpretation [37]. The "abcd" model is a nonlinear watershed model that accepts monthly precipitation, air temperature and potential evapotranspiration as inputs and produces streamflow as an output [38]. This hydrologic model was originally introduced by Thomas in the 1980s and is applied using an annual interval. Some scholars found that a monthly "abcd" model compared favorably with several other monthly water balance models [37–39]. A detailed description of the model has been summarized by Alley (1984) [38] and Fernandez et al. (2000) [37].

The model has four parameters, a, b, c, and d, each of which is presumed to have some degree of physical interpretation. The parameter a ($0 < a \leq 1$) reflects the propensity of runoff to occur before the soil is fully saturated. Values of parameter a less than 1.0 result in runoff when $W_i < b$. The parameter b is an upper limit on the sum of evapotranspiration and soil moisture storage. The parameter c is related to the fraction of mean runoff that comes from groundwater. Finally, parameter d is the reciprocal of the groundwater residence time.

In this study, two criteria, Nash-Sutcliffe Efficiency (NSE) and Water Balance Error (WBE), are used to evaluate the performance of the "abcd" model. The model was calibrated by maximizing the NSE of monthly streamflow [7]. NSE is defined as follows:

$$\mathrm{NSE} = 1 - \frac{\sum_{i=1}^{n}\left(Q_{obs,i} - Q_{sim,i}\right)^2}{\sum_{i=1}^{n}\left(Q_{obs,i} - \overline{Q_{obs}}\right)^2} \tag{11}$$

The WBE in percentage is considered the linear inequality constraint that compels the total simulated streamflow to be within 5% of the total recorded streamflow [7]. WBE is calculated as follows:

$$\mathrm{WBE} = 100\left(\frac{\sum_{i=1}^{n} Q_{sim,i} - \sum_{i=1}^{n} Q_{obs,i}}{\sum_{i=1}^{n} Q_{obs,i}}\right) \tag{12}$$

where Q_{sim} and Q_{obs} are the simulated and observed streamflow, respectively. $\overline{Q_{obs}}$ is the arithmetic mean of the observed runoff.

4 Results

4.1 The Change in Annual Runoff, Precipitation and E_0

Historical trends of hydro-meteorology factors can help identify the effects of climate change on water resource systems. Figures 2 and 3 show the long-term variation of the observed annual streamflow, precipitation and potential evapotranspiration from 1956 to 2012. The results indicated that an abrupt change point in approximately 1997 was detected by the MK test. Therefore, the streamflow series was divided into two periods: "natural period (period I: 1956–1996)" and "impacted period (period II: 1997–2012)." In period I, the change in streamflow is primarily controlled by climatic factors, and the effects of human activities can be ignored. On the contrary, human activities play an important role in streamflow change in period II and must be considered. The results of the trend analysis show that the annual runoff presented a remarkable negative trend ($P<0.05$) by 6.05 mm per decade. The observed annual streamflow was 41.7 mm in period I, whereas it decreased to 19.3 mm in period II. The relative rate of change reached −53.7% (Table 1). Figure 3 shows the change in annual precipitation, the potential evapotranspiration, the temperature and the runoff coefficient. Compared with period I, the precipitation decreased by 40.67 mm in period II, whereas the temperature and E_0 increased by 1.54 °C and 59.7 mm, respectively (Table 1). Because of the impacts of human activities, there is a significant change in the relationship between precipitation and runoff. As observed in Fig. 3a, b, the runoff coefficient in period II is clearly lower than that observed in period I. This finding indicates that the same annual precipitation in period II generated less runoff than it did in period I.

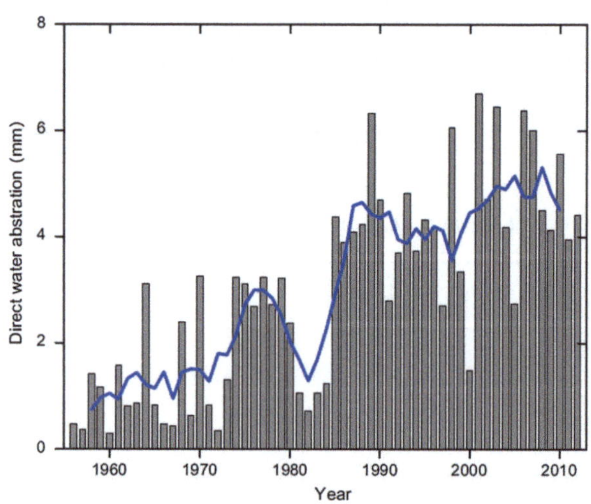

Fig. 2 Water diversion in the JRB from 1956 to 2012. The blue line represents 5-years moving average

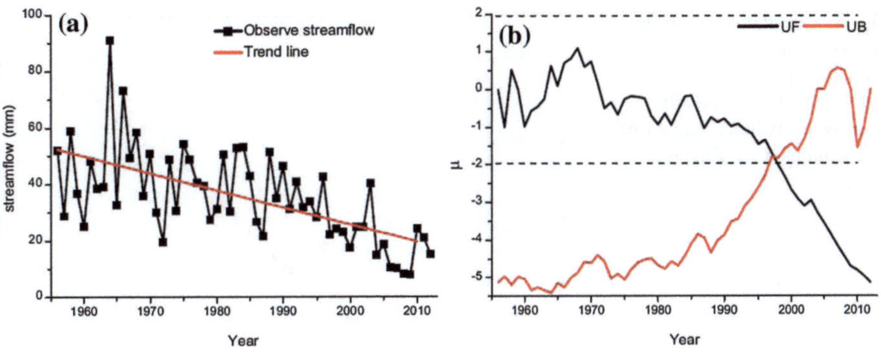

Fig. 3 **a** Variation of annual observed stream flow in the JRB. **b** The Mann-Kendall analysis of annual observed stream flow from 1956 to 2012

Table 1 Hydrological variables statistics during the two periods

Hydrological variables	Period I	Period II	Relative change	
			(mm)	(%)
Observed stream flow (mm)	41.7	19.3	−22.4	−53.7
Q_{Direct} (mm)	2.35	6.94	4.59	195
Precipitation (mm)	513.62	472.95	−40.67	−8.0
Temperature (°C)	8.43	9.97	1.54	18.3
E_0 (mm)	909.8	969.5	59.7	6.6

4.2 *Attribution of Change in Streamflow Using the Climate Elasticity Method*

Through parameter calibration, when the plant-available coefficient $\omega = 0.89$, the best agreement between the calculated Eq. (10) and observed (annual precipitation minus runoff) actual evapotranspiration occurred (Fig. 4). Following Eq. (9), the elasticity index of the annual streamflow to precipitation (εP) and ET_0 (εET_0) was 2.15 and −1.15, respectively. This result indicated that a 10% increase in precipitation would result in a 21.5% increase in streamflow, whereas a 10% increase in ET_0 would result in an 11.5% decrease instreamflow. The annual streamflow was more sensitive to the change in precipitation than the change in ET_0. The 40.67 mm decrease in precipitation led to a 7.58 mm decrease in streamflow, whereas the 59.7 mm increase in ET_0 resulted in a decrease in annual streamflow of 3.34 mm. The changes in precipitation and ET_0 together led to a decrease in streamflow by 10.92 mm in period II, accounting for 48.8% (10.92/22.40 mm) of the total observed decrease in annual streamflow. Correspondingly, human activities resulted in a decrease in annual streamflow of 11.48 mm, accounting for 51.2% of the decrease in streamflow. Among the various human activities, the direct impact

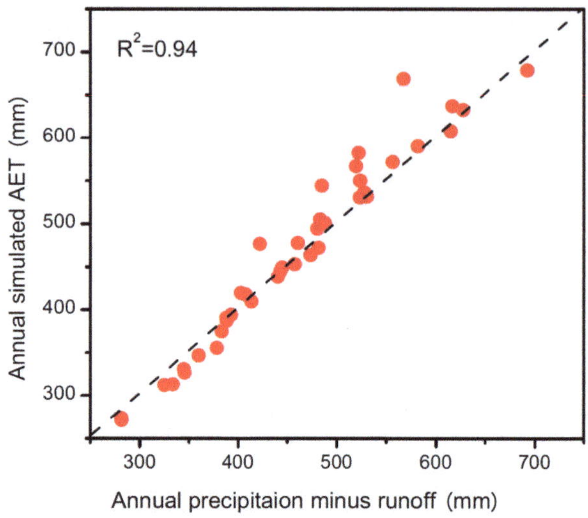

Fig. 4 Correlation relations between annual actual evapotranspiration (AET) calculated directly from water balance equation and simulated by Zhang's curve

(water diversion) was responsible for the 20.5% decrease in annual streamflow; the remainder of the decrease came from indirect impacts, such as land-use change, soil and water conservation practices and dam construction.

4.3 Attribution of Change in Streamflow Using the Hydrological Model

The "abcd" model was calibrated using the monthly series from 1956 to 1980 and validated using data from 1981 to 1996. Model performance during calibration and the validation periods can be found in Fig. 5 and Table 2. The NSE of the simulated monthly streamflow in the calibration and validation period was 0.86 and 0.80, respectively. The WBE was 2.41% and 3.92% in the two periods, respectively. This result indicates that the model can achieve good performance in simulating long-term streamflow. To estimate the impact of climatic variation on the streamflow, the model parameters obtained during the calibration period were used to simulate the natural streamflow in period II. The difference between the simulated and observed streamflow was the streamflow change caused by human activities. According to the results of hydrological modeling, the average annual streamflow at the Zhangjiashan station was 36.10 mm in period II under the natural scenario (Fig. 6), whereas the naturalized streamflow in period II was 26.24 (19.3 + 6.94) mm (Table 1). The difference between the simulated and naturalized streamflow was 9.86 mm, which was the impact of climatic variation on streamflow. Therefore, climatic variation accounted for 44.0% (9.86/22.4 mm) of the decrease in streamflow in period II, and human activities led to 56.0% of the decrease in annual runoff.

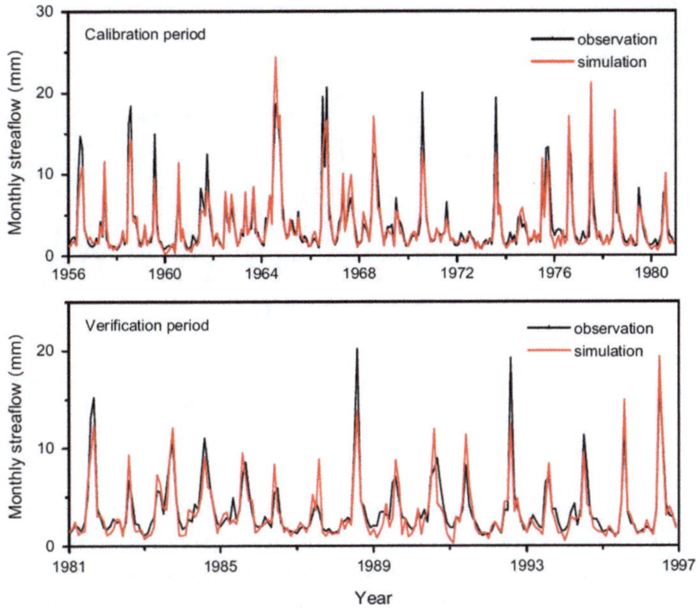

Fig. 5 Model performances in calibration and validation period

Table 2 The contributions of climatic variation and human activities to the stream flow decrease

Climate elasticity method	ε_P	ε_{ET}	ΔQ_C (mm)	ΔQ_H (mm)	ΔQ_C (%)	ΔQ_H (%)	
						ΔQ_{Direct}	$\Delta Q_{Indirect}$
	2.15	−1.15	−10.92	−11.48	48.8	20.5	30.7
Hydrological modeling	NSE	MAE	ΔQ_C (mm)	ΔQ_H (mm)	ΔQ_C (%)	ΔQ_H (%)	
						ΔQ_{Direct}	$\Delta Q_{Indirect}$
	0.86[a]	2.41%[a]	−9.85	−12.55	44.0	20.5	35.5
	0.80[b]	3.92%[b]					

[a,b]Data series from 1956 to 1980 and validated using data from 1981 to 1996
ε_P and ε_{ET} are climate elasticity of runoff to precipitation and potential evapotranspiration, respectively. ΔQ_C and ΔQ_C are the contribution of climatic variation and human activities, respectively. ΔQ_{Direct} and $\Delta Q_{Indirect}$ are the direct and indirect influence in Eq. (7)

5 Discussions

5.1 Uncertainties

This paper performs a quantitative assessment of the impacts of climate change and human activities on the decrease in runoff in the JRB using the climate elasticity method and a hydrological model. Some uncertainties are inevitable in the input data, model parameters and assessment methods. The distribution of precipitation in the basin is uneven. Thus, the precipitation data from 11 hydro-meteorological

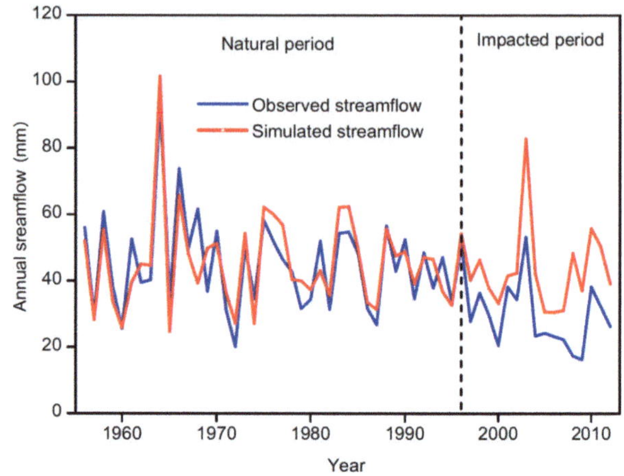

Fig. 6 Simulated and observed annual stream flow in "natural period" and "impacted period"

stations can hardly represent actual regional precipitation. In addition, due to the inhomogeneity in land surface and the "equifinality of parameter sets" (hydrological models of many different parameter sets are equally good at reproducing an output signal), the model parameters obtained via automatic optimization cannot represent the real model parameters in the basin. Thus, the uncertainty of model parameters also affects the assessment results. Moreover, uncertainties exist in the climate elasticity method. The impacts of precipitation and potential evapotranspiration are separated under the assumption that they are independent in Eq. (8). However, they do, in fact, impact each other [35]. Finally, there is also large uncertainty in the water diversion data. Any error of the water diversion data can affect the assessment results to a certain degree.

5.2 *Impact of Human Activities in the Jinghe River*

Water diversion is primarily used to meet the demands of domestic, industrial and agricultural water use. With economic development and the increasing population, water diversion in the JRB increased clearly from 1956 to 2012. The average annual amount of water diversion was 2.35 mm in period I, whereas it increased by 4.59 mm and reached 6.94 mm in period II. The increase in water diversion accounted for 20.5% of the total decrease in streamflow in period II (Table 2). Water diversion is thus an important driving factor and can lead to a decrease in streamflow at the outlet of the watershed.

In addition, the JRB is part of the Loess Plateau, which is characterized by thick and highly erodible loess, sparse vegetation, unevenly distributed rainfall, and a relatively high intensity of rainstorms [40]. To reduce soil erosion and maintain ecosystem health, massive soil and water conservation measures have been implemented since the 1970s. These conservation measures include building

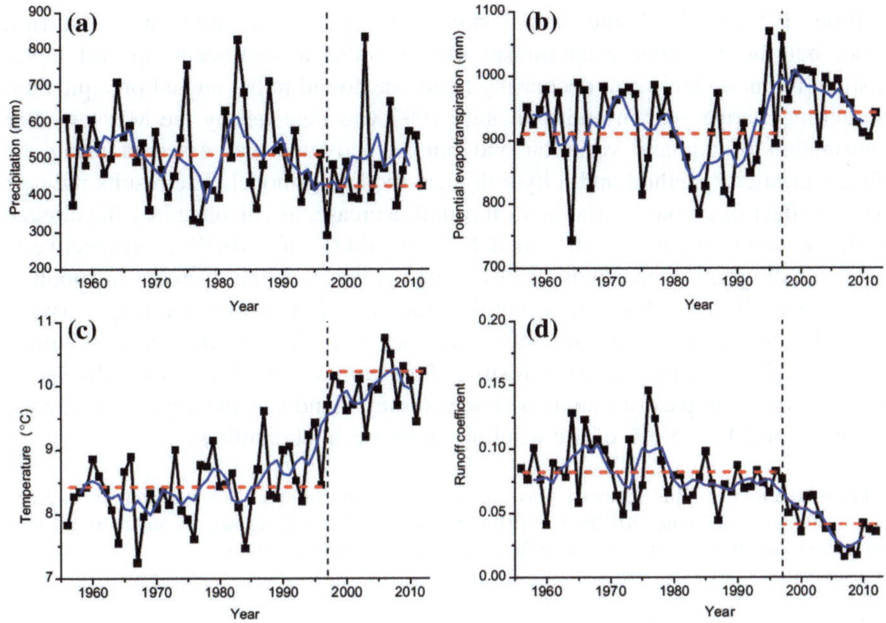

Fig. 7 Variation of annual precipitation (**a**), potential evapotranspiration (**b**), temperature (**c**) and runoff coefficient (**d**). The blue line shows the five-year moving average and the red horizontal dotted lines represent the averages of the corresponding period

terraces, planting trees, and constructing sediment-trappingdams. The areas of terraces, forest, and sediment-trapping dams in 1999 reached up to 1445, 6751, and 672 hm^2, respectively. Especially in the late 1990s, a national ecological restoration project named the Grain for Green Project was implemented over the Loess Plateau. As a result, there has been an increase in vegetation coverage and significant land use change in the JRB [40], leading to an increase in canopy interception, plant transpiration and soil moisture storage capacity. There will eventually be a decrease in direct runoff and an increase in actual evapotranspiration [41]. In this way, soil and water conservation measures have changed the underlying surface gradually and impacted the relationship between precipitation and streamflow (Fig. 7d).

6 Conclusions

In this paper, the trend in streamflow and driving factors for the JRB were investigated. The MK test was used to detect the trend and the abrupt change of climatic and hydrological variables. Two methods, the climatic elasticity method and the hydrological model, were used to quantitatively assess the impacts of climate change and human activities on streamflow.

From 1956 to 2012, the annual streamflow showed a significant downward trend, but the potential evapotranspiration showed a significant upward trend. A slight but non-significant decreasing trend was found in the annual precipitation. An abrupt change year in approximately 1997 was detected by the MK test. The contribution of climatic variation and human activities was assessed using the climate elasticity method and a hydrological modeling model. The results indicate that the effect of climate variability on runoff decrease, as calculated by the climate elasticity and hydrological model, are 44.0 to 48.8%, respectively. Correspondingly, the contribution of human activities on the decrease in runoff is 51.2–56.0%. Both methods indicated that human activities were the major driving factor for the decrease in streamflow in the JRB. Among the various human activities, 20.5% of the decrease in streamflow comes from direct water diversion, and the indirect impact of human activities through land use and vegetation change accounts for 30.7–35.5% of the resulting decrease in streamflow.

Acknowledgements This research was supported by the National Science and Technology Support Program of China (2012BAC09B05), the National Natural Sciences Foundation of China (90302009) and the Ministry of Water Resources of China (201101047).

References

1. Gardner, L.R.: Assessing the effect of climate change on mean annual runoff. J. Hydrol. **379** (3–4), 351–359 (2009). https://doi.org/10.1016/j.jhydrol.2009.10.021
2. Fu, G., Okoren, C., Yao, C., Chen, S.: Impact of climate change on runoff of Yellow River. Yellow River (11), 77 (2005)
3. Lan, Y.C., Zhao, G.H., Zhang, Y.N., Wen, J., Liu, J.Q., Hu, X.L.: Response of runoff in the source region of the Yellow River to climate warming. Quatern. Int. **226**(1–2), 60–65 (2010). https://doi.org/10.1016/j.quaint.2010.03.006
4. Lyu, J.Q., Zhang, Z.Z., Shen, B.: Impacts of climate change on runoff in the Yellow River. J. Am. Water Works Assoc. **106**(5), 87–88 (2014). http://dx.doi.org/10.5942/jawwa.2014.106.0025
5. Jiang, S.H., Renm, L.L., Yong, B., Fu, C.B., Yang, X.L.: Analyzing the effects of climate variability and human activities on runoff from the Laohahe basin in northern China. Hydrol. Res. **43**(1–2), 3–13 (2012). https://doi.org/10.2166/nh.2011.133
6. Liu, Q., McVicar, T.R.: Assessing climate change induced modification of Penman potential evaporation and runoff sensitivity in a large water-limited basin. J. Hydrol. **464**, 352–362 (2012). https://doi.org/10.1016/j.jhydrol.2012.07.032
7. Liu, X.M., Dai, X.Q., Zhong, Y.D., Li, J.J., Wang, P.: Analysis of changes in the relationship between precipitation and streamflow in the Yiluo River, China. Theor. Appl. Climatol. **114** (1–2), 183–191 (2013). https://doi.org/10.1007/s00704-013-0833-0
8. Peng, H.Y., Jia, W., Qiu, Y.Q., Niu, C.W., Ding, X.Y.: Assessing climate change impacts on the ecohydrology of the Jinghe River basin in the Loess Plateau, China. Hydrol. Sci. J. Des. Sci. Hydrol. **58**(3), 651–670 (2013). https://doi.org/10.1080/02626667.2013.774457
9. Tang, Q., Oki, T., Kanae, S., Hu, H.: Hydrological cycles change in the Yellow River basin during the last half of the twentieth century. J. Clim. **21**, 1790–1806 (2008). https://doi.org/10.1175/2007jcli1854.1

10. Chen, Z., Chen, Y., Li, B.: Quantifying the effects of climate variability and human activities on runoff for Kaidu River basin in arid region of northwest China. Theor. Appl. Climatol. (2012). https://doi.org/10.1007/s00704-012-0680-4
11. Bao, Z., Zhang, J., Wang, G., Fu, G., He, R., Yan, X., Jin, J., Liu, Y., Zhang, A.: Attribution for decreasing streamflow of the Haihe River basin, northern China: climate variability or human activities? J. Hydrol. **460–461**, 117–129 (2012). https://doi.org/10.1016/j.jhydrol.2012.06.054
12. Zhang, Z.X., et al.: Influence of the three gorges dam on streamflow for the Poyang lake. In: International Conference on Earth and Environmental Science, pp. 19–25 (2013)
13. Dunne, K., Milly, P.: Macroscale water fluxes 2. Water and energy supply control of their inter-annual variability. Water Resour. Res. **38**, 1206 (2002)
14. Du, M.Y., Kawashima, S., Yonemura, S., Zhang, X.Z., Chen, S.B.: Mutual influence between human activities and climate change in the Tibetan Plateau during recent years. Glob. Planet. Change **41**(3–4), 241–249 (2004). https://doi.org/10.1016/j.gloplacha.2004.01.010
15. Liu, X., Dai, X., Zhong, Y., Li, J., Wang, P.: Analysis of changes in the relationship between precipitation and streamflow in the Yiluo River. Theor. Appl. Climatol., China (2013). https://doi.org/10.1007/s00704-013-0833-0
16. Wang, W., Shao, Q., Yang, T., Peng, S., Yu, Z., Taylor, J., Xing, W., Zhao, C., Sun, F.: Changes in daily temperature and precipitation extremes in the Yellow River basin, China. Stoch. Env. Res. Risk Assess. **27**(2), 401–421 (2012). https://doi.org/10.1007/s00477-012-0615-8
17. Zhang, X.P., Zhang, L., Zhao, J., Rustomji, P., Hairsine, P.: Responses of streamflow to changes in climate and land use/cover in the Loess Plateau, China. Water Resour. Res. **44** (2004). https://doi.org/10.1029/2007wr006711
18. Huang, M., Zhang, L.: Hydrological responses to conservation practices in a catchment of the Loess Plateau, China. Hydrol. Process. **18**(10), 1885–1898 (2004). https://doi.org/10.1002/hyp.1454
19. Wang, H., Yang, Z., Saito, Y., Liu, J.P., Sun, X.: Interannual and seasonal variation of the Huanghe (Yellow River) water discharge over the past 50 years: connections to impacts from ENSO events and dams. Glob. Planet. Change **50**(3–4), 212–225 (2006). https://doi.org/10.1016/j.gloplacha.2006.01.005
20. Yang, T., Zhang, Q., Chen, Y.D., Tao, X., Xu, C.-Y., Chen, X.: A spatial assessment of hydrologic alteration caused by dam construction in the middle and lower Yellow River, China. Hydrol. Process. **22**(18), 3829–3843 (2008). https://doi.org/10.1002/hyp.6993
21. Fu, G., Chen, S., Liu, C., Shepard, D.: Hydro-climatic trends of the Yellow River basin for the last 50 years. Clim. Change **65**(1), 149–178 (2004). https://doi.org/10.1023/b:clim.0000037491.95395.bb
22. Xu, J., Li, X., Chen, J., Gao, Y., Li, M.: The Influence of Water Conservancy Projects on Storm-Flood-Sediment in the Middle Reaches of Yellow River Basin. Water Conservancy Press, Zhenzhou, China (2009)
23. Dong, Q., Chen, X., Chen, T.: Characteristics and changes of extreme precipitation in the Yellow-Huaihe and Yangtze-Huaihe rivers basins, China. J. Clim. **24**(14), 3781–3795 (2011). https://doi.org/10.1175/2010JCLI3653.1
24. Yue, S., Wang, C.Y.: Applicability of prewhitening to eliminate the influence of serial correlation on the Mann-Kendall test. Water Resour. Res. **28**(6) (2007)
25. Kendall, M.: Rank Correlation Methods (Charles Griffin Book Series). Oxford University Press (1990)
26. Helsel, D.R., Hirsch, R.M.: Statistical Methods in Water Resources. Elsevier, Amsterdam (1992)
27. Li, L.J., Zhang, L., Wang, H., Wang, J., Yang, J.W., Jiang, D.J., Li, J.Y., Qin, D.Y.: Assessing the impact of climate variability and human activities on streamflow from the Wuding River basin in China. Hydrol. Process. **21**(25), 3485–3491 (2007). https://doi.org/10.1002/hyp.6485

28. Zhang, X.P., Zhang, L., Zhao, J., Rustomji, P., Hairsine, P.: Responses of streamflow to changes in climate and land use/cover in the Loess Plateau, China. Water Resour. Res. **44** (2008) https://doi.org/10.1029/2007wr006711
29. Wang, W.G., Shao, Q.X., Peng, S.Z., Zhang, Z.X., Xing, W.Q., An, G.Y., Yong, B.: Spatial and temporal characteristics of changes in precipitation during 1957–2007 in the Haihe River basin, China. Stoch. Env. Res. Risk Assess. **25**(7), 881–895 (2011). https://doi.org/10.1007/s00477-011-0469-5
30. Yue, S., Wang, C.Y.: Applicability of pre-whitening to eliminate the influence of serial correlation on the Mann-Kendall test. Water Resour. Res. **38**(6) (2002). https://doi.org/10.1029/2001wr000861
31. Arora, V.K.: The use of the aridity index to assess climate change effect on annual runoff. J. Hydrol. **265**(1–4), 164–177 (2002). https://doi.org/10.1016/S0022-1694(02)00101-4
32. Dooge, J.C.I., Bruen, M., Parmentier, B.: A simple model for estimating the sensitivity of runoff to long-term changes in precipitation without a change in vegetation. Adv. Water Resour. **23**(2), 153–163 (1999). https://doi.org/10.1016/S0309-1708(99)00019-6
33. Pike, J.: The estimation of annual runoff from meteorological data in a tropical climate. J. Hydrol. **2**, 116–123 (1964). https://doi.org/10.1016/0022-1694(64)90022-8
34. Fu, B.: On the calculation of the evaporation from land surface. Sci. Atmos. Sin. **5**, 23–31 (1981)
35. Zhang, L., Dawes, W.R., Walker, G.R.: Response of mean annual evapotranspiration to vegetation changes at catchment scale. Water Resour. Res. **37**(3), 701–708 (2001). https://doi.org/10.1029/2000wr900325
36. Milly, P.C.D., Betancourt, J., Falkenmark, M., Hirsch, R.M., Kundzewicz, Z.W., Lettenmaier, D.P., Stouffer, R.J.: Climate change—stationarity is dead: whither water management? Science **319**(5863), 573–574 (2008). https://doi.org/10.1126/science.1151915
37. Fernandez, W., Vogel, R.M., Sankarasubramanian, A.: Regional calibration of a watershed model. Hydrol. Sci. J. Des. Sci. Hydrol. **45**(5), 689–707 (2000). https://doi.org/10.1080/02626660009492371
38. Alley, W.M.: On the treatment of evapotranspiration, soil-moisture accounting, and aquifer recharge in monthly water-balance models. Water Resour. Res. **20**(8), 1137–1149 (1984). https://doi.org/10.1029/WR020i008p01137
39. Nasseri, M., Zahraie, B., Ansari, A., Solomatine, D.P.: Uncertainty assessment of monthly water balance models based on incremental modified fuzzy extension principle method. J. Hydroinformatics **15**(4), 1340–1360 (2013). https://doi.org/10.2166/hydro.2013.159
40. Dang, S.Z., Dong, G.T., Jiang, X.H., He, H.M., Zhang, W.G.: Impacts of climate change and human activities on runoff in the upper reaches of Jinghe river. S. N. Water Transf. Water Sci. Technol. **04** (2014). https://doi.org/10.13476/j.cnki.nsbdqk.2014.04.007
41. Wu, Z.T., Zhang, H.J., Krause, C.M., Cobb, N.S.: Climate change and human activities: a case study in Xinjiang, China. Clim. Change **99**(3–4), 457–472 (2010). https://doi.org/10.1007/s10584-009-9760-6

Research on Water Resources Carrying Capacity Based on ET

Su Li and Bin Liu

Abstract With the progress of society, the human demand for water resources is increasing. However, due to the shortage of water resources and the serious water pollution, the contradiction between people and water is becoming more prominent. Therefore, it is necessary to study the water resources carrying capacity. Based on the previous studies, the water resources carrying capacity is defined as a region whose water resources is not beyond the allowable transpiration evaporation and can bear the largest population and the corresponding arable land area in the condition of sustainable development and ensuring the normal material level in this paper, using Weixian as an example. The method of ET was used and the relationship between subject and object was considered; meanwhile, the regional evapotranspiration was defined as the subject and the largest population and the corresponding cultivated land area was the object; and the water resources carrying capacity of different levels in Weixian was calculated.

Keywords Water resources carrying capacity · ET · Weixian

1 Introduction

Water resources carrying capacity is an extension and expansion of various natural resources carrying capacity in the sustainable development of a country or region. It has a vital influence on the social and economic development of a country or region [1]. Because it concerns different water resources and social economic systems of different countries, it has uncertainty and ambiguity.

The first time that the concept of water resources carrying capacity was explicitly put forward was in the study of the carrying capacity of water resources in Urumqi

S. Li · B. Liu (✉)
College of Hydropower, Hebei University of Engineering, Handan 056038, China
e-mail: liubin820104@163.com

S. Li
e-mail: 2522954666@qq.com

W. Dong et al. (eds.), *Sustainable Development of Water Resources and Hydraulic Engineering in China*, Environmental Earth Sciences,
https://doi.org/10.1007/978-3-319-61630-8_31

and the river basin in 1992 by Shi and Qu [2]. It is defined as a region of water resources, in a certain social history and the development of the science and technology stage, without destroying the social and ecological system, the largest ability bearing industry, agriculture, city scale and population capacity. It is a comprehensive index with the social, economic, scientific and technological development. After much research [3–5], China has made great progress in the direction of water resources carrying capacity. However, the definition of water resources carrying capacity is still not a unified concept. To comprehensively analyze its definition, it is generally divided into three views: the first view is the largest water resource support capacity; the second one is the largest water resources support scale, both of which are called the sustainable theory of water resources; and the third view is the greatest development capacity of water resources, which is called the scale of water resource development [6, 7]. Although the sustainable theory of water resources and the scale of water resource development are from different views, they both define water resources as the main body, emphasize the importance of the ecological environment and make optimal allocation of water resources in the condition of sustainable development. In the long run, the sustainable theory of water resources is more in line with the shortage of water resources.

To summarize previous research on the carrying capacity of water resources, the carrying subject and object and the relationship between both should be included. It is generally believed that the main body is the water resources system, the object is the human impact and survival of the social economic system and the environmental system [8]. However, the analysis of water resources carrying capacity is often focused on the amount of water resources so far and carrying objects on the production and life of the economic use and economic benefits. Although it has been recognized that the ecological environment is also the carrying object, the course of the study either ignores or roughly estimates or relates to ecological environment of computations and neglects the full utility value of water resources in the ecosystem level [9]. Because the factors affecting the water resources carrying capacity is more and the research content is complex, there is not an accurate response evaluation index system of the relationship between the subject and object. The construction of the index system tends to simplify and is artificial, and also the selection of indicators and complete standard is often not enough [10, 11]. Water resources carrying capacity in a certain sense has become a "particular solution under the background", which means that the water resources carrying capacity can be improved by some ways and means. This will lead to the emergence of "optimization" or "maximum" carrying capacity in this period. It is hard to determine which is the water resources carrying capacity and which is contrary to the limit of water resources carrying capacity [9]. At the same time, there is also a great gap between the bearing scale obtained by various quantitative research methods. Even when the same method is used, corresponding to a different scheme of bearing scale also has a large gap.

Based on the previous studies, according to the problems existing in the development and utilization of water resources in Weixian, the research method of water resources carrying capacity based on ET is proposed. Different from the traditional meaning of ET, the concept of ET is extended and expanded. It refers to the area of real water consumption. The shortage of water resources, the serious overexploitation of groundwater and the phenomenon of irrational use of water resources is general in Weixian. Therefore, it has a certain practical significance to calculate the carrying capacity of water resources in Weixian from the point of view of water consumption.

2 General Situation

Weixian is located at the southern tip of Hebei Province, Southeast of Handan city, between the north latitude 36°03′–36°26′36″00″, east longitude 114°43′42″–115° 07′24″. The east of Weixian connects Da Ming county, north Guang Ping, and the south borders on Qingfeng, Neihuang, and Nanle counties which are three counties of Henan Province; the west is adjacent to Cheng' An and Lin Zhang counties. Weixian covers the area of 862 km^2. It belongs to the warm temperate zone, semi-humid and semi-arid continental monsoon climate zone; and its main rivers are Zhang he and the Wei River. By the end of 2012, its population had reached 976.5 thousand and 912 thousand mu of the cultivated area. Rainfall is mainly concentrated in 6–9 months; the average annual precipitation is 550.7 mm; and the average annual surface water resources is 5.629 million m^3. The average years of groundwater resources is 77.74 million m^3 and the gross amount of water resources is 83.37 million m^3 when recharge methods were used.

3 Research Methods

In this paper, the ET analysis method was used to evaluate the water resources carrying capacity in Weixian. The main body of water resources carrying capacity is the regional transpiration evaporation. It includes not only the regional evapotranspiration of produced water, but also the consumption of the entry of water outside the region with the largest population and the corresponding cultivated land area as the object. According to the regional average water balance equation, the average years of target ET was calculated. Meanwhile, according to the dynamic principle, and to keep the dynamic balance of groundwater, there was a high flow year of remaining water stored in the reservoir and groundwater was appropriately in dry years. Based on this, the target ET of the wet year, normal year and dry year was calculated. Through the establishment of the population, the cultivated land area, ET-cultivated land area, and ET-non-arable land, the water resources carrying capacity of the target ET at different level years was determined.

4 Analysis of Carrying Capacity Based on ET

4.1 Analysis and Calculation of the Main Body

Regional evapotranspiration is a process of moving water from the earth's surface to the atmosphere, and it is an important part of the water cycle in nature. In the natural state, the water balance equation of a closed basin is expressed as follows:

$$P = R + ET \pm \Delta V \tag{1}$$

where P is the annual precipitation; R is the annual runoff (the surface runoff and underground runoff); ET is the basin annual evapotranspiration; and ΔV is the annual water storage variation.

For many years on average, ΔV is zero. In this way, the equation of water balance equation is expressed as:

$$P = R + ET \tag{2}$$

For a region, in view of the regional water quantity and the external exchange, the average regional water balance equation is:

$$P + I = ET + O \tag{3}$$

where I is the regional mean annual inflow; O is the regional mean annual outflow (the surface runoff and underground runoff).

Regional evapotranspiration is the sum of the regional interior surface evaporation, water surface evaporation, plant emission, and the social and economic activities of human beings. Regional evapotranspiration not only contains the transpiration and evaporation in the natural water cycle, but also covers the evaporation of the water in the artificial water cycle to meet the needs of life, production and ecology.

From (3), it is known that the average annual rainfall is 550.7 mm in Weixian; the average inflow minus outflow is 59.8 mm. By calculation, ET is 610.5 mm, that is 52,624.3 m^3. Evapotranspiration of the township is mainly determined by the current situation of water resources and water supply. Table 1 indicates the distribution of the average ET of the townships in Weixian.

To truly reflect the regional water resources carrying capacity, the target ET of wet year, normal year and dry year should be determined, respectively. In the premise of ensuring the groundwater dynamic balance, the dynamic principle to determine the target ET of different level years was used. By calculation, the target ET of a wet year is 620.2 mm; the target ET of a normal year is 607.9 mm; and the target ET of a dry year is 598.0 mm.

Table 1 The statistical table of ET of towns in Weixian

Name	Area (km^2)	Precipitation (mm)	Precipitation ($\times 10^4$ m^3)	Inflow minus outflow ($\times 10^4$ m^3)	Target ET	
					($\times 10^4$ m^3)	(mm)
Weicheng	65	547.5	3558.6	465.4	4024	619.1
Dondaigu	29	548.5	1590.6	165.9	1756.5	605.7
Jizhenzhia	28	545.9	1528.5	382.7	1911.1	682.6
Dezheng	24	551.1	1322.6	192	1514.6	631.1
Shakouji	62	556.7	3451.4	350.9	3802.3	613.3
Yehuguai	26	553.5	1439	284.9	1723.9	663
Shiwangji	24	549.9	1319.7	107	1426.7	594.5
Beigao	66	535.8	3536.1	234.6	3770.7	571.3
Qiandamo	37	544.9	2016	127.5	2143.5	579.3
Yuanbao	21	539.7	1133.3	72.8	1206.1	574.3
Shuangjing	49	553.8	2713.5	374.5	3088	630.2
Nanshuangmiao	44	554.1	2437.9	254.4	2692.3	611.9
Damacun	22	540.2	1188.4	90.9	1279.3	581.5
Daixnhzuang	46	548.8	2524.4	238.3	2762.7	600.6
Bianma	51	553.4	2822.2	190.3	3012.5	590.7
Yali	48	560.9	2692.2	376	3068.2	639.2
Zhangerzhuang	61	564.9	3445.7	638.5	4084.2	669.5
Chewang	45	544.5	2450.1	154.8	2604.9	578.9
Builong	44	550.2	2420.8	150.5	2571.3	584.4
Beitaitou	27	549.9	1484.7	92.3	1576.9	584.1
Bokou	43	556.9	2394.6	210	2604.5	605.7
Total	862	550.7	47,470.3	5154	52,624.3	610.5

4.2 Analysis and Calculation of the Bearing Objects

In this paper, the object of water resources carrying capacity is the population that water resources can bear and the corresponding cultivated land area. Wheat and corn are mainly grown in Weixian; the plant system is to double crop a year. In this paper, two kinds of grain crops of wheat and corn are used as the basis for the calculation of grain yield in Weixian. Through the statistics, the wheat production of Weixian is about 450 kg/acres, and corn production is about 550 kg/acres.

According to the actual situation of Weixian, grain output was 588 thousand tons and per capita grain production was 602 kg in 2012. On the basis of the present level of per capita consumption of grain, ensuring the self-sufficiency of grain, meeting the improvement of people's living standards and preserving a certain amount of wealth, the per capita consumption of grain in Weixian was defined as 600 kg/acre.

By establishing the correlation between population to cultivated land area, population to grain consumption, grain to population, cultivated land area to ET consumption, and the area of non-cultivated land to ET consumption, the regional carrying capacity level of Weixian under different levels of annual target ET was determined.

The relationship between population and cultivated land area is expressed as:

$$(P_{wheat} + P_{corn}) \times A_{cultivate} = 600 \times R \tag{4}$$

where P_{wheat} is wheat yield, in kg/acre; the value of P_{wheat} is 450 kg/acre; P_{corn} is wheat yield, in kg/acre; the value of P_{corn} is 550 kg/acre. $A_{cultivate}$ is the cultivated area, in acre; R is population, in million.

Comparing the per capita water consumption quota in the developed area comprehensively, the comprehensive quota of small city life is 220 L person-1 d − 1; comprehensive quota of rural life is 200 L person-1 d − 1. Taking into account the actual situation of Weixian, the comprehensive water consumption quota of residents as 240 L person-1 d − 1 was defined:

$$ET_{target} - ET_{life} \times R = ET_{cultivate} + ET_{non} \tag{5}$$

where ET_{target} is the target of ET; ET_{life} is residents living ET; $ET_{cultivate}$ is the cultivated land ET; ET_{non} is the non-cultivated land ET.

$$(ET_{wheat} + ET_{corn}) \times A_{cultivate} = ET_{non} \tag{6}$$

In the Hai River Plain, it is estimated that ET_{non} is 0.6 × $ET_{cultivate}$.

$$0.6 \times (ET_{wheat} + ET_{corn}) \times (A - A_{cultivate}) = ET_{non} \tag{7}$$

where ET_{wheat} is the ET of wheat, the value of ET_{wheat} is 387 mm/acre; ET_{corn} is the ET of corn, the value of ET_{corn} is 313 mm/acre; A is the total area.

According to the measured ET values of various crops in Weixian in 2010 refers to ET data of the evaporation experiment station in the North China Plain. After a comprehensive analysis, the ET of various crops in Weixian was determined. The ET of wheat is 387 mm/acre; The ET of corn is 313 mm/acre.

From Eqs. (4)–(7), the population can be known that water resources can bear and the corresponding cultivated land area are, respectively, 1116.7 thousand people and 670 thousand acres for many years on average. The population that water resources can bear in a wet year, normal year and dry year are, respectively, 1173.7 thousand people, 1101.6 thousand people, and 1043.3 thousand people; the corresponding cultivated land areas are 704.2 thousand acres, 661 thousand acres, and 626 thousand acres.

The carrying capacity in different level years are listed in the Table 2.

Table 2 The statistical table of carrying capacity in different level years in Weixian

Level year	Target ET (mm)	Population ($\times 10^4$ people)	Cultivated area ($\times$ 10^4 people)	Population ($\times 10^4$ people)		Cultivated area ($\times 10^4$ acres)	
				2012 year	Difference	Status	Difference
Average	610.5	111.67	67.00	97.65	14.02	92.19	−25.19
Wet	620.2	117.37	70.42	97.65	19.72	92.19	−21.77
Normal	607.9	110.16	66.10	97.65	12.51	92.19	−26.09
Dry	598.0	104.33	62.60	97.65	6.68	92.19	−29.59

4.3 Result Analysis of Water Resources Carrying Capacity

Through the above analysis of water resources carrying capacity of subject and object in Weixian, the target of ET in different level years can be obtained and the population and corresponding cultivated land area calculated that water resources can bear. On the whole, the actual area of cultivated land has been beyond the scope of bearing in Weixian. The main reason is the level of agricultural water consumption which has been beyond the carrying capacity. However, the population of some villages and towns also exceeded the carrying capacity. These towns mainly rely on the ultra deep groundwater to meet the current water scale. Based on this, water-saving measures should be taken to alleviate this phenomenon, such as the adjustment of planting structure, the regional water resources allocation, regional water diversion and so on.

5 Analysis of Carrying Capacity After South North Water Diversion Water

The South North Water Diversion Water would increase the water consumption and alleviate the water resources shortage pressure. The Yellow River began to supply water to Handan in 2011 and the south north water transfer would be scheduled to supply to Handan in 2015. Through the data analysis, evapotranspiration can increase 34015 thousand m^3. Therefore, the allowable transpiration evaporation in Weixian will add to 560258 thousand m^3, which is 650 mm.

From Eqs. (4)–(7), it could obtain the population that water resources could bear and the corresponding cultivated land area, which are 1348.1 thousand people and 808.8 thousand acres, respectively, after the south north water transfer. It increased by 231.4 thousand people and 138.8 thousand acres more than before; however, it still cannot carry the existing cultivated land area.

6 Conclusion

1. Based on ET, the summarizing of the previous research on water resources carrying capacity, and the research method of water resources carrying capacity is proposed. Meanwhile, the ultimate bearing capacity can be reflected with the regional evapotranspiration, the largest population and the corresponding land area evaluation index.
2. In the study of water resources carrying capacity of Weixian, the carrying objects have been beyond the scope of bearing more or less because of the shortage of water resources, the serious groundwater overdraft and exceeding water consumption.
3. The proportion of agricultural water consumption is far greater than the domestic water and industrial water in Weixian. It is necessary to take water-saving measures, for instance, the adjustment of planting structure, and the increase of irrigation water coefficient.
4. Although it alleviated the contradiction between supply and demand of water resources to some extent after South North Water Diversion Water, there are still some problems.
5. Analysis of water resources carrying capacity based on ET is just a preliminary study; there are still many deficiencies. for example, incomplete evaluation index and the ignoring of the calculation on water consumption in the ecological environment.

References

1. Yuan, Y., Gan, H., Wang, Z.J., Bao, W.F.: Discussion on progress and development trend of the research on water resources carrying capacity. J. China Inst. Water Resour. Hydropower Res. **01**, 62–67 (2006). http://dx.doi.org/10.13244/j.cnki.jiwhr.2006.01.012. (in Chinese)
2. Shi, Y.F., Qu, Y.G.: Carrying Capacity of Water Resources in Urumqi River Basin and Its Rational Utilization, pp. 94–111. Science Press, Beijing (1992). (in Chinese)
3. Xu, Y.P.: A study of comprehensive evaluation of the water resources carrying capacity in the arid area—a case study in the Hetian river basin of Xinjiang. J. Nat. Resour. **03**, 229–237 (1993). (in Chinese)
4. Ruan, B.Q., Shen, J.: Study on calculation model of moderate carrying capacity of regional water resources. J. Soil Eros. Soil Water Conservation **03**, 58–62+86 (1998). (in Chinese)
5. Hui, Y.H., Jiang, X.H., Huang, Q., Xuan, X.J.: Study on evaluation index system of water resources carrying capacity. J. Bull. Soil Water Conservation **01**, 30–34 (2001). http://dx.doi.org/10.13961/j.cnki.stbctb.2001.01.008. (in Chinese)
6. Duan, C.Q. Liu. C.M., Chen, X.N., Liu, W.H., Zhen, H.X.: Discussion on the concept and research method of regional water resources carrying capacity. J. Geogr. **01**, 82–90 (2010). (in Chinese)
7. Zhang, L., Dong, Z.C.: Analysis on carrying capacity of water resources in River Basin. J. China Water Resour. **10**, 100–104 (2002). (in Chinese)

8. Zuo, Q.T., Zhang, P.J., Ma, J.X.: Calculation model and key problems of water resources carrying capacity. J. Water Conservancy Hydropower Eng. **02**, 5–8+11+94 (2004). http://dx.doi.org/10.13928/j.cnki.wrahe.2004.02.002. (in Chinese)
9. Long, T.R., Jiang, W.C., He, Q.: New understanding of the connotation of water resources carrying capacity. J. Hydraul. Eng. **03**, 116–120 (2004). (in Chinese)
10. Luan, F.F., Xia, J.X.: Comparison of theory and method of regional water resources carrying capacity. J. Water Resour. Water Eng. **03**, 116–120 (2013). (in Chinese)
11. Liu, D., Wang, L.P., Li, R.B., Zhang, Y.K.: Evaluation of regional water resources carrying capacity based on health of water resources system. J. South North Water Transf. Water Conservancy Science Technol. **02**, 214–21 (2015). https://dx.doi.org/10.13476/j.cnki.nsbdqk.2015.02.001. (in Chinese)

Estimation of Environmental Water Requirements via an Ecological Approach: A Case Study of Yongnian Wetland, Haihe Basin, China

Lingyue Wang and Xiaoliu Yang

Abstract Due to natural change and special human destruction, the area of wetland is greatly decreasing. To recover the wetland ecological system has become an urgent task that must be faced. Water plays an important role in the course of wetland recovery. With Yongnian wetland as a case study, the ecological function method was used, based on structures and functions of the wetland ecological system, the ecological water requirement was divided into four parts: water requirement of wetland vegetation, water requirement of wetland soil, water requirement of wildlife habitats and water requirement of recharging ground water. According to the indexes that were presented in the paper, each part was classified to three grades, respectively. According to the formulas that were given in the paper, all kinds of ecological water requirement of Yongnian wetland were calculated; the results showed that the least ecological water requirement is 1.256×10^8 m^3. The proper ecological water requirement is between 3.041×10^8 m^3 and 4.301×10^8 m^3, and the most ecological water requirement is 6.350×10^8 m^3.

Keywords Wetland ecosystem health · Ecological water requirement
Haihe River basin · Yongnian wetland

1 Introduction

Wetlands, pivotal components of our environment and the cradles of biodiversity, cover approximately 5–8% of the world's land surface (7–10 million km^2) and must be preserved for their significant functions as natural habitats and global carbon cycling. A wetland has important resources, the unique ecosystem formed by the

L. Wang · X. Yang (✉)
College of Urban and Environmental Science, Peking University,
Beijing 100871, China
e-mail: xlyang@urban.pku.edu.cn

L. Wang
e-mail: wanglingyue@pku.edu.cn

W. Dong et al. (eds.), *Sustainable Development of Water Resources and Hydraulic Engineering in China*, Environmental Earth Sciences,
https://doi.org/10.1007/978-3-319-61630-8_32

surface-land interaction, which plays a very important role in flood storage drought, regulating climate, controlling soil erosion, promoting silt-making and degrading environmental pollution, etc. [1]. Water is the dominant factor of a wetland's formation, development, succession, extinction and regeneration; and it is the key to the formation of the soil and the maintenance of biological species in the wetland ecosystem [2]. With the increasing tension of water resources, the research of ecological water demand has been paid close attention to by scholars. In this paper, the ecological function method [3] is used to study the ecological environment water demand of Yongnian Wetland in Hebei province.

2 Study Area

Ziya water system is located in the sea of Henan province, from the Taihang Mountains of Fuyang and from the northern slope of Wutai Mountain, Hutuo River. The catchment area of the Haihe River basin is 25%, and the population accounted for 16%. Yongnian wetland is located in the eastern part of Hanhan Yongnian County, north latitude 36°40′60″–36°41′06″, east of the 114°41′15″–114°45′00″, and the central lake is located in the ancient city; the south and east are adjacent to the Fuyang left dike; the west and the northern is a dike (Fig. 1). In recent years, the surface area of Yongnian wetland has been reduced seriously; its water resources and water environment are one of the most serious river systems in Haihe River basin, which is a typical epitome of the ecological environment of Haihe River basin.

3 Method

A wetland ecosystem is a complicated continuous system, and the existing calculation method lacks the cause and mechanism of wetland whole ecological evolution; therefore, the calculation results cannot be applied well to wetland protection and management. In the past two years, according to the theory of wetland eco-hydrological structure, the area of wetland central area is determined by hydrologic connection degree, and the research of wetland water demand analysis has made great progress. The research on ecological environmental water demand in foreign countries mainly concentrates on the river [4–6], which has less research on the wetland eco-environmental water demand. However, due to the lack of relevant data in most wetland research areas, the application range of hydrological connectivity has been limited. There is no specific quantitative analysis on how to carry out effective wetland water replenishment and how to rationally plan and design wetland water replenishment plan. There are two main methods to calculate the water demand of wetland ecological environment: wet-week method and Habitat method, which is costly and requires long-term observation and experiment.

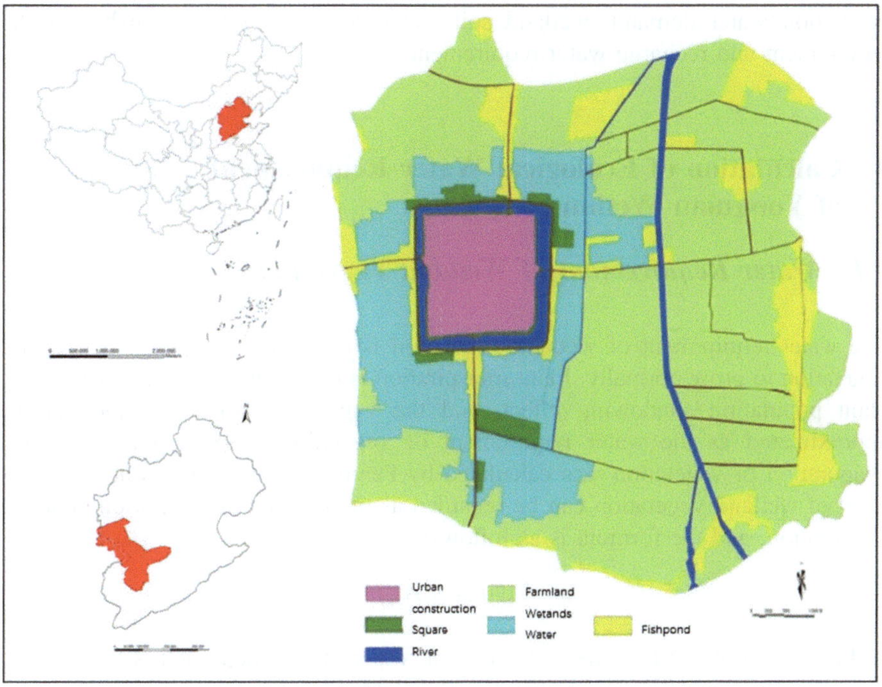

Fig. 1 Yongnian wetland and Ziya River basin

Domestic scholars have carried out much work in the theory and practical application of wetland eco-environmental water demand [7–13]; from a disciplinary point of view, there are hydrology and ecological methods. The former is mainly based on a series of data, through comparison, selection, modeling and so on; the latter is mainly according to the wetlands that are facing the key problems and ecological function value of water demand assessment. Because the research area is small and there is poor continuity of hydrology, the ecological function method is used to calculate.

Ecological water demand has a certain objective and maintains different ecological functions. To facilitate the study, the ecological water demand is divided into the largest, smallest and most suitable ecological water requirement. The maximum ecological water demand is the maximum amount of water that the system can withstand, and if it exceeds this threshold system, it may mutate; the minimum ecological water demand is the minimum amount of water needed for the ecosystem to maintain its own development; and the lower than the minimum water ecosystem will gradually shrink, degenerate or disappear. The most suitable ecological water requirement is the optimum water quantity for sustaining the development of the ecosystem, and the system is in the optimal state of development. According to the definition of wetland ecological water demand, the ecological water demand of Yongnian depression can be decomposed into: wetland vegetation water demand, wetland soil water demand, biological habitat water requirement and recharge water requirement.

4 Calculation of Ecological Water Requirement of Yongnian Wetland

4.1 Water Requirement of Wetland Vegetation

The water requirement of wetland vegetation is the amount of water needed for vegetation to grow normally. Plant transpiration is much larger than that caused by plant population shadowing effect, and the vegetation water demand can be approximated as the water requirement of vegetation evaporation. The actual evaporation of vegetation was calculated by Penman formula. The water requirement of wetland vegetation can be regarded as the product of vegetation area and evaporation, and the formula is as follows:

$$Wp = a(t)ET \tag{1}$$

In the formula, WP is the wetland plant water requirement; a(t) is the wetland area; E is the vegetation coverage; T is the steam emission.

Referring to historical documents, the surface area of Yongnian WA in 1957 was 7 km^2; The research institute uses remote sensing imagery Landsat TM7, TM4–5, MSS image data to download the Earth explorer image retrieval engine from the US Geological Survey (USGS). The time period is for years 1980, 1990, 2000, 2010, 2015; each year is 6-August data, which guaranteed the wetland is in the abundant water period. The interpretation of remote sensing image is carried out by Envi software. To improve the resolution of water and vegetation, 4, 3 and 2 bands of false color synthetic images were used, and the results of the Earth high-resolution image and field investigation were shown in Table 1. After on-the-spot investigation, the wild plant community in Yongnian WA wetland is the dominant species, which is more than 90%. Table 2 shows that the community has Moss, Lotus,

Table 1 The surface area of Yongnian wetland during 1957–2015

Year	Area/km^3
1957	15.7
1980	5.31
1990	3.57
2000	10.12
2005	8.68
2010	2.61
2015	3.86

Table 2 Water requirement of wetland vegetation

	Reed coverage/%	Annual steam emission/mm	Wetland area/km^2	Plant water requirement/ million m^3
Min	40	900	15.7	565.2
Suitable	60–70	1400–1600	15.7	1313.8–1758.4
Max	>90	1900	15.7	2331.45

Calamus and so on. To calculate the water requirement, the level of water requirement is divided into reed coverage degree.

4.2 Wetland Soil Water Demand

At present, the wetland soil water demand has not been clearly defined [14]; in the author's case, that is the wetland ecosystem corresponding to the reserves. In the calculation, the soil moisture content is approximate, so the soil is special. And the two large water constants often appear in the calculation. First, the field holding water, indicating that if the water level is relatively deep, the soil layer can maintain the maximum water content. Because of the different characteristics of chemistry and physics, there are different water demands in all kinds of wetland soils. Combined with the requirement, the calculation can be completed by saturation and field water-holding capacity, which is the concrete formula as follows:

$$Qt = \alpha HtAt \tag{2}$$

The above formulas correspond to the water requirement, area, thickness, saturation, or volume water content of the wetland soil.

The soil moisture content test and wetland soil investigation in Time Domain Reflector (TDR) were carried out in August 2016, and the soil types involved in the region included peat soil, meadow and humus soil. Topsoil, grass-roots layer, is the thickness of about 10 cm. The soil layer of humus is 30 cm deep, and the peat layer is about 20 cm thick. The peat soil grass layer and peat layer thickness are 20, 40 cm, respectively. Because the peat soil has extremely strong water absorption, there exists the perennial stagnant water in which the soil is more saturated. Combined with the proportion of all kinds of soil and water characteristics, with saturation, field water-holding capacity is a concrete basis, according to the peat Swamp soil standards; the level of water requirements is delineated. The results are given in Table 3.

Table 3 Water requirement of wetland soil

Type of water holding	Volume moisture/%	Soil thickness/cm	Wetland area/km^2	Soil water demand/million m^3
Water-holding capacity in the field	30	30	15.7	141.3
Saturated water-holding capacity	50–60	30	15.7	235.5–282.6
Capacity for saturated water storage	>80	30	15.7	376.8

4.3 Biological Habitats

Birds, fish and other breeding habitats are wetland habitat water requirements as the basic water demand. The wetland system is composed of swamp vegetation and water surface, which can provide habitats for waterfowl, and can determine how the species richness is affected by the ratio of surface water and swamp vegetation area to a large extent [15]. Therefore, in general, in the calculation of this water demand, according to the water depth of the habitat, the proportion of the surface area to the level of water requirements can be divided and calculated; the specific formula is as follows:

$$Wq = A(t)CH(t) \tag{3}$$

In the formula, Wq is a biological habitat water requirement; C is the flooded area as a percentage of the total area; and $H(t)$ is water depth.

In recent years, due to the reduction of water area in sediment deposition, the wetland storage is often not more than 20 million m^3, which is a large exposed land. Grain planting activity appeared in the periphery of the WA Lake, and the number and species of fishes decreased significantly. In the light of the 1983–1986 related survey, the natural fishing volume continues to decrease. Residents built about a hundred acres of aquarium, more for catfish, grass carp and other aquaculture; therefore, production has improved. In the continuous development of artificial aquaculture, fishery production has been increasing continuously. As a result of the local more pollution-resistant fish and waterfowl for minimum water requirement, the survival of waterfowl space can be ensured, in which the water flooded area accounted for the minimum value of 25%; and for the appropriate water index set, normal living space to fish, waterfowl can be provided. For the maximum water demand index, even if the optimal survival of fish meets the needs of water depth, Chen et al. [16] and other scholars give the results of the study, and the ideal depth is 2 m. The results of water requirement for biological habitat are given in Table 4.

Table 4 Water requirement of wildlife habitats

	Flooded area/%	Depth/m	Wetland area/km^2	Water requirement of wildlife habitats/million m^3
Min	25	0.5	15.7	196.25
Suitable	50–60	1.0–1.5	15.7	785–1413
Max	>80	2	15.7	2512

4.4 Water Requirement for Recharge Groundwater

Generally, combined with the leakage pathway, the wetland recharge groundwater can be completed because the wetland itself is related and more complex. The calculation makes the assumption that the recharge of groundwater is only partially involved in the water surface of the wetlands; and that if the swamp is not water-free, it only needs to be maintained for its own requirements, as follows:

$$Wq = kIAT \tag{4}$$

In the formula, Wq is the recharge water requirement; k is the permeability coefficient; I is hydraulic gradient; and A is the seepage section area, which is the calculation period length.

In 1985–2015, the monthly precipitation data of Lotus Estuary Hydrologic station were selected and analyzed; the precipitation appeared in May–October reached 438.2 mm, about 90% of the year. The replenishment time is calculated according to 180 days. In combination with the local soil type, the number of osmotic systems is determined, i.e. 0.005 m/d. Table 5 shows the result of water requirement of recharging groundwater in Yongnian wetland.

Table 5 Water requirement of recharging groundwater

	Flooded area/%	Permeability coefficient/ (m d^{-1})	Wetland area/km^2	Replenishment time/d	Recharge groundwater requirement/ million m^3
Min	25	0.005	15.7	180	353.3
Suitable	50–60	0.005	15.7	180	706.5–847.8
Max	>80	0.005	15.7	180	1130.4

Table 6 Wetland ecological water requirement

	Min/million m^3	Suitable/million m^3	Max/million m^3
Plant water requirement	565.2	1313.8–1758.4	2331.45
Soil water requirement	141.3	235.5–282.6	376.8
Water requirement for biological habitats	196.25	785–1413	2512
Water requirement for recharge groundwater	353.25	706.5–847.8	1130.4
Total	1256	3040.8–4301.8	6350.45

5 Results

After calculating the water demand (Table 6), it is found that the local ecological water demand is minimum, maximum and suitable values of 1256 million m^3, 6350.45 million m^3, 3040.8–4301.8 million m^3.

5.1 *Analysis on the Rationality of Maximum Value of Wetland Ecological Water Demand*

In the Ziya Basin, Yongnian wetland belongs to one of the flood detention areas with design area of 16 km^2, and a lag of flood up to 54 million m^3. According to the definition of maximum water requirement, it can be found that when the amount of water is exceeded, there will be floods and the calculation of the maximum ecological water demand is more scientific.

5.2 *Analysis of the Proportions of Each Type*

According to the different types of water requirement, the most important proportion is the water demand of plants, which is 45.86%, 42.75%–45.65% and 41.61%, respectively, in the minimum, the suitable and the maximum water requirement. With the level of increase, the proportion continued to decrease. The corresponding proportion of wetland soil water requirement is 11.25, 7.7–6.6 and 5.9%. With the level of increase, the proportion continued to decrease. The corresponding proportion of water demand for habitats is 15.6, 25.8–32.8 and 39.6%. With the level of increase, the proportion continues to increase. Groundwater recharge supplies corresponding to 28.1, 23.2–19.7 and 17.8%. The ratio continues to decrease as the level increases.

It can be found that the water demand in this area is mainly the evapotranspiration of vegetation, and the increase of water quantity will be more beneficial to the conservation of habitats.

5.3 Countermeasures for Ecological Protection of Yongnian Wetland

To protect the wetland ecological environment, water problems should be solved, and ecological water supplement work should be carried out. According to the local amount of water, for the problems faced, the corresponding seasonal characteristics and the implementation of concrete measures include the following:

(1) This area uses Fuyang River as the main source of water and provides protection for urban and factory water supply. To ensure the ecological water replenishment, it is necessary to improve water efficiency and ensure the downstream ecological water.
(2) Local government should unify the management and dispatching of water resources so that the utilization efficiency of water resources can be improved and the water environment can be improved so as to achieve the purpose of sustainable utilization. In the meantime, water conservancy projects should be dispatched reasonably and gradually transformed from open management to closed ones. Through the use of engineering measures, local water resources should be optimally allocated.

Acknowledgements This research was supported by the Major Science and Technology Program for Water Pollution Control and Treatment of China, 2012.

References

1. Pearsell, W.G., Mulamoottil, G.: Toward the integration of wetland functional boundaries into suburban landscapes. In: Mulamoottil, G., Warner, B.G., McBean, E.A.: Wetlands Environmental Gradients, Boundaries and Buffers, pp. 25–37. CRC Press, Inc. (1996)
2. Hughes, D.A.: Towards the incorporation of magnitude-frequency concepts into the building block methodology used for quantifying ecological flow requirements of South African rivers. Water SA **25**(3), 279–284 (1999)
3. Feng, X., Chang, K.-N.: Research progress of wetland ecological water demand. J. Ecol. **27** (12), 2228–2234 (2008)
4. Jowet, T.I.G.: Instream stream flow methods: a comparison of approaches. Regul. Rivers Res. Manag. **13**, 115–127 (1997). http://dx.doi.org/10.1002/(SICI)1099-1646(199703)13:2<115::AID-RRR440>3.0.CO;2-6
5. Mosley, M.P.: Flow requirements for recreation and wildlife in New Zealand rivers—a review. J. Hydrol. **22** (1983)

6. King, J., Brown, C., Sabet, H.: A scenario-based holistic approach to environmental flow assessments for rivers. River Res. Appl. **19**, 619–639 (2003). https://doi.org/10.1002/rra.709
7. Wang, X., Liu, C., Yang, Z.: Research progress and prospect of ecological and environmental water demand. Adv. Water Sci. **04**, 507–514 (2002). https://doi.org/10.3321/j.issn:1001-6791.2002.04.020
8. Liu, J., Yang, Z.: Lake study on calculation method of ecological water demand. J. Nat. Resour. **05**, 604–609 (2002). https://doi.org/10.3321/j.issn:1000-3037.2002.05.011
9. Shi, W., Wang, G.: Estimation of ecological water demand in the lower reaches of the Yellow River. Acta Geogr. Sin. **05**, 595–602 (2002). https://doi.org/10.3321/j.issn:0375-5444.2002.05.012
10. Cui, B., Yang, Z.: Study on eco-environmental water requirement of wetland. J. Environ. Sci. **02**, 219–224 (2002). https://doi.org/10.3321/j.issn:0253-2468.2002.02.018
11. Cui, S.: Discussion about some problems of eco-environmental water demand. China **08**, 71–74 (2001). https://doi.org/10.3969/j.issn.1000-1123.2001.08.030
12. Lei, C., Wu, Y.: Study on present situation, problems and countermeasures of water resources in China. Water Sav. Irrig. **04**, 41–43 (2007). https://doi.org/10.3969/j.issn.1007-4929.2007.04.014
13. Jiang, D., Wang, H.-X., Lau: Summary of classification and calculation method of eco-environmental water demand. Progress Geogr. **04**, 369–378 (2003). https://doi.org/10.3969/j.issn.1007-6301.2003.04.005
14. Cui, B., Yang, Z.: Analysis on eco-environmental water requirement of wetland classification and examples. Science **01**, 21–28 (2003). https://doi.org/10.3321/j.issn:1007-7588.2003.01.004
15. Yang, A., Tang, K., Wang, H., Liu, X.: The basic theory and calculation method of ecological water. J. Hydraul. Eng. **12**, 39–45 (2004). https://doi.org/10.3321/j.issn:0559-9350.2004.12.007
16. Chen, M., Wang, L., Feng, H.: Wetland hydrology structure theory and the analysis. Acta Ecol. Sin. **6**, 2887–2893 (2008). https://doi.org/10.3321/j.issn:1000-0933.2008.06.054

Establishment and Application of Evaluation Index System of Soil Salinization in Plain Reservoir: An Example for Shifosi Reservoir

Lijun Xu

Abstract Plain reservoir surrounding soil salinization assessment method for the small scope of application, low evaluation precision of the common problems proposed for underground water level elevation, soil water salinity, evaporation and precipitation ratio were used as evaluation factors of the soil salinization assessment index system. Shifosi Reservoir, as an example, in the field water test obtained evaluation parameters based on using AHP fuzzy comprehensive evaluation method of the reservoir after impoundment and the degree of soil salinization was evaluated. The results show that the most serious salinization area is in the left bank of Shifosi Reservoir, after the impoundment of Shifosi reservoir; and closer to the main salinization is more serious, and away from the main dam area the salinization degree is gradually reduced. The right bank of the terrain is relatively low in soil salinization degree due to a serious problem in most parts; less parts presented mild salinization. According to the result of the evaluation, the measures of prevention and control of the salinization of precipitation and water supply were determined, and the rational development and utilization of water resources were realized. This study provides a new idea for the evaluation of soil salinization and provides a scientific basis for the control and prevention of environmental geological hazards in the reservoir area.

Keywords Plain reservoir · Soil salinization · Evaluation index system
Water storage test · Fuzzy comprehensive evaluation

1 Introduction

Plain reservoir water storage easily led to a problem of soil salinization in the reservoir and bank of the reservoir. It could cause a variety of environmental geological disasters and bring huge economic losses; and it is the important content

L. Xu (✉)
Liaoning Province Water Resources Management Company,
Shenyang 110103, People's Republic of China
e-mail: xulijun007@126.com

W. Dong et al. (eds.), *Sustainable Development of Water Resources and Hydraulic Engineering in China*, Environmental Earth Sciences,
https://doi.org/10.1007/978-3-319-61630-8_33

of an engineering survey and evaluation of the reservoir, which is also one of the four main engineering geological problems [1]. Therefore, soil salinization degree and evaluation caused by plain reservoir water storage is deeply studied by domestic and foreign scholars. The evaluation method of soil salinization can only forecast according to limited groundwater level data of the research area. In terms of landform, geologic parameters of an applicable area also have more constraints. The accuracy of the evaluation is low, and the applicable scope is small [2, 3]. Therefore, to establish a scientific evaluation index system and evaluation method of soil salinization degree in the reservoir area is worth deep exploration.

Soil salinization refers to the phenomenon of salt accumulation in the surface soil caused by the salt in the subsoil, and groundwater rises to the surface with the capillary water, for which the phenomenon caused by groundwater level around the reservoir is higher than the ground as a result of the supply of surface water to groundwater that increased rapidly after reservoir filling [4, 5]. Groundwater elevation of the reservoir area is the most intuitive index of salinization degree. Soil salinity should be used as an evaluation index of salinization degree because it is also an important factor of soil salinization [6]. Besides, evaporation-precipitation ratio has a certain influence on soil salinity. Evaporation and infiltration are important drivers of salt movement in the soil; the distribution and existing state of salts in the soil is directly controlled [7, 8]. Therefore, evaporation-precipitation ratio is also an important index of salinization degree. Hence, groundwater elevation, soil salinity and evaporation-precipitation ratio as evaluation index of soil salinization was proposed in this study. The standard of soil salinization degree was divided, which will provide a new research method and idea for the quantitative evaluation of soil salinization degree in a reservoir.

Using the mainstream of Liao River dominant water project using Shifosi reservoir as an example, fuzzy comprehensive evaluation methods of soil salinization degree were studied on the basis of the establishment of the evaluation index system of soil salinization in Shifosi reservoir. Reasonable control measures of soil salinization were proposed according to the actual situation of the reservoir. It overcame the limitations of traditional evaluation methods of soil salinization, provided scientific basis for environmental geologic calamity and rational development and utilization of water resources in the reservoir.

2 Establishment of Evaluation Index System

2.1 Groundwater Elevation

There are two conditions of soil salinization. One is a certain amount of salt in groundwater, where either groundwater elevation or capillary water rises to the height of the root zone; and the salinity detained in the soil with the evaporation of water leads to an increase in soil salinity. The other is that in low-lying ground, the water cannot be discharged in time and detained salinity of the ground leads to an increase in soil salinity after evaporation [9]. Hence, groundwater elevation is selected as the evaluation index. Groundwater elevation is divided into groundwater depth and the rising height of capillary water; these two evaluation indexes were used as the impact of soil salinization degree. Height of the capillary water was calculated according to Appendix C of Code for Water Resources and Hydropower Engineering Geological Investigation Standard (GB50287-99) [10].

The limit depth of evaporation in Shenyang is 4.0 m without salinization, which was defined as the range of groundwater level below 4.0 m. Most of the soil is silty clay in the study area. Average height of capillary water is 1.5 m. Therefore, serious salinization was defined as the rising height of capillary water below 4.0 m.

2.2 *Soil Salinity*

Soil salinity is the direct factor of influence on soil salinization. Soil salinity is divided into the content of chlorine ion and salinity. The relationship between salinity and critical depth of groundwater is shown in Table 1 [11]. The total content of salt and sodium chloride has a great influence on crops. As an example of rice, critical concentration of salt tolerance in each reproductive stage of rice is shown in Table 2. The relationship between soil salinity and salinization degree has been defined through analysis and summary on different spatial variation of soil salinity and the actual situation of Shifosi reservoir [12].

2.3 *Evaporation-Precipitation Ratio*

Soil salinization is the gradual accumulation process of soluble salts in the soil surface. Evaporation and infiltration are important drivers of salt movement in the soil; the distribution and existing state of salts in the soil is directly controlled. Evaporation-precipitation ratio has a certain influence on soil salinity [13]. The relationship between the evaporation-precipitation ratio and soil salinization is shown in Table 3. In conclusion, the soil salinization evaluation index and standards are shown in Table 4.

Table 1 Relationship between salinity and critical depth of groundwater

Salinity (g/L)	2	4	6	10
Critical depth of groundwater (m)	2	2.5	3	3.3

Table 2 Critical concentration of salt tolerance in each reproductive stage of rice

Reproductive stage	Growth	Soil salinization degree	
		Total content of salt (%)	Sodium chloride (%)
Seedling stage–returning green stage	Growth death	<0.212	<0.088
	Gradual death	>0.265	>0.117
Returning green stage–tillering stage	Growth death	<0.242	<0.088
	gradual death	>0.348	>0.208
Tillering stage–heading stage	Growth death	<0.250	<0.117
	Gradual death	>0.400	>0.280
Heading stage–maturation stage	Growth death	<0.280	<0.126
	Gradual death	>0.469	>0.322

Table 3 Relationship between of evaporation-precipitation ratio and soil salinization

Evaporation-precipitation ratio	Soil salinization
<1	Salinization or little salinization
5–10	Segmental salinization
10–15	Saline soil area < 5%
15–20	Saline soil area 5–10%
>20	Saline soil area > 20%

Table 4 Soil salinization evaluation index and standards

Index soil salinization	Groundwater elevation (m)		Soil salinity		
	Ground-water depth	Height of capillary water	Salinity	Sodium chloride	Evaporation-precipitation ratio
Serious salinization	≤ 1.5	2.2–2.5	7–10	>0.322	10–20
Moderate salinization	1.5–3.0	1.8–2.2	5–6	0.117–0.28	6–10
Mild salinization	3.0–4.0	0.8–1.8	3–4	0.088–0.117	2–5
Without salinization	≥ 4.0	0.5–0.8	≤ 2	<0.088	1

3 Application Example

3.1 Study Area Profile

Shifosi reservoir is located in Huangjia Town and Faku County, XinChengZi District, Liaoning Province. It is only a dominant water project in the mainstream of Liao River; the study area is shown in Fig. 1. It is located in the middle of an alluvial fan of Liao River in Lower Liao River alluvial plain. It belongs to the temperate continental monsoon climate. Average annual rainfall and evaporation is 665.5 mm and 1300–1800 mm, respectively [14]. Most of the aquifer rock groups are medium and fine-grained sand and gravel in the dykes. The thickness of medium and fine-grained sand is 2–9.5 m, and the thickness of gravel is 35–45 m. A continuous water-resisting layer does not exist. Most of the aquifer rock group is mainly clayey soil of Quaternary loose sediments; aquifer rock groups are medium and fine-grained sand and gravel, and the thickness is 5–10 m.

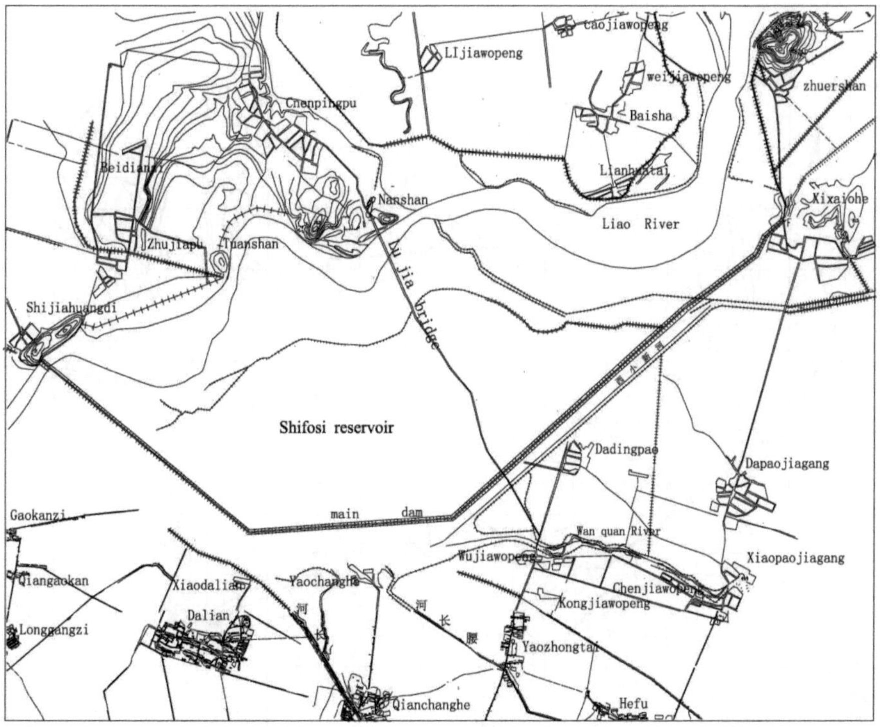

Fig. 1 Scope of the study area

3.2 Evaluation of Salinization Degree

Division of Evaluation Area. The division method of the evaluation area used the grid points; it divides the study area into many same size grids. The sampling points set in the center of the grid, grid division, is shown in Fig. 2.

Grid division is based on the immersion scope of beginning judgment. The range was divided into 500 m × 500 m same size grids. Each grid is an evaluation grid; a total of grid cell is 280. The grid center points were selected as the grid feature points; evaluation data in the grid area is based on the value of the grid point.

Calculated Parameters. To determine the relationship between groundwater level with ground elevation and soil physical property parameters into and surrounding the reservoir, the wild water storage test was proceeded. Long-term monitoring of groundwater level in the study area was obtained through groundwater level observation and a survey well. The rising height of the capillary water, natural void ratio and saturation were obtained and are shown in Tables 5 and 6.

Evaluation Analysis and Results. (1) Evaluation index system of soil salinization degree.

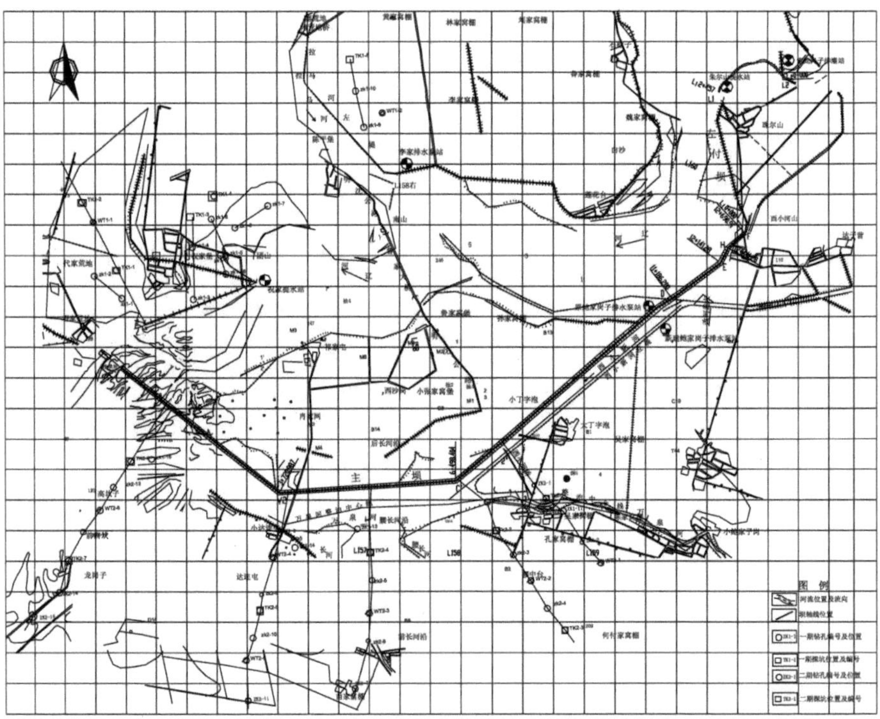

Fig. 2 Grid division of evaluation area

Table 5 Statistical chart of rising capillary water in test pit

Left bank			Right bank		
Test pit number	Stratum lithology	Rising height of capillary water(m)	Test pit number	Stratum lithology	Rising height of capillary water (m)
TK2-1	Silty clay	1.23	TK1-1	Silty clay	1.24
TK2-2	Silty clay	1.20	TK1-2	Silty clay	1.2
TK2-3	Silty clay	0.89	TK1-3	Silty clay	0.94
TK2-4	Silty clay	1.10	TK1-4	Silty clay	0.7
TK2-5	Silty clay	0.70	TK1-5	Silty clay	0.89
TK2-6	Silty clay	0.89			
TK2-7	Silty clay	0.99			

Table 6 Statistical chart of natural void ratio and saturation level for different depth pit

Soil sample number	Sample depth (m)	Natural void ratio (e_o)	Saturation (S_r)	Soil sample number	Sample depth (m)	Natural void ratio (e_o)	Saturation (S_r)
ZK1-1-1	2–2.3	0.82	0.892	ZK1-6-1	1.5–1.8	0.943	0.897
ZK1-1-2	4–4.3	0.89	0.949	ZK1-6-2	3.8–4	1.147	0.902
ZK1-1-3	6–6.3	0.906	0.965	ZK1-6-3	6–6.3	1.235	0.971
ZK1-1-6	11–11.3	0.663	0.979	ZK1-6-4	8–8.3	0.734	0.928
ZK1-2-1	1.3–1.6	0.668	0.882	ZK1-8-1	1.8–2	0.770	0.908
ZK1-2-5	9–9.3	0.753	0.957	ZK1-8-2	2.8–3	0.834	0.974
ZK1-2-6	12.5–12.8	0.676	0.955	ZK1-8-8	15–15.2	0.707	0.894
ZK1-3-1	1.8–2.1	1.104	0.955	ZK1-9-1	2–2.2	0.611	0.878
ZK1-4-1	2–2.3	0.815	0.931	ZK1-10-1	1–1.3	0.726	0.86
ZK1-4-2	4–4.3	0.886	0.933	ZK1-11-1	2–2.2	0.97	0.801
ZK1-5-1	1.5–1.8	1	0.91	ZK1-11-2	2.3–2.5	0.831	0.984
ZK1-5-2	3–3.3	0.850	0.937	ZK1-12-1	1.5–1.7	0.825	0.922
ZK1-5-3	6–6.3	1.381	0.971	ZK1-12-2	1.7–1.9	0.873	0.878
ZK1-5-4	9–9.3	0.84	0.894	ZK1-13-1	1.5–1.8	1.107	0.931
ZK1-5-5	11–11.3	1.410	0.967				

U (soil salinization degree) = {groundwater elevation, soil salinity, evaporation-precipitation ratio}; each factor was divided into N factors according to its nature and degree. Shown by the following: groundwater elevation = {groundwater depth, rising height of capillary water}; soil salinity = {salinity, sodium chloride}.

According to monitoring data of groundwater elevation and soil salinity in the study area and parameters of water storage test, using two-level fuzzy comprehensive evaluation method, evaluated the soil salinization degree in the reservoir by the evaluation index system of soil salinization degree. The result is shown in Fig. 3.

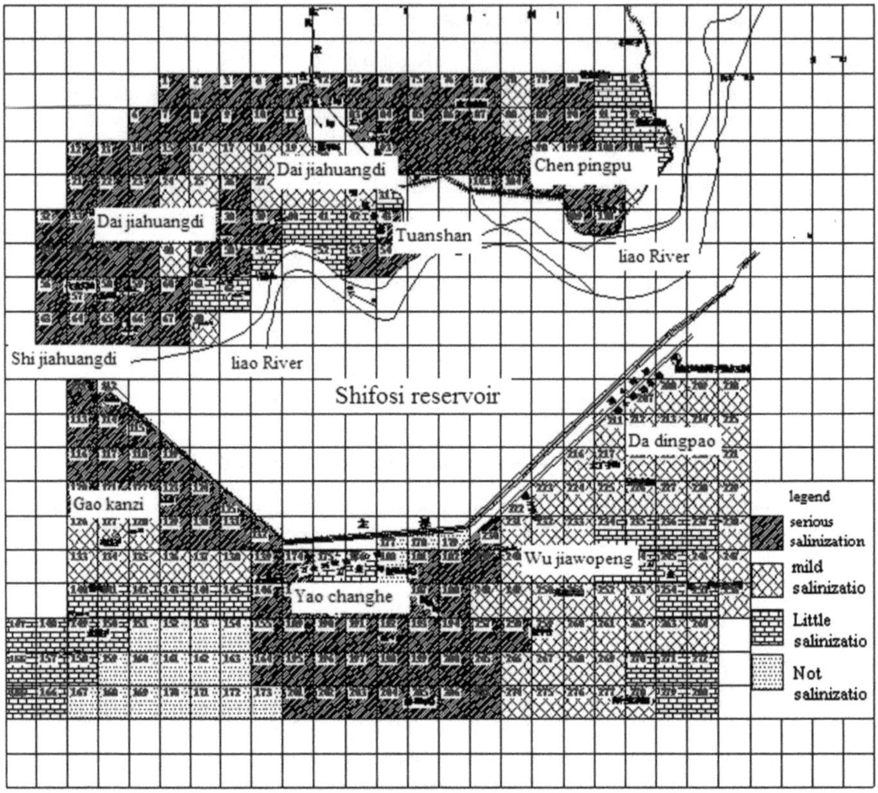

Fig. 3 Evaluation result of soil salinization degree in study area

Figure 3 shows that most areas of Gaokanzi District, Yaochanghe District and Dadingzipao District, respectively, located at the southwest, south and southeast of the main dam downstream of the left bank in the reservoir appear to have serious salinization. Part of area of southwest Longgangzi District and southeast Xiaobaojiazigang District, far away from main dam, appear to have mild salinization. A small part of south of Longgangzi District appears without salinization. Most areas of Daijia wasteland, Shijia wasteland, Chenpingpu and Lijiawopeng District, located at right bank in the reservoir, appear to have serious salinization. Part of the areas of Lijiawopeng, Weijiawopeng and Tuanshan District, near the right bank of Liao River, appear to have mild salinization.

The evaluation results of soil salinization degree show that reservoir water storage would lead to serious environmental geological disasters around most of the Shifosi reservoir with serious salinization and have a certain influence on local residents.

4 Solution of Prevention and Control for Soil Salinization

To eliminate geological disasters of soil salinization due to reservoir water storage, designed intake programs of deep-and-shallow well were used to deploy and control soil salinization around the reservoir. Groundwater extraction is a water source supply for Shenbei New District, Shenyang. A lower groundwater level was used to reduce environmental geological disasters; and at the same time, full use was made of water resources from the reservoir. According to the environmental condition of Shifosi reservoir, Mingshen Road as an important transportation road connects Shenyang city and Mingshen city of Heilongjiang Province. Ground elevation of Mingshen Road as a limiting height of water storage is 46.5 m. Numerical simulation was used to intake programs by Visual Modflow software; the result is shown in Fig. 4.

Figure 4 shows that until the main dam downstream on the left bank in the reservoir for most of Gaokanzi District appears to have mild salinization. Part of area of Southwest Shijia wasteland, Lijiawopeng and Chenpingpu on the right bank in the reservoir appears to have mild salinization. The rest of the area appears without salinization. In the comparison of Fig. 3 with Fig. 4 it can be seen that the scope of serious salinization and mild salinization greatly reduces. The problem of immerged soil in Shifosi reservoir is basically solved by a deploying and controlling program of deep-and-shallow wells.

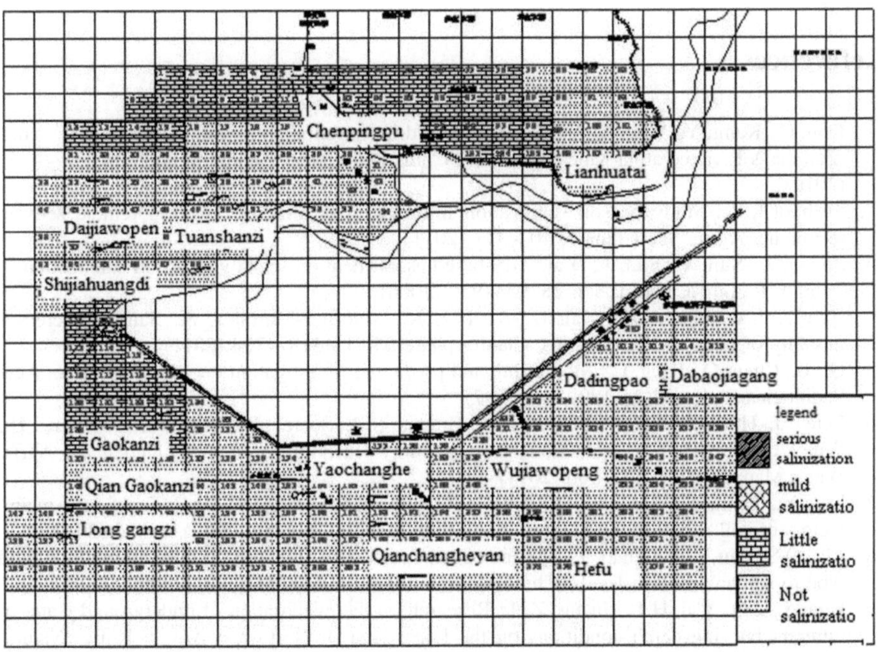

Fig. 4 Prediction result of prevention and control for soil salinization

5 Conclusion

(1) Plain reservoir surrounding soil salinization assessment method of the common problems proposed groundwater level elevation, soil salinity, evaporation—and precipitation ratio as evaluation factors of the soil salinization assessment index system. The evaluation standard of soil salinization was proposed according to the actual situation of the reservoir. This study provides a new idea for the evaluation of soil salinization.
(2) According to the evaluation index system of soil salinization degree, the soil salinization degree of the reservoir after water storage was evaluated using the AHP fuzzy comprehensive evaluation method. The results show that there are different degrees of soil salinization around the reservoir. Above all the main dam downstream on the left bank and Chenpingpu District on the right bank in the reservoir there appears to be serious salinization. According to the result of the evaluation, soil salinization must be controlled at the same time of the source supply for Shenbei New District. The evaluation realizes the rational development and utilization of water resources and provides a scientific basis for soil salinization.

Acknowledgements This research was financially supported by the Environmental Protection Public Welfare Industry Research special funds for major projects (201009009).

References

1. Dou, C., Kang, Y., Wan, S., Lu, G.: Effects of drip irrigation with saline water on way maize (Zea mays L.var.ceratina Kulesh) in North China Plain. Agric. Water Manag. **26**(03), 44–51 (2010)
2. Tashpolat, T.: Study on soil salinization information in arid region using remote sensing technique. Agric. Sci. China **3**, 404–411 (2011)
3. Zhang, Z., Sun, Y., Sun, J., et al.: GIS-based assessment on sensitivity to soil salinization in Tianjin. J. Agro-Environ. Sci. **25**, 954–957 (2006)
4. Acosta, J.A., Jansen, B., Kalbitz, A., Faz, K., Martínez-Martínez, S.: Salinity increases mobility of heavy metals in soils. Chemosphere **85**(8), 1318–1324 (2011)
5. Qi, L., Chen, Q., Pang, C.: Monitoring and predicting investigation on soil salinisation. Acta Pedologica Sinica **34**(2), 189–197 (1997)
6. Letey, J., Hoffman, G.J., Hopmans, J.W., Grattan, S.R., Suarez, D., Cowin, D.L., Oster, J.D., Wu, L., Amrhein, C.: Evaluation of soil salinity leaching requirement guidelines. Agric. Water Manag. **98**(4), 502–506 (2011)
7. Liu, Y., Feng, C., Yinglan, O., Zheng, P.: Information statistical analysis of influencing factors of soil salinization in Yanqi Basin. Environ. Sci. Technol. **33**(3), 23–26 (2010)
8. Liu, Q.S., Liu, G.H., Zhao, J.: The indication function of soil type and soil texture and land type to soil salinization levels. Chin. Agric. Sci. Bull. **24**(1), 297–300 (2008)
9. Huang, M.Q., Cai, H.J., Huang, Z.H.: Research on the evaporation of underground water in different burying depth conditions on the Loess area. J. Northwest A & F Univ. (Natural Science Edition). **35**(3), 233–237 (2007)

10. GB50287-99: Appendix C of Code for Water Resources and Hydropower Engineering Geological Investigation Standard. 49–51
11. Hu, K., Chen, H., Zhang, Y., et al.: Spatial distribution of shallow groundwater depth, total dissolved solid and nitrate pollution. Trans. CSAE **25**, 21–25 (2009)
12. Liu, J.J, Cheng, D.J., Yu, Z.H.: Effect of soil sodium chloride content on soil capillary water rise characteristics. Water Sav. Irrig. **2**, 8–14 (2011)
13. He, Y.J., Jin, M.G., Liu, Y., et al.: Spatial variability of soil salt on the surface of cotton fields in the south of Xinjiang. Geol. Sci. Technol. Inf. **29**(6), 89–93 (2010)
14. Investigation and Design Institute of Water Resources and Hydropower Liaoning Province: Geological exploration and impact assessment report on immersion of Shifosi reservoir water storage test (first phase) (2010)

Effects of Discharge of Dripper on Soil Water and Heat Energy Movements Under Water Storage Pit Irrigation

Fei Li, Chao Ma, Lihua Hao, Shubin Li, Qiang Ma, Ming Xu, Yunxin Zhang, Lishu Wang and Yunpu Zheng

Abstract Water storage pit irrigation was combined with drip irrigation, and the influence of discharge of the dripper on soil water and heat movements was examined to explore a better water-saving irrigation. The results showed that the maximum soil moisture content appeared below the drop point after irrigation. In addition, the soil moisture content near the drop point has a greater change than soil moisture content near the wetting front. There was only a little trend after 24 h of irrigation because of water evaporation. It was also found that increased discharge of the dripper not only increased the volume of the wetted soil, but also increased the soil moisture content of each layer. In addition, a different dripper discharge pit has an effect on soil temperature about water storage pit; the temperature is higher with greater flow than with a small flow.

Keywords Water storage pit irrigation · Drip irrigation · Wetting front
Soil temperature

F. Li · C. Ma · L. Hao · Y. Zhang · L. Wang · Y. Zheng (✉)
School of Water Conservancy and Hydropower, Hebei University of Engineering,
178 S. Zhonghua Street, Handan 056021, Hebei, China
e-mail: zhengyunpu_000@sina.com

S. Li
Office of Fu-Zhang River Irrigation and Water Supply Management,
4 Fuxi South Street, Handan 056021, Hebei, China

Q. Ma
Yahoo! Inc., 701 1st Ave, Sunnyvale, CA 94089, USA

M. Xu (✉)
Center for Remote Sensing and Spatial Analysis, Department of Ecology,
Evolution and Natural Resources, Rutgers University, 14 College Farm Road,
New Brunswick, NJ 08901, USA
e-mail: mingxu@crssa.rutgers.edu

W. Dong et al. (eds.), *Sustainable Development of Water Resources and Hydraulic Engineering in China*, Environmental Earth Sciences,
https://doi.org/10.1007/978-3-319-61630-8_34

1 Introduction

China has a vast territory and different natural characteristics, and there are also differences in water conservancy conditions for developing agriculture. North of Huaihe, generally referred to as the north, the annual rainfall is generally less than 800 mm, which belongs to arid or semi-arid areas. Most of the irrigation methods are surface irrigation, and the field engineering facilities are simple and without energy and easy to implement; however, it is easy to cause a soil surface knot and decrease water use efficiency because of evaporation [1]. Water storage pit irrigation is a method in which the irrigation water can infiltrate directly into the deep soil. To maintain the dry soil surface and reduce the evaporation, the irrigation water can directly infiltrate into the middle and deep layer soil by the water storage pit. In addition, the method of the water storage pit can greatly improve the irrigation water use efficiency and water retention property [2]. Drip irrigation is developed by underground irrigation. The irrigation water is transported directly to the root and matches the root distribution of the crop. Many studies have reported that the shape of wetted soil by drip irrigation has an important influence on root water uptake of crops. For example, Li et al. [3] studied the relationship between discharge of the dripper and the volume of wetted soil and found that the discharge of the dripper irrigation has a great influence on the shape and size of wetted soil. Zhang et al. [4] found that the relationship between the horizontal or vertical moving distance of the wetting front with irrigation time appeared as a power function.

Most researches mainly focus on indoor experimental research and a physical model that are based on water storage pit irrigation and drip irrigation [5]. However, it should be noted in future application of drip irrigation and water storage pit irrigation. The simulation experiment was examined and the effect of the discharge of the dripper on the characteristic of wetted soil in was analyzed in a greenhouse. In other words, the discharge of the dripper, measured soil moisture content and temperature in different depths and locations were controlled.

2 Materials and Methods

2.1 Test Soil Parameters

Field experiments were conducted at the experiment field of Hebei University of Engineering in 2016. The greenhouse experiment was started when the temperature and humidity were suitable in late spring and early summer.

2.2 Experimental Design

The hole with the same depth and the same diameter according to the requirements of the experiment was excavated, and the diameter of each water storage pit is 20 cm and 15 cm in depth. Three discharges of the dripper were controlled for experiments which were 1.4 L/h, 2.16 L/h and 2.8 L/h, respectively. In addition, for the experiment 10.8 g nitrogen fertilization was applied with the same method and the same amount of water, about 2 L for each treatment, and meanwhile soil temperature and irrigation time were recorded.

The drip irrigation device is a simple piece of equipment made up of a Markov bottle, a rubber hose, a syringe needle and three tripods. The drip irrigation dripper was fixed on the three tripods, just above the selected suitable drop point. The discharge of the dripper was controlled by the altitude difference between the outlet of the Markov bottle and the outlet of the emitter; and the greater the height, the greater the discharge of the dripper. Soil samples were taken at 0, 24 and 72 h after irrigation with a small soil sampler according to the design and test requirements.

The pit center was the sampling center, and then along the 4 o' clock direction, the 8 o'clock direction and the 12-point direction radial sampling was performed, respectively. The first sampling point is A (distance center 5 cm), and the interval is 5 cm as the second sampling point B, and three sampling points were set up (Fig. 1). The vertical direction is downward from the soil surface, spaced at 5 cm, until the wetted soil front is the wetted front boundary (Fig. 1). Soil tamped pits were used due to sampling to prevent pit collapse.

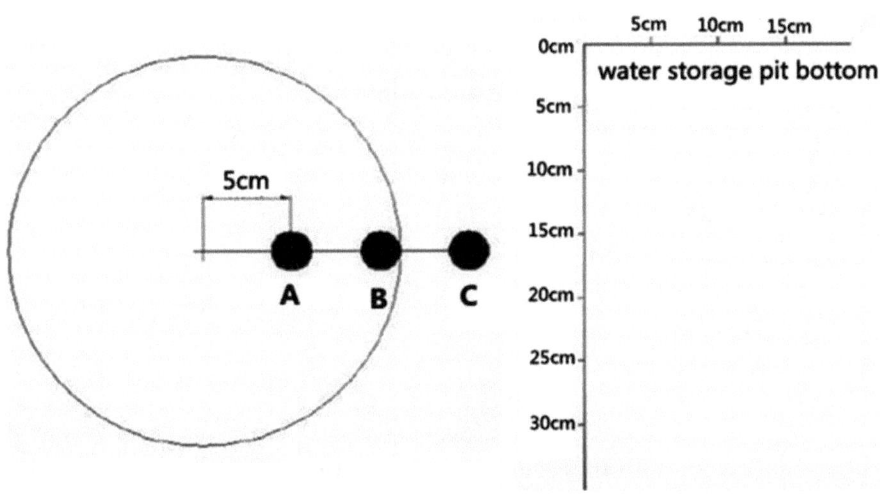

Fig. 1 Distribution of sampling points

2.3 *Determination of Soil Moisture Content*

The oven-drying method was used to measure the soil moisture content for soil samples immediately. The temperature was 106 °C for oven-drying and the time was set for 5 h. The soil initial water content is 3.9%; bulk density is 1.38 g/cm^3; and field capacity is 27.8%. The calculation of soil water content is as follows: soil moisture content = (soil wet weight-soil dry weight)/soil dry weight × 100%.

3 Results

3.1 *Effect of Dripper Discharge and Irrigation Time on Soil Moisture Content Distribution*

The results showed that the contour distribution was from sparse to dense from the drip irrigation point to the wetting front direction. Near the drop point, the soil moisture content had little change, while the soil moisture content changed significantly near the wetting front. In addition, the maximum soil moisture content occurred at the drip point just below the emitter (Figs. 2a, 3a and 4a); and the soil moisture content decreased gradually with the increase in the distance from the drop point. The distribution of soil water in 24 and 72 h after irrigation appears as a similar rule (Figs. 2b, c, 3b, c and 4b, c). When the dripper discharge was 1.44 L/h in 72 h after irrigation soil moisture distribution, the soil moisture content increased with the increase of 0–10 cm soil depth. This trend was obvious in the soil moisture contour map after 72 h for each flow irrigation, but not obvious after 24 h irrigation.

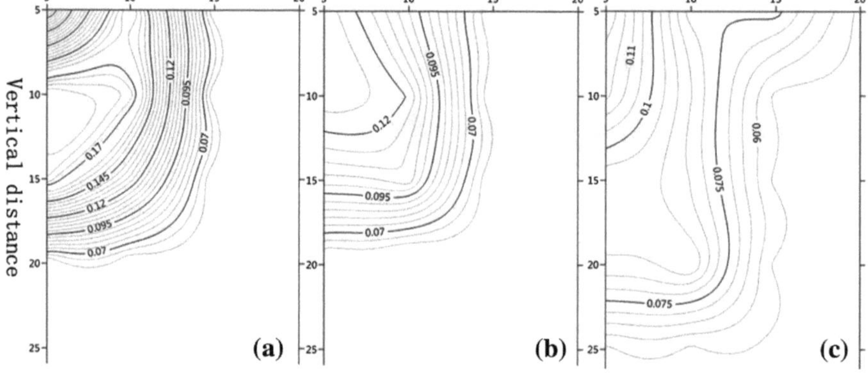

Fig. 2 Soil moisture content at different time with dripper flow rate of 1.44 L/h

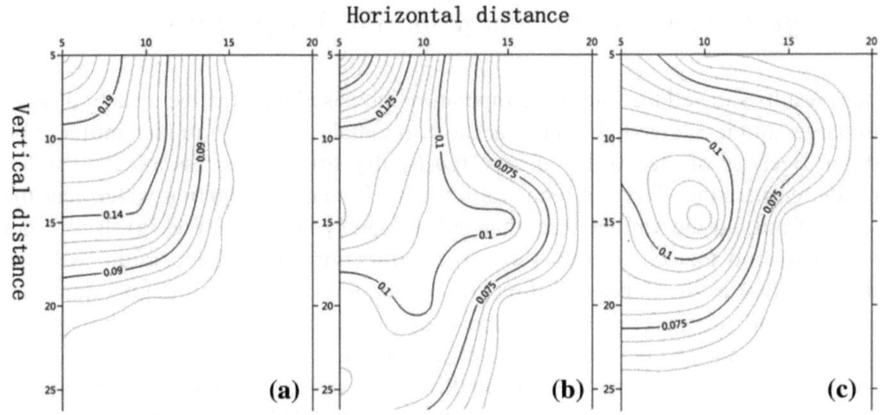

Fig. 3 Soil moisture content at different time with dripper flow rate of 2.16 L/h

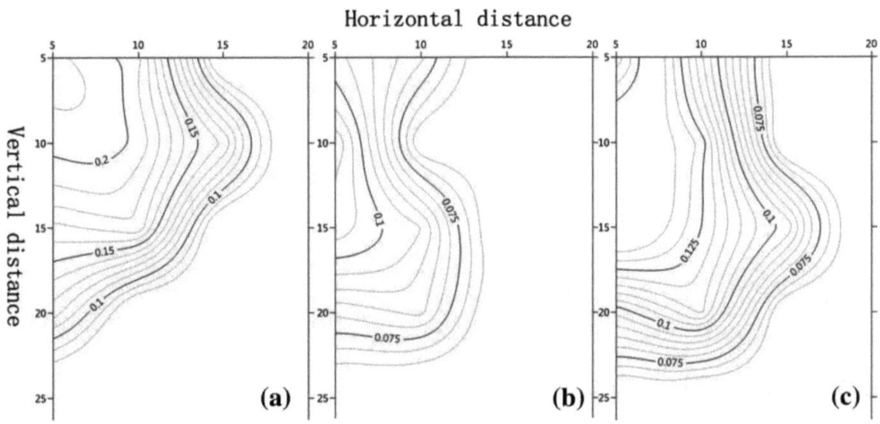

Fig. 4 Soil moisture content at different time with dripper flow rate of 2.80 L/h

It was also found that the soil moisture content increased with the increase of dripper discharge after 0 h irrigation at the same sampling point (Figs. 2, 3 and 4). However, this rule was not satisfied at the wet boundary. The distribution of water content in soil after 24 and 72 h redistribution still met this rule (Figs. 2b, c, 3b, c and 4b, c). These results also showed that the vertical wetting front was 25 cm after 72 h irrigation, and the soil water movement is no longer obvious after 72 h of irrigation by the above discussion. Therefore, when the dripper discharge was in 1.44–2.8 L/h and the irrigation amount was 2 L, soil moisture changed significantly only at 25 cm soil depth above; below 25 cm was not influenced by the experiment of drip irrigation.

3.2 *Effect of Dripper Discharge on Soil Temperature*

These results showed that the soil temperature increased with the increase of dripper discharge, and it changed obviously at the horizontal distance from the emitter at 0–10 cm. Because of the 10–14 cm covered by dry soil, it was less affected by the outside temperature. The soil temperature was affected by irrigation water near the wetting front, so the temperature changed obviously (Figs. 5a, 6a, 7a). It was not difficult to find that changes in the vertical direction of the temperature was the same direction as the horizontal direction, which was that the soil temperature first increased and then decreased with the increase in distance from the drop point, which corresponds to the trend of water migration.

Comparing the three pictures, it could clearly be seen that the maximum soil temperature gradually moved down as time went on. When the emitter flow was

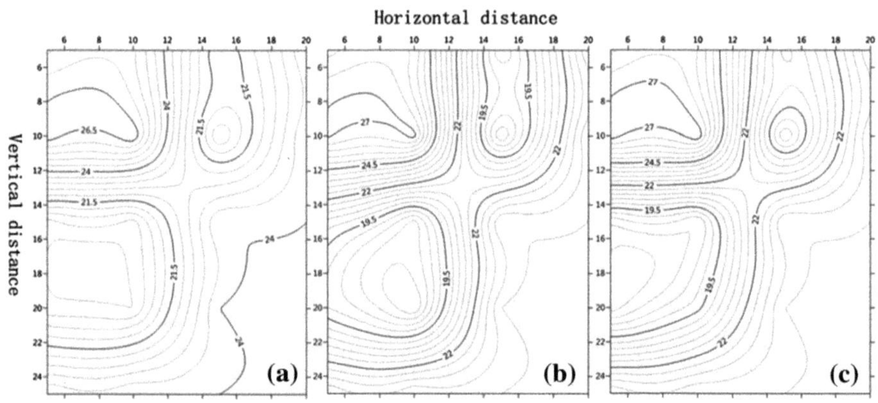

Fig. 5 Soil temperature at 9 a.m. under different head flow

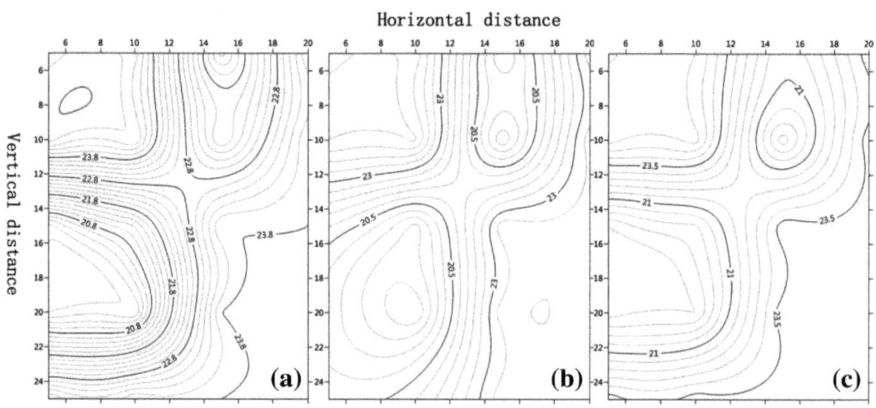

Fig. 6 Soil temperature at 14 noon under different head flow

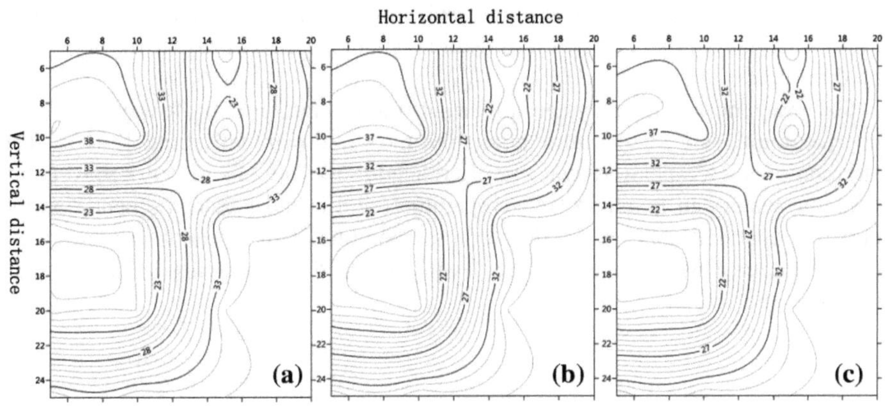

Fig. 7 Soil temperature at 19 p.m. under different head flow

1.44 L/h, it could clearly be seen that at 9 in the morning, the highest temperature appeared at 10 cm layers and 14 at noon; the highest temperature appears at 10–11 cm layers, and 19 in the evening; the highest temperature appeared at 11–12 cm layers.

The variation of soil temperature was closely related to the law of soil water movement. Irrigation water can maintain soil temperature, and it had greater influence on topsoil temperature than on the deep soil temperature (Figs. 2, 3 and 4).

4 Discussion

As is known, one of the purposes of drip irrigation is to enable crops to obtain suitable moisture in their root active layer [6, 7]. It is very important according to the depth of crop roots and reasonable selection of dripper discharge and irrigation quota. To avoid wetting depth is not enough to limit crop root growth or wetting depth beyond the crop root layer which causes loss of soil water and nutrient. The results showed that the maximum soil moisture content occurred at the drip point just below the emitter, and the soil moisture content decreased gradually with the increase of the distance from the drop point. The distribution of soil water in 24 and 72 h after irrigation appears to be a similar law (Figs. 2b, c, 3b, c and 4b, c). Maybe this situation had a great relationship with the surface soil moisture evaporation, at 10 cm depth, soil water evaporation for 24 h after irrigation, soil moisture content decreased in 0–5 cm layer, while the 5–10 cm layer was not affected; therefore, soil moisture content would be increased from top to bottom. After 72 h irrigation, the surface evaporation was more serious; the surface water content is very low, and the evaporation depth also increased; therefore, the trend was more obvious (Figs. 2c, 3c and 4c). The soil moisture content was almost unaffected by evaporation under the 10 cm layer, and the distribution of soil moisture content is the same as that at 0 h after irrigation; that is, the soil moisture content increases with the distance from the drop point.

Soil temperature also plays an important role in plant growth. It was found that the maximum soil temperature gradually moved down with time [7–10]. The variation of soil temperature was closely related to the law of soil water movement. Irrigation water can maintain soil temperature, and it had greater influence on the topsoil temperature than on the deep soil temperature. This research may provide a good theoretical basis for the better development of water-saving irrigation technology in the future.

References

1. Cheng, D.J., Wang, L.J., Hou, Y.K., Fang, Y.: Experiment study on impact from soil initial water content on water movement distribution characteristics of soil under injection irrigation. Water Resour. Hydropower Eng. **43**(7), 91–94 (2012)
2. Booth, R.J., Lightfoot, C.J.: The reticulation of ethanol stillage through irrigation systems and its use for fertilisation of sugarcane in Zimbabwe. Agric. Water Manag. **17**(13), 49–58 (1990). https://doi.org/10.1016/0378-3774(90)90055-4
3. Li, M.S., Kang, S.Z., Yang, H.M.: Effects of plastic film mulch on the soil wetting pattern, water consumption and growth of cotton under drip irrigation. Trans. CSAE **23**(6), 49–54 (2007)
4. Zhang, R., Cheng, Z.Y., Zhang, J.X.: Sandy loam soil wetting patterns of drip irrigation: a comparison of point and line sources. Procedia Eng. **28**, 506–511 (2012)
5. Cheng, D.J., Fei, L.J., Lei, Y.B., et al.: Study on the rule of nitric nitrogen movement and distribution under film hole irrigation with fertilizer. Agric. Res. Arid Areas **1**(26), 237–240 (2008)
6. Zhang, J., Wenquan, N., Zhang, L.L., et al.: Experimental study on characters of wetted soil in moisture irrigation. J. Drain. Irrig. Mach. Eng. **10**(6), 32–38 (2012)
7. Bhatnagar, P.R., Chauhan, H.S.: Soil water movement under a single surface trickle source. Agric. Water Manag. **95**(7), 799–808 (2008). https://doi.org/10.1016/j.agwat.2008.02.003
8. Poulovassilis, A.: The influence of the initial water content on the redistribution of soil water about infiltration. Soil Sci. **5**, 275–281 (1983)
9. Patel, N., Rajput, T.B.S.: Dynamics and modeling of soil water under subsurface drip irrigated. Agric. Water Manag. **95**(12), 335–1349 (2007). https://doi.org/10.1016/j.agwat.2008.06.002
10. Li, X.B., Kang, Y.H., Wan, S.Q., et al.: Reclamation of very heavy coastal saline soil using drip-irrigation with saline water on salt-sensitive plants. Soil Tillage Res. **146**, 159–173 (2015). https://doi.org/10.1016/j.still.2014.10.005

The Relationship and Influencing Factors Water-Gas Interface Mercury Emission Flux and Water Suspended Mercury of a Gold Mining Area River

Rui-Ping Liu, You-Ning Xu and Jiang-Hua Zhang

Abstract In order to study the coupling relationship and its influencing factors river water-air interface emission flux of mercury with the rivers of mercury, based on the small Qin Lingjin mining area shuangqiao river as the research object, with field investigation, monitoring and indoor suspended adsorption desorption experiments means. Obtaining the datas about winter and summer season water, suspended solids, gaseous mercury and hydrological and meteorological conditions of river. The result show that water velocity and Quantity of flow, relative pressure were positively correlated with the river-gas interface mercury exchange flux; humidity and rivers of mercury volatilize were positively correlated in summer, the winter not no monitoring data shows that correlation; river volatile mercury is restricted in air and water temperatures, Water temperature affect the suspended solids of mercury adsorption desorption, the higher the temperature, the lower adsorption. Mercury not easily soluble in river water, suspended matter mercury concentration as positively related with river-gas interface mercury emission flux. Research results will enrich the river mercury migration transformation rule and pollution prevention and control work.

Keywords Gold mining area · River · Water-gas interface mercury emission flux Water suspended mercury

---

The original version of this chapter was revised: A new chapter has been included. The erratum to this chapter is available at https://doi.org/10.1007/978-3-319-61630-8_36

R.-P. Liu (✉)
Key Laboratory of Loess Geological Disaster, Key Laboratory of Subsurface Hydrology and Ecological Effect in Region of Ministry of Education, China Geological Survey Geological Survey Center of Xi'an, Chang'an University, Xi'an, China
e-mail: lrp1331@163.com; 164632469@qq.com

Y.-N. Xu · J.-H. Zhang
China Geological Survey Geological Survey Center of Xi'an, Xi'an, China
e-mail: 948477575@qq.com

W. Dong et al. (eds.), *Sustainable Development of Water Resources and Hydraulic Engineering in China*, Environmental Earth Sciences,
https://doi.org/10.1007/978-3-319-61630-8_35

1 Introduction

Mercury is one of the most toxic heavy metals in water environment, due to special physical and chemical properties of elemental mercury by atmospheric transmission of cross-border global pollution [1–3]. After entering rivers of mercury under certain conditions will translate into greater toxicity and biological validity more methyl mercury, and through the way such as wastewater irrigation, breathing into the food chain, constitute the harm to human health [4]. 50–60 s of the 20th century, Japan kyushu minamata bay nearby residents due to consumption of methyl mercury contamination of seafood, thousands of people appear nervous system disease, the disease was known as the "minamata disease", since then, mercury pollution become the research focus of the world's attention in 1990, 1992 and 1996, respectively, in Sweden, the United States, Canada and Germany held a mercury as a global pollutant international academic conferences. In 2013 the United Nations environment programme (unep) through the minamata convention, again on the agenda will be global mercury pollution abatement [5].

Mercury mine development brings a lot of characteristics of pollutants, and to be the only can exist as a liquid at normal temperature and has a volatile metals, it exists in the environment in a variety of chemical form, mercury is the only and in continuous movement between water, air, soil and biological transformation. Among them, the water and air are mercury "source entrance" or "discharge", on the one hand, the release of water into the atmosphere from mercury reduces the load of mercury in water. On the other hand, also increase the content of mercury in the atmosphere, and vice versa. The exchange of water and atmospheric mercury on the interface between flux affects rivers of mercury in water migration transformation way. Was spread to the surrounding environment into the atmosphere, on the other hand, in the river further into methyl mercury and with flow in the mutual transformation between dissolved and suspended state, sedimentary state, eventually pose a threat to human health, the water is released into the atmosphere every year accounted for about 32–77% of the natural sources of atmospheric mercury mercury. Mercury exchange flux between water and air interface, therefore, the study received extensive attention of the researchers at home and abroad [6–9]. A large number of studies have suggested that the thermodynamics, the interface of water-gas mercury exchange function controlled by Henry law ($[Hg^0]/[Hg^0]$ water = H), $[Hg^0]$ the content of water, the water dissolved gaseous mercury (Hg^0) gaseous elemental mercury content in the atmosphere. Day and night, seasons, and mercury exchange flux exists in the fine rain change rule [10–16]. About river mercury vapour interface exchange research, under the condition of rainfall, flood, dry gas input and output of mercury and rivers of mercury research [17, 18], also have Hg^0 experts to study the seasonal change in the atmosphere—transformation of river system [19]. Mercury exists mainly in the forms of water soluble state, suspension type, sedimentary state in river, How the coupling relationship and the influence factors of river of the water soluble state, suspension state of mercury with in the atmosphere, there is few people discussed. The author expect to provide reference basis for water environment management in gold mining area studying this subiect.

2 Organization of the Text

2.1 *Sample Collection and Test Methods*

(1) The Sample Collection

The study area lie in our country's second largest gold producing region—small Qin Lingjin mining area. The shuangqiao river flow into the Yellow River, length of 29.73 km, water catchment area of 171.64 km^2, the average annual runoff 37.67 million m^3. Its contain seven tertiary water. Because mercury is brought about by the human activities, this article all sample distribute in Tai valley river near to the pollutant, the river flow cross over the alluvial plain, two kinds of geomorphic units. In order to compare the winter and summer for rivers, the effect of mercury in the atmosphere, the sampling respectively on August 27 and December 21, 2015. We collected in the same place at the time from 8 in the morning to 15 in the afternoon, water, suspended solids, water-gas mercury exchange flux of river surface and the corresponding meteorological parameters, the physics properties of rivers, hydraulic parameters, and so on, the summer once every 1 h, sampling, sampling winter every 2 h. Water-gas mercury exchange flux using Zeeman effect between mercury analyzer (LUMEX Zeeman RA915+) field measurement as shown in Fig. 1, the continuous determination of each point 3 min, repeat 3 times, averaging mercury concentrations at various points as per sample.

According to the law of conservation of mass, Hg emission flux calculation formula is:

$$F = (C_{out} - C_{in} - C_0)Q \times A$$

F for river/gas interface between mercury emission flux, ng/(m^2/h);
C_{in} Cout of flux respectively for import and export of total gaseous mercury concentrations (ng/m^3);
C_0 for flux tank wall adsorption or release of mercury(ng/m^3);
Q for flux in the air flow rate (m^3/h);
A is flux box at the bottom of the area, the m^2

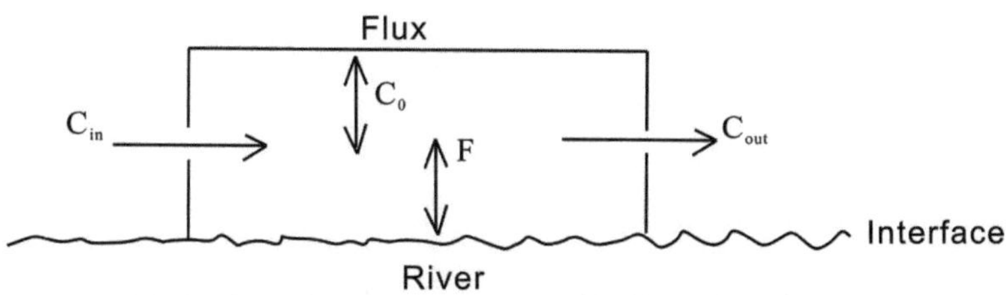

Fig. 1 The river-gas interface mercury exchange flux diagram

(2) Suspension of Mercury Adsorption Test Method

The suspension adsorption mercury experiment, objective to study the suspended matter on the biggest mercury adsorption quantity: weigh and dry sediment sample 5 g in the number of points in dissolved oxygen bottles (in view of the volatile mercury, mercury dissolved oxygen bottle sealed with black cloth), respectively, to join the background electrolyte concentration of $NaNO_3$ is 0.01 mol L^{-1}, several groups of mercury concentration (e.g., in the form of Hg $(NO_3)_2$) (0.01, 0.05, 0.01, 0.05, 1, 5, 10, 50 mg/L) 50 ml of the solution. With 0.5 mol L^{-1} HNO_3 or NaOH to adjust pH value of soil sample suspension, with suspended solids of the pH = 8 value of the sample. In constant temperature water bath oscillation on continuous oscillation of not less than 4 h later, with the speed 4000 r. min^{-1} for 15 min, take the supernatant water, analysis of mercury. using atomic spectrophotometer determination. Change the water bath temperature (15, 25, 35 °C), repeat the experiment two times.

2.2 *The Results and Analysis*

(1) Influence Factors Analysis

The river-gas interface mercury exchange flux and relevant parameters collect at time on August 27, 2015, and December 31, 2015 at 8:30–15:30. From the Table 1 and Fig. 2: gas mercury decrease migrating with time, the highest value at 9:30 in the morning, at noon time (11–13:30) keep a certain concentration range, in the afternoon the gas mercury concentrations to minimize (13:30–15:30). The daily variation of gas mercury with humidity, relative pressure, velocity, and rate of flow

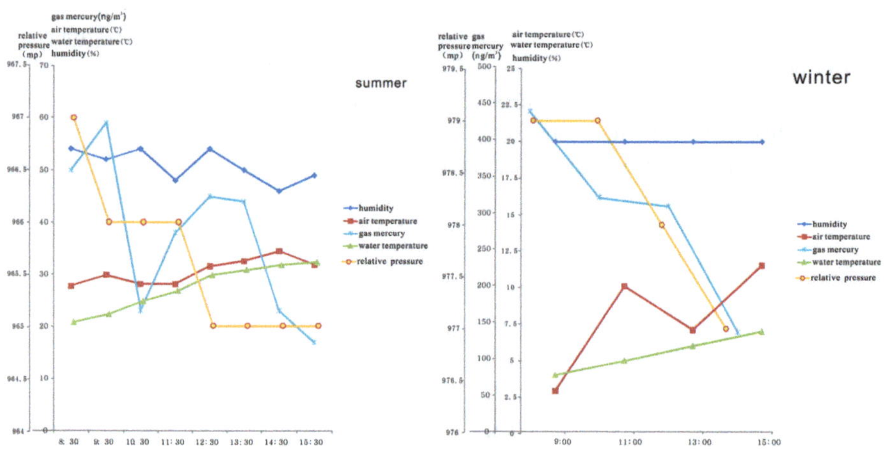

Fig. 2 The river-gas interface mercury exchange flux and relevant parameters of a day on summer and winter river

Table 1 The river-gas interface mercury exchange flux and environmental factors at three-river section in the gold mining area river

Season	Time	pH	V (m/s)	Q (m^3/h)	River width (m)	deep (cm)	Water tem. (°C)	Gas humidity (%)	Gas tem. (°C)	Relativity atmospheric press (mp)	Gas mercury (ng/m^3)	Water mercury contents (mg/L)
Summer	8:30	7.97	0.54	903.25	5.7	8	20	54	27.7	967	50	–
	9:30	8.01	0.53	655.55	5.7	6	21.5	52	29.8	966	59	–
	10:30	8.04	0.49	403.14	5.7	4	24	54	28.1	966	23	–
	11:30	7.98	0.42	519.67	5.7	6	26	48	28.1	966	38	–
	12:30	7.97	0.48	587.63	5.6	6	29	54	31.5	965	45	–
	13:30	8.01	0.44	363.33	5.5	4	30	50	32.5	965	44	–
	14:30	8.06	0.50	587.93	5.4	6	31	46	34.4	965	23	–
	15:30	8.1	0.21	225.61	5.4	6	31.5	49	31.8	965	17	–
Winter	9:00	8.1	0.53	519.45	5.2	6	4	20	2.9	979	438	–
	11:00	7.9	0.42	426.17	5.1	6	5	20	10.1	979	321	–
	13:00	7.9	0.44	357.46	5.0	4.9	6	20	7.1	978	309	–
	15:00	7.9	0.2	191.73	4.9	5.9	7	20	11.5	977	136	–

Table 2 The correlation coefficient between the river-gas interface mercury exchange flux and physical parameters

Parameters	Summer air mercury	Gas mercury in winter
Water tem. (°C)	−0.60	−0.95
Gas humidity (%)	0.67	–
Gas tem. (°C)	−0.31	−0.87
Relativity atmospheric press (mp)	0.43	0.90
V (m/s)	0.60	0.57
Q (m^3/h)	0.65	0.68

is positive correlation, the negative correlation with air and water temperatures (see Table 2).

With seasonal change rule: the temperature, humidity, absolute pressure is lower in summer, the relative air pressure is higher in summer than in winter, the gas-water interface mercury exchange flux is high in winter rather than in summer.

In some river sections, the velocity of speed affects the river of turbulence, the velocity of fast promote the collision between mercury molecular in river; And in general, the higher surface temperature and humidity, the greater the pressure is lower, the smaller the density of the gas, so lead to the smaller the mercury content in the gas.

b. Andersson Maria E study of spring and summer season season (warm) water interface mercury emission flux is higher than the autumn and winter festival (or cold season) [20], why this paper point of view is different from his?

Andersson Maria describes mercury vapour interface to exchange information about a region types of lake or coastal. In our study area is a small river system, and the difference between the lakes, is the water area of the lake is broad, under the intense sun radiation factors control, increasing temperature of water, so as to show the water flux of the release into the atmosphere of mercury. And the river section is not wide rivers, and rivers water relatively large liquidity, environmental factor and velocity of speed, suspended adsorption (Fig. 3) are all important. Mercury is not soluble in water (Table 1), this paper think that as water temperature is higher rivers suspended solids on mercury adsorption quantity is smaller. Suspended matter not easy adsorption of heavy metals in river mercury, mercury to escape the water surface and deposition in the sediment. The higher the gas temperature and the faster velocity of speed, gaseous mercury in the gas escape very faster to higher sky, The solubility of mercury in air, Which lead the gas mercury values is lower than in winter by LUMEX Zeeman RA915+.

(2) The River Surface Gas Relationship with Rivers

From Table 1 shows river-gas interface mercury exchange flux have no correlation with mercury concentration of river water after filtering suspended solids, if the river-gas interface mercury exchange flux have a relation with the rivers? This paper discusses the river-gas interface mercury exchange flux and mercury

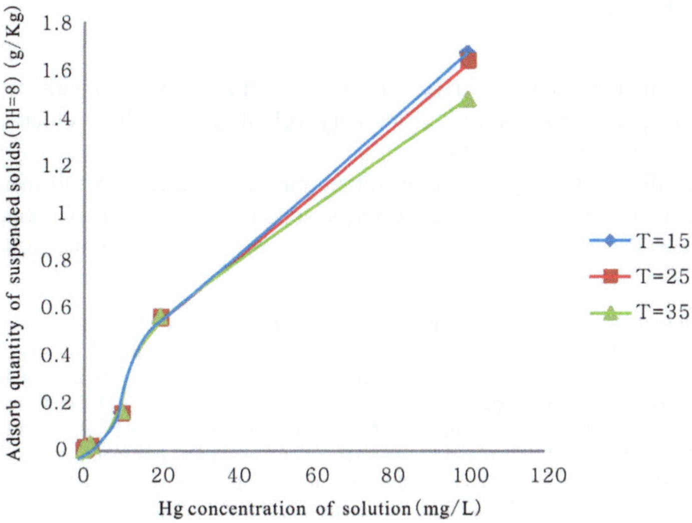

Fig. 3 Rivers adsorption quantity and temperature of the suspended solids

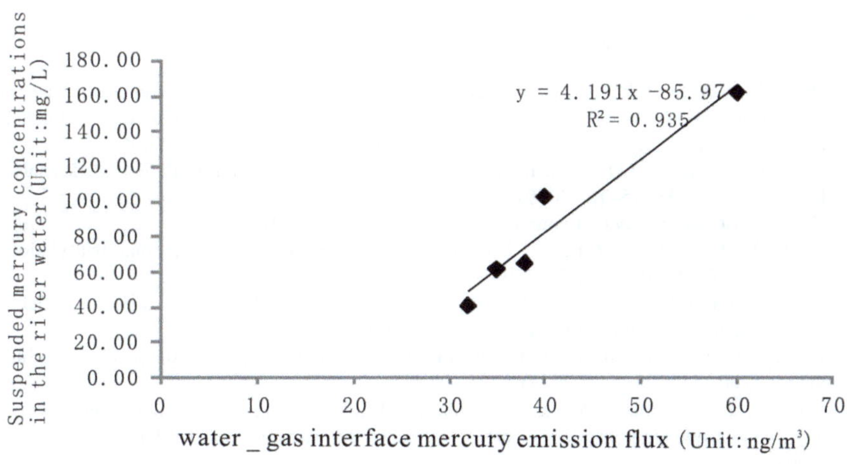

Fig. 4 Water-gas interface exchange flux of mercury and mercury in water containing suspended solids

concentration in water containing suspended solids, as shown in Fig. 4, river-gas interface mercury exchange flux and mercury concentration in water containing suspended matter relationship in a straight line in the gold mine, the correlation coefficient of 0.935.

3 Conclusion

(1) The influence factors of river mercury volatilize the water velocity of flow, relative pressure and rivers of mercury volatilize, humidity, Volatile mercury in air and water temperatures.
(2) Water flow is mainly have transmitted and disturbance effect to mercury, and mercury not easily soluble in water, suspended matter mercury concentration was positively related with the river-gas interface mercury emission flux.

Acknowledgements This research was financially supported by the Land and Resources Survey Project (12120101100015002 2 1212010741003); Shaanxi Province Natural Science Foundation Research Project (2015 jm4129), Central Business Expenses for Basic Research of University's Scientific Research Program (open fund) (310829161128), the Ministry of Land and Resources Research Projects (2008334061), "the Ministry of Land and Resources Public Welfare Industry Scientific Research (201111020121011140)"; Mine Geological Environment, Shaanxi Tongguan Field Base. At the same time thanks to the data analysis from Xiu-hua CHENG of Xi'an test center, professor Juan XIE of Chang'an university provide laborator. Dong GAO of Chang'an university field sampling.

References

1. Feng, X.B., Wang, S.F., Qiu, G.L., et al.: Total gaseous mercury exchange between water and air during cloudy weather conditions over Hongfeng Rservoir, Guizhou, China [J]. J. Geophys. Res. **113**, D15309. https://doi.org/10.1029/2007JD009600 (2008)
2. Qin, D.-L., Slang, J.Q., Yousheng, F.: Global mercury pollution review and analysis. J. Environ. Sci. **35**, 75–78 (2009)
3. Ye, C.: Arid environmental sources of mercury research (2013)
4. Hu, H.-Y., Feng, X.-B., Zeng, Y.-P., et al.: The research progress of microbial methylation of mercury. J. Ecol. **30**, 874–882 (2011)
5. Lin, D., Wu, J.-F., Xiaoming, R.: Atmospheric mercury pollution and emission reduction policy research in China. Ecol. Econ. **30**, 153–156 (2014)
6. Wasserman, C.J., Hacon, S., et al.: The amazon river mercury biogeochemical studies. J. R. Swed. Acad. **32**, 336–342 (2013)
7. Sommar, J., Andersson, M.E., Gardfeldt, K., Lindqvist, O.: Enhanced concentrations of dissolved gaseous mercury in the surface waters of the Arctic Ocean. J. Mar. Chem. 190–194 (2008)
8. Zheng, D.-M., Winniy, W.Q.: Different pollution types sediment the formation distribution of mercury. J. Environ. Sci. Technol. 44–56 (2010)
9. Gang, Z., Ning, W., Chao, Y.J., et al.: Continuous precipitation weather conditions between soil/atmospheric mercury flux characteristics. China Environ. Sci. 409–415 (2013)
10. Muresan, B., Cossa, D., Richard, S.: The mercury speciation and exchanges at the air—water interface of a tropical artificial reservoir, French Guiana. Sci. Total Environ. 132–132 (2007)
11. Zhu, J.S., Wang, D.Y., Ma, M.: Mercury release flux and its influencing factors at the air—water interface in paddy field in Chongqing, China. Chin. Sci. Bull. 58 (2013): 266–266
12. Andersson, M.E., Gårdfeldt, K., Wängberg, I., et al.: The seasonal and daily variation of mercury evasion at coastal and off shore sites from the Mediterranean sea. J. Mar. Chem. **104**, 214–226 (2007)

13. Ci, Z., Zhang, X., Wang, Z.: The elemental mercury in coastal seawater of Yellow Sea, China: temporal variation and air-sea exchange. J. Atmos. Environ. 183–190 (2011)
14. White, E.M., Landis, M.S., Keeler, G.J., et al.: Study of mercury wet deposition physicochemistry in the Ohio River Valley through automated sequential from. Sci. Total Environ. **448**, 107–448 (2013)
15. Van Metre, P.C.: Increased atmospheric deposition of mercury in the reference 'lakes near major urban areas. Environ. Pollut. **162**, 109–215 (2012)
16. Christianson, E.M., Keeler, G.J., Landis, M.S., Norris, G.A., Dvonch, J.T.: Sources of mercury wet deposition in eastern, Ohio, USA. Environ. Sci. Technol. **40**, 5874–5881 (2006)
17. Menz, F.C., Seip, H.M.: Acid rain in Europe and the United States: an update. Environ. Sci. Policy 253–265 (2004)
18. Poissant, L., Pilote, M., Constant, P., et al.: The mercury gas exchanges over selected bare soil and flooded sites in the bay St. Francois wetlands (Québec, Canada). J. Atmos. Environ. **38**, 4205–4214 (2004)
19. Lindberg, S.E., Dong, W., Tilden Meyers, T.: Transpiration of gaseous elemental mercury through vegetation in a subtropical wetland in Florida. Atmos. Environ. 5207–5219 (2002)
20. Muresan, B., Cossa, D., Richard, S., Burban, B.: Mercury speciation and exchanges at the air—water interface of a tropical artificial reservoir, French Guiana. Sci. Total Environ. **385**, 132–145 (2007)

Erratum to: Sustainable Development of Water Resources and Hydraulic Engineering in China

Wei Dong, Yanqing Lian and Yong Zhang

Erratum to:
W. Dong et al. (eds.), *Sustainable Development of Water Resources and Hydraulic Engineering in China*, Environmental Earth Sciences, https://doi.org/10.1007/978-3-319-61630-8

In the original version of the book, a new chapter has to be included as the last chapter, which is a belated correction. The erratum book has been updated with the change.

The updated online version of this chapter can be found at
https://doi.org/10.1007/978-3-319-61630-8_35
https://doi.org/10.1007/978-3-319-61630-8

W. Dong et al. (eds.), *Sustainable Development of Water Resources and Hydraulic Engineering in China*, Environmental Earth Sciences,
https://doi.org/10.1007/978-3-319-61630-8_36

MIX
Papier aus verantwortungsvollen Quellen
Paper from responsible sources
FSC® C105338

If you have any concerns about our products,
you can contact us on
ProductSafety@springernature.com

In case Publisher is established outside the EU,
the EU authorized representative is:
Springer Nature Customer Service Center GmbH
Europaplatz 3, 69115 Heidelberg, Germany

Printed by Libri Plureos GmbH
in Hamburg, Germany